# Selected Titles in This Series

191 **Shun-ichi Amari and Hiroshi Nagaoka,** Methods of information geometry, 2000

190 **Alexander N. Starkov,** Dynamical systems on homogeneous spaces, 2000

189 **Mitsuru Ikawa,** Hyperbolic partial differential equations and wave phenomena, 2000

188 **V. V. Buldygin and Yu. V. Kozachenko,** Metric characterization of random variables and random processes, 2000

187 **A. V. Fursikov,** Optimal control of distributed systems. Theory and applications, 2000

186 **Kazuya Kato, Nobushige Kurokawa, and Takeshi Saito,** Number theory 1: Fermat's dream, 2000

185 **Kenji Ueno,** Algebraic Geometry 1: From algebraic varieties to schemes, 1999

184 **A. V. Mel'nikov,** Financial markets, 1999

183 **Hajime Sato,** Algebraic topology: an intuitive approach, 1999

182 **I. S. Krasil'shchik and A. M. Vinogradov, Editors,** Symmetries and conservation laws for differential equations of mathematical physics, 1999

181 **Ya. G. Berkovich and E. M. Zhmud',** Characters of finite groups. Part 2, 1999

180 **A. A. Milyutin and N. P. Osmolovskii,** Calculus of variations and optimal control, 1998

179 **V. E. Voskresenskii,** Algebraic groups and their birational invariants, 1998

178 **Mitsuo Morimoto,** Analytic functionals on the sphere, 1998

177 **Satoru Igari,** Real analysis—with an introduction to wavelet theory, 1998

176 **L. M. Lerman and Ya. L. Umanskiy,** Four-dimensional integrable Hamiltonian systems with simple singular points (topological aspects), 1998

175 **S. K. Godunov,** Modern aspects of linear algebra, 1998

174 **Ya-Zhe Chen and Lan-Cheng Wu,** Second order elliptic equations and elliptic systems, 1998

173 **Yu. A. Davydov, M. A. Lifshits, and N. V. Smorodina,** Local properties of distributions of stochastic functionals, 1998

172 **Ya. G. Berkovich and E. M. Zhmud',** Characters of finite groups. Part 1, 1998

171 **E. M. Landis,** Second order equations of elliptic and parabolic type, 1998

170 **Viktor Prasolov and Yuri Solovyev,** Elliptic functions and elliptic integrals, 1997

169 **S. K. Godunov,** Ordinary differential equations with constant coefficient, 1997

168 **Junjiro Noguchi,** Introduction to complex analysis, 1998

167 **Masaya Yamaguti, Masayoshi Hata, and Jun Kigami,** Mathematics of fractals, 1997

166 **Kenji Ueno,** An introduction to algebraic geometry, 1997

165 **V. V. Ishkhanov, B. B. Lur'e, and D. K. Faddeev,** The embedding problem in Galois theory, 1997

164 **E. I. Gordon,** Nonstandard methods in commutative harmonic analysis, 1997

163 **A. Ya. Dorogovtsev, D. S. Silvestrov, A. V. Skorokhod, and M. I. Yadrenko,** Probability theory: Collection of problems, 1997

162 **M. V. Boldin, G. I. Simonova, and Yu. N. Tyurin,** Sign-based methods in linear statistical models, 1997

161 **Michael Blank,** Discreteness and continuity in problems of chaotic dynamics, 1997

160 **V. G. Osmolovskii,** Linear and nonlinear perturbations of the operator div, 1997

159 **S. Ya. Khavinson,** Best approximation by linear superpositions (approximate nomography), 1997

158 **Hideki Omori,** Infinite-dimensional Lie groups, 1997

157 **V. B. Kolmanovskii and L. E. Shaikhet,** Control of systems with aftereffect, 1996

156 **V. N. Shevchenko,** Qualitative topics in integer linear programming, 1997

Translations of
# MATHEMATICAL MONOGRAPHS

Volume 191

# Methods of Information Geometry

Shun-ichi Amari
Hiroshi Nagaoka

Translated by
Daishi Harada

**Editorial Board**

Shoshichi Kobayashi (Chair)
Masamichi Takesaki

# 情報幾何の方法

JOHO KIKA NO HOHO
(Methods of Information Geometry)
by Shun-ichi Amari and Hiroshi Nagaoka

Copyright © 1993 by Shun-ichi Amari and Hiroshi Nagaoka
Originally published in Japanese by Iwanami Shoten, Publishers, Tokyo, 1993

The authors, Oxford University Press, and the American Mathematical Society gratefully acknowledge the financial support provided by the Daido Life Foundation for the editing of this work.

Translated from the Japanese by Daishi Harada

2000 *Mathematics Subject Classification.* Primary 00A69, 53–02, 53B05, 53A15, 62–02, 62F05, 62F12, 93C05, 81Q70, 94A15.

---

**Library of Congress Cataloging-in-Publication Data**
Amari, Shun'ichi.
 [Joho kika no hoho. English]
 Methods of information geometry / Shun-ichi Amari, Hiroshi Nagaoka ; [translated from the Japanese by Daishi Harada].
  p. cm. — (Translations of mathematical monographs, ISSN 0065-9282 ; v. 191)
 Includes bibliographical references and index.
 ISBN 0-8218-0531-2 (alk. paper)
  1. Mathematical statistics.  2. Geometry, Differential.  I. Nagaoka, Hiroshi, 1955–  II. Title. III. Series.
QA276.A56313  2000
519.5—dc21                                                                                                    00-059362

---

**Copying and reprinting.** Individual readers of this publication, and nonprofit libraries acting for them, are permitted to make fair use of the material, such as to copy a chapter for use in teaching or research. Permission is granted to quote brief passages from this publication in reviews, provided the customary acknowledgment of the source is given.

Republication, systematic copying, or multiple reproduction of any material in this publication is permitted only under license from the American Mathematical Society. Requests for such permission should be addressed to the Assistant to the Publisher, American Mathematical Society, P. O. Box 6248, Providence, Rhode Island 02940-6248. Requests can also be made by e-mail to reprint-permission@ams.org.

© 2000 by the American Mathematical Society. All rights reserved.
The American Mathematical Society retains all rights
except those granted to the United States Government.
Printed in the United States of America.
∞ The paper used in this book is acid-free and falls within the guidelines
established to ensure permanence and durability.
Visit the AMS home page at URL: http://www.ams.org/
10 9 8 7 6 5 4 3 2 1    05 04 03 02 01 00

# Contents

**Preface**     vii

**Preface to the English edition**     ix

**1 Elementary differential geometry**     1
    1.1 Differentiable manifolds . . . . . . . . . . . . . . . . . . . . . . 1
    1.2 Tangent vectors and tangent spaces . . . . . . . . . . . . . . . . 5
    1.3 Vector fields and tensor fields . . . . . . . . . . . . . . . . . . . 8
    1.4 Submanifolds . . . . . . . . . . . . . . . . . . . . . . . . . . . . 10
    1.5 Riemannian metrics . . . . . . . . . . . . . . . . . . . . . . . . 11
    1.6 Affine connections and covariant derivatives . . . . . . . . . . . 13
    1.7 Flatness . . . . . . . . . . . . . . . . . . . . . . . . . . . . . . . 17
    1.8 Autoparallel submanifolds . . . . . . . . . . . . . . . . . . . . . 19
    1.9 Projection of connections and embedding curvature . . . . . . . 22
    1.10 Riemannian connection . . . . . . . . . . . . . . . . . . . . . . 23

**2 The geometric structure of statistical models**     25
    2.1 Statistical models . . . . . . . . . . . . . . . . . . . . . . . . . . 25
    2.2 The Fisher metric . . . . . . . . . . . . . . . . . . . . . . . . . 28
    2.3 The $\alpha$-connection . . . . . . . . . . . . . . . . . . . . . . . . . 32
    2.4 Chentsov's theorem and some historical remarks . . . . . . . . 37
    2.5 The geometry of $\mathcal{P}(\mathcal{X})$ . . . . . . . . . . . . . . . . . . . . . . . 40
    2.6 $\alpha$-affine manifolds and $\alpha$-families . . . . . . . . . . . . . . . . . 45

**3 Dual connections**     51
    3.1 Duality of connections . . . . . . . . . . . . . . . . . . . . . . . 51
    3.2 Divergences: general contrast functions . . . . . . . . . . . . . . 53
    3.3 Dually flat spaces . . . . . . . . . . . . . . . . . . . . . . . . . . 58
    3.4 Canonical divergence . . . . . . . . . . . . . . . . . . . . . . . . 61
    3.5 The dualistic structure of exponential families . . . . . . . . . . 65
    3.6 The dualistic structure of $\alpha$-affine manifolds and $\alpha$-families . . . 70
    3.7 Mutually dual foliations . . . . . . . . . . . . . . . . . . . . . . 75
    3.8 A further look at the triangular relation . . . . . . . . . . . . . 77

## 4 Statistical inference and differential geometry — 81
- 4.1 Estimation based on independent observations — 81
- 4.2 Exponential families and observed points — 85
- 4.3 Curved exponential families — 87
- 4.4 Consistency and first-order efficiency — 89
- 4.5 Higher-order asymptotic theory of estimation — 94
- 4.6 Asymptotics of Fisher information — 97
- 4.7 Higher-order asymptotic theory of tests — 100
- 4.8 The theory of estimating functions and fiber bundles — 107
  - 4.8.1 The fiber bundle of local exponential families — 107
  - 4.8.2 Hilbert bundles and estimating functions — 109

## 5 The geometry of time series and linear systems — 115
- 5.1 The space of systems and time series — 115
- 5.2 The Fisher metric and the $\alpha$-connection on the system space — 118
- 5.3 The geometry of finite-dimensional models — 123
- 5.4 Stable systems and stable feedback — 127

## 6 Multiterminal information theory and statistical inference — 133
- 6.1 Statistical inference for multiterminal information — 133
- 6.2 0-rate testing — 136
- 6.3 0-rate estimation — 140
- 6.4 Inference for general multiterminal information — 142

## 7 Information geometry for quantum systems — 145
- 7.1 The quantum state space — 145
- 7.2 The geometric structure induced from a quantum divergence — 150
- 7.3 The geometric structure induced from a generalized covariance — 154
- 7.4 Applications to quantum estimation theory — 159

## 8 Miscellaneous topics — 167
- 8.1 The geometry of convex analysis, linear programming and gradient flows — 167
- 8.2 Neuro-manifolds and nonlinear systems — 170
- 8.3 Lie groups and transformation models in information geometry — 172
- 8.4 Mathematical problems posed by information geometry — 175

## Guide to the Bibliography — 181

## Bibliography — 187

## Index — 203

# Preface

Information geometry provides the mathematical sciences with a new framework for analysis. This framework is relevant to a wide variety of domains, and it has already been usefully applied to several of these, providing them with a new perspective from which to view the structure of the systems which they investigate. Nevertheless, the development of the field of information geometry can only be said to have just begun.

Information geometry began as an investigation of the natural differential geometric structure possessed by families of probability distributions. As a rather simple example, consider the set $S$ of normal distributions with mean $\mu$ and variance $\sigma^2$:

$$p(x; \mu, \sigma) = \frac{1}{\sqrt{2\pi}\sigma} \exp\left\{-\frac{(x-\mu)^2}{2\sigma^2}\right\}.$$

By specifying $(\mu, \sigma)$ we determine a particular normal distribution, and hence $S$ may be viewed as a 2-dimensional space (manifold) which has $(\mu, \sigma)$ as a coordinate system. However, this is not a Euclidean space, but rather a Riemannian space with a metric which naturally follows from the underlying properties of probability distributions. In particular, when $S$ is a family of normal distributions, it is a space of constant negative curvature. The underlying characteristics of probability distributions lead not only to this Riemannian structure; an investigation of the structure of probability distributions leads to a new concept within differential geometry: that of mutually dual affine connections. In addition, the structure of dual affine connections naturally arises in the framework of affine differential geometry, and has begun to attract the attention of mathematicians researching differential geometry.

Probability distributions are the fundamental element over which fields such as statistics, stochastic processes, and information theory are developed. Hence not only is the natural dualistic differential geometric structure of the space of probability distributions beautiful, but it must also play a fundamental role in these information sciences. In fact, considering statistical estimation from a differential geometric viewpoint has provided statistics with a new analytic tool which has allowed several previously open problems to be solved; information geometry has already established itself within the field of statistics. In the fields of information theory, stochastic processes, and systems, information geometry is being currently applied to allow the investigation of hitherto unexplored possibilities.

The utility of information geometry, however, is not limited to these fields. It has, for example, been productively applied to areas such as statistical physics and the mathematical theory underlying neural networks. Further, dualistic differential geometric structure is a general concept not inherently tied to probability distributions. For example, the interior method for linear programming may be analyzed from this point of view, and this suggests its relation to completely integrable dynamical systems. Finally, the investigation of the information geometry of quantum systems may lead to even further developments.

This book presents for the first time the entirety of the emerging field of information geometry. To do this requires an understanding of at least the fundamental concepts in differential geometry. Hence the first three chapters contain an introduction to differential geometry and the recently developed theory of dual connections. An attempt has been made to develop the fundamental framework of differential geometry as concisely and intuitively as possible. It is hoped that this book may serve generally as an introduction to differential geometry. Although differential geometry is said to be a difficult field to understand, this is true only of those texts written by mathematicians for other mathematicians, and it is not the case that the principal ideas in differential geometry are hard. Nevertheless, this book introduces only the amount of differential geometry necessary for the remaining chapters, and endeavors to do so in a manner which, while consistent with the conventional definitions in mathematical texts, allows the intuition underlying the concepts to be comprehended most immediately.

On the other hand, a comprehensive treatment of statistics, system theory, and information theory, among others, from the point of view of information geometry is for each distinct, relying on properties unique to that particular theory. It was beyond the scope of this book to include a thorough description of these fields, and inevitably, many of the relevant topics from these areas are rather hastily introduced in the latter half of the book. It is hoped that within these sections the reader will simply gather the flavor of the research being done, and for a more complete analysis refer to the corresponding papers. To complement this approach, many topics which are still incomplete and perhaps consist only of vague ideas have been included.

Nothing would make us happier than if this book could serve as an invitation for other researches to join in the development of information geometry.

# Preface to the English Edition

Information geometry provides a new method applicable to various areas including information sciences and physical sciences. It has emerged from investigating the geometrical structures of the manifold of probability distributions, and has applied successfully to statistical inference problems. However, it has been proved that information geometry opens a new paradigm useful for elucidation of information systems, intelligent systems, control systems, physical systems, mathematical systems, and so on.

There have been remarkable progresses recently in information geometry. For example, in the field of neurocomputing, a set of neural networks forms a neuro-manifold. Information geometry has become one of fundamental methods for analyzing neurocomputing and related areas. Its usefulness has also been recognized in multiterminal information theory and portfolio, in nonlinear systems and nonlinear prediction, in mathematical programming, in statistical inference and information theory of quantum mechanical systems, and so on. Its mathematical foundations have also shown a remarkable progress.

In spite of these developments, there were no expository textbooks covering the methods and applications of information geometry except for statistical ones. Although we published a booklet to show the wide scope of information geometry in 1993, it was unfortunately written in Japanese. It is our great pleasure to see its English translation. Mr. Daishi Harada has achieved an excellent work of translation.

In addition to correction of many misprints and errors found in the Japanese edition, we have made revision and rearrangement throughout the manuscript to make it as readable as possible. Also we have added several new topics, and even new sections and a new chapter such as §2.5, §3.2, §3.5, §3.8 and Chapter 7. The bibliography and the guide to it have largely been extended as well. These works were done by the authors after receiving the original translation, and it is the authors, not the translator, who should be responsible for the English writing of these parts.

This is a small booklet, however. We have presented a concise but comprehensive introduction to the mathematical foundation of information geometry in the first three chapters, while the other chapters are devoted to an overview of wide areas of applications. Even though we could not show detailed and comprehensive explanations for many topics, we expect that the readers feel its flavor and prosperity from the description. It is our pleasure if the book would play a key role for further developments of information geometry.

Year 2000

Shun-ichi Amari
Hiroshi Nagaoka

# Chapter 1

# Elementary differential geometry

Differential geometry is a mature field of mathematics and has many introductory texts; still, it is not an easy field to master. However, in this book we shall require only the fundamental ideas and methodologies of differential geometry. The main theme of modern differential geometry has been to characterize the global properties of manifolds, and much theory has been developed towards this end. At this time, the field of information geometry (mostly) requires only the theory of the locally characterizable properties of manifolds.

For information geometry the most important aspects of differential geometry are those which allow us to take problems from a variety of fields: statistics, information theory, and control theory; visualize them geometrically; and from this develop novel tools with which to extend and advance these fields. In this chapter we present an introduction to differential geometry from this point of view.

## 1.1 Differentiable manifolds

A **differentiable manifold** is a mathematical concept denoting a generalization/abstraction of geometric objects such as smooth curves and surfaces in an $n$-dimensional space. Intuitively, a manifold $S$ is a "set with a coordinate system." Since $S$ is a set, it has elements. It does not matter what these elements are (these elements are also called the **points** of $S$.) For example, in this book, we shall introduce manifolds whose points are probability distributions and also those whose points are linear systems. $S$ must also have a **coordinate system**. By this we mean a one-to-one mapping from $S$ (or its subset) to $\mathbb{R}^n$, which allows us to specify each point in $S$ using a vector of $n$ real numbers (this vector is called the **coordinates** of the corresponding point). We call the natural number $n$ the **dimension** of $S$, and write $n = \dim S$.

We call a coordinate system that has $S$ as its domain a global coordinate

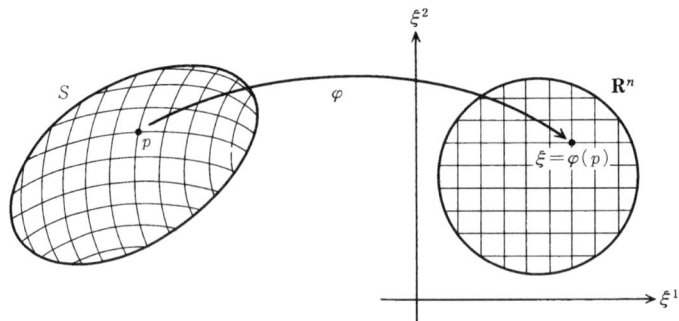

Figure 1.1: A coordinate system for $S$.

system. In our analysis below, we shall consider only the case where there exists a global coordinate system. However, in general there are many manifolds which do not have global coordinate systems. Examples of such a manifold include the surface of a sphere and the torus (the surface of a donut). These manifolds have only local coordinate systems. This may be viewed informally in the following way. Consider an open subset $U$ of $S$, and suppose that $U$ has a coordinate system. This provides a local coordinate system for those points contained in $U$. For a point not contained in $U$, consider another open subset $V$ containing that point which also has a coordinate system. Repeat this process until the original set $S$ is covered, so that each point in $S$ is contained in an open subset which has a coordinate system. Then this collection of open subsets of $S$ and their corresponding coordinate systems would allow us to express any point in $S$ using coordinates. However, as mentioned above, in this chapter we shall consider only the case when there exists a global coordinate system. This will suffice to prepare us for the later chapters. Indeed, since in this chapter we principally develop the local theory of manifolds, this assumption does not typically affect the generality of the analysis.

Let $S$ be a manifold and $\varphi : S \to \mathbb{R}^n$ be a coordinate system for $S$. Then $\varphi$ maps each point $p$ in $S$ to $n$ real numbers: $\varphi(p) = [\xi^1(p), \cdots, \xi^n(p)] = [\xi^1, \cdots, \xi^n]$. These are the coordinates of the point $p$. Each $\xi^i$ may be viewed as a function $p \to \xi^i(p)$ which maps a point $p$ to its $i^{\text{th}}$ coordinate; we call these $n$ functions $\xi^i : S \to \mathbb{R}$ ($i = 1, \cdots, n$) the **coordinate functions**.[1] We shall write the coordinate system $\varphi$ in ways such as $\varphi = [\xi^1, \cdots, \xi^n] = [\xi^i]$ (Figure 1.1).

Let $\psi = [\rho^i]$ be another coordinate system for $S$. Then the same point $p \in S$ has both the coordinates $[\xi^i(p)] = [\xi^i] \in \mathbb{R}^n$ with respect to the coordinate system $\varphi$, and the coordinates $[\rho^i(p)] = [\rho^i] \in \mathbb{R}^n$ with respect to the coordinate system $\psi$. The coordinates $[\rho^i]$ may be obtained from $[\xi^i]$ in the following way. First apply the inverse mapping $\varphi^{-1}$ to $[\xi^i]$; this gives us a point $p$ in $S$. Then apply $\psi$ to this point; this result is $[\rho^i]$. In other words, we apply the

---

[1] We shall use $\xi^i$, $\rho^i$ to denote both (the variable representing) the $i^{\text{th}}$ coordinate of a point and a coordinate function. This is similar to writing "the function $y = y(x)$."

## 1.1. DIFFERENTIABLE MANIFOLDS

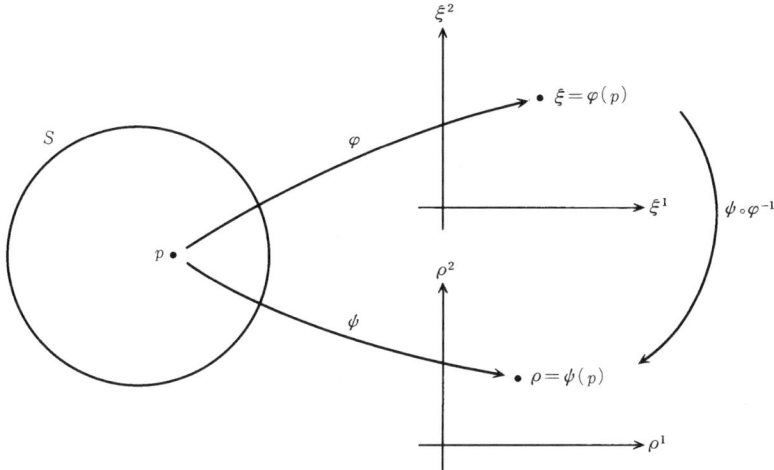

Figure 1.2: Coordinate transformation.

transformation on $\mathbb{R}^n$ given by

$$\psi \circ \varphi^{-1} : [\xi^1, \cdots, \xi^n] \mapsto [\rho^1, \cdots, \rho^n]. \tag{1.1}$$

This is called the coordinate transformation from $\varphi = [\xi^i]$ to $\psi = [\rho^i]$ (Figure 1.2).

To consider $S$ as a manifold means that one is interested in investigating those properties of $S$ which are invariant under coordinate transformations. In particular, differential geometry analyzes the geometry of objects using differential operators with respect to a variety of functions on $S$, and it would be problematic if these operators depended fundamentally on the choice of coordinates. Hence it is necessary to restrict the coordinate systems to those which allow smooth transformations between each other.

In order to properly formalize the concepts described above, let us now formally define manifolds for which there exists a global coordinate system.

Let $S$ be a set. If there exists a set of coordinate systems $\mathcal{A}$ for $S$ which satisfies the conditions (i) and (ii) below, we call $S$ (more properly, $(S, \mathcal{A})$) an $n$-dimensional $C^\infty$ **differentiable manifold**, or more simply, a **manifold**.

(i) Each element $\varphi$ of $\mathcal{A}$ is a one-to-one mapping from $S$ to some open subset of $\mathbb{R}^n$.

(ii) For all $\varphi \in \mathcal{A}$, given any one-to-one mapping $\psi$ from $S$ to $\mathbb{R}^n$, the following holds:

$$\psi \in \mathcal{A} \iff \psi \circ \varphi^{-1} \text{ is a } C^\infty \text{ diffeomorphism.}$$

Here, by a $C^\infty$ **diffeomorphism** we mean that $\psi \circ \varphi^{-1}$ and its inverse $\varphi \circ \psi^{-1}$ are both $C^\infty$ (infinitely many times differentiable). From these conditions, and given the coordinate transformation described in Equation (1.1), it follows that we may take the partial derivative of the function $\rho^i = \rho^i(\xi^1, \cdots, \xi^n)$ with respect to its variable arguments as many times as needed, and that the same holds for $\xi^i = \xi^i(\rho^1, \cdots, \rho^n)$. In this book, the condition $C^\infty$ is used a number of times, but in fact it is usually not necessary; it would suffice for the relevant functions to be differentiable some appropriate number of times. Intuitively, then, we may consider $C^\infty$ to simply mean "sufficiently smooth".

Let $S$ be a manifold and $\varphi$ be a coordinate system for $S$. Let $U$ be a subset of $S$. If the image $\varphi(U)$ is an open subset of $\mathbb{R}^n$, then we say that $U$ is an open subset of $S$. From condition (ii) above, we see that this property is invariant over the choice of coordinate system $\varphi$. This allows us to consider $S$ as a topological space. For any non-empty open subset $U$ of $S$, we may restrict $\varphi$, the coordinate system of $S$, to obtain $\varphi|_U$ (the mapping $U \to \mathbb{R}^n$ obtained by restricting the domain of $\varphi$ to $U$), which may be taken as a coordinate system for $U$. Hence we see that $U$ is a manifold whose dimension is the same as that of $S$.

Let $f : S \to \mathbb{R}$ be a function on a manifold $S$. Then if we select a coordinate system $\varphi = [\xi^i]$ for $S$, this function may be rewritten as a function of the coordinates; i.e., letting $[\xi^i]$ denote the coordinates of the point $p$, we have $f(p) = \bar{f}(\xi^1, \cdots, \xi^n)$, where $\bar{f} = f \circ \varphi^{-1}$. Note that $\bar{f}$ is a real-valued function whose domain is $\varphi(S)$, an open subset of $\mathbb{R}^n$. Now suppose that $\bar{f}(\xi^1, \cdots, \xi^n)$ is partially differentiable at each point in $\varphi(S)$. Then the partial derivative $\frac{\partial}{\partial \xi^i} \bar{f}(\xi^1, \cdots, \xi^n)$ is also a function on $\varphi(S)$. By transforming the domain back to $S$, we may define the partial derivatives of $f$ to be $\frac{\partial f}{\partial \xi^i} \stackrel{\text{def}}{=} \frac{\partial \bar{f}}{\partial \xi^i} \circ \varphi : S \to \mathbb{R}$. We write $\left( \frac{\partial f}{\partial \xi^i} \right)_p$ to denote the value of this function at point $p$ (the partial derivative at point $p$).

When $\bar{f} = f \circ \varphi^{-1}$ is $C^\infty$, in other words when $\bar{f}(\xi^1, \cdots, \xi^n)$ can be partially differentiated with respect to its variables an unbounded number of times, we call $f$ a $C^\infty$ **function** on $S$. This definition does not depend on the choice of coordinate system $\varphi$. The partial derivatives $\frac{\partial f}{\partial \xi^i}$ of a $C^\infty$ function $f$ are also $C^\infty$ functions. We may similarly define the higher-order partial derivatives, e.g. $\frac{\partial^2 f}{\partial \xi^j \partial \xi^i} = \frac{\partial}{\partial \xi^j} \frac{\partial f}{\partial \xi^i}$. These will also be $C^\infty$. As with the case of $C^\infty$ functions on $\mathbb{R}^n$, $\frac{\partial^2 f}{\partial \xi^j \partial \xi^i} = \frac{\partial}{\partial \xi^j} \frac{\partial f}{\partial \xi^i}$ holds.

Let us denote the class of $C^\infty$ functions on $S$ by $\mathcal{F}(S)$, or simply $\mathcal{F}$. For all $f$ and $g$ in $\mathcal{F}$ and a real number $c$, we define the sum $f + g$ as $(f+g)(p) = f(p) + g(p)$, the scaling $cf$ as $(cf)(p) = cf(p)$, and the product $f \cdot g$ as $(f \cdot g)(p) = f(p) \cdot g(p)$; these functions are also members of $\mathcal{F}$.

Let $[\xi^i]$ and $[\rho^j]$ be two coordinate systems. Since the coordinate functions $\xi^i$ and $\rho^j$ are clearly $C^\infty$, the partial derivatives $\frac{\partial \xi^i}{\partial \rho^j}$ and $\frac{\partial \rho^j}{\partial \xi^i}$ are well defined, and they satisfy

$$\sum_{j=1}^n \frac{\partial \xi^i}{\partial \rho^j} \frac{\partial \rho^j}{\partial \xi^k} = \sum_{j=1}^n \frac{\partial \rho^i}{\partial \xi^j} \frac{\partial \xi^j}{\partial \rho^k} = \delta_k^i, \tag{1.2}$$

where $\delta_k^i$ is 1 if $k = i$, and 0 otherwise (the Kronecker delta). In addition, for

## 1.2. TANGENT VECTORS AND TANGENT SPACES

any $C^\infty$ function $f$, we have

$$\frac{\partial f}{\partial \rho^j} = \sum_{i=1}^n \frac{\partial \xi^i}{\partial \rho^j} \frac{\partial f}{\partial \xi^i} \quad \text{and} \quad \frac{\partial f}{\partial \xi^i} = \sum_{j=1}^n \frac{\partial \rho^j}{\partial \xi^i} \frac{\partial f}{\partial \rho^j}. \tag{1.3}$$

**Note:** In this book there often appear equations which contain indices such as $i, j, \cdots$, and are to be summed over those indices that are both super and subscripted. For these equations we shall abbreviate by omitting the summation sign $\sum$ corresponding to these indices. For example, Equations (1.2) and (1.3) above would be written as

$$\frac{\partial \xi^i}{\partial \rho^j} \frac{\partial \rho^j}{\partial \xi^k} = \frac{\partial \rho^i}{\partial \xi^j} \frac{\partial \xi^j}{\partial \rho^k} = \delta_k^i$$

$$\frac{\partial f}{\partial \rho^j} = \frac{\partial \xi^i}{\partial \rho^j} \frac{\partial f}{\partial \xi^i}, \qquad \frac{\partial f}{\partial \xi^i} = \frac{\partial \rho^j}{\partial \xi^i} \frac{\partial f}{\partial \rho^j}.$$

We shall also abbreviate $\sum_{i=1}^n \sum_{j=1}^n A_{jk}^{ij} B_i^h$ as $A_{jk}^{ij} B_i^h$. Hence (unless there is ambiguity), whenever there appears such an equation we shall assume that there is an implicit $\sum$ (i.e., there is a summation over the relevant indices). Note therefore that $A_j^i X^j = A_k^i X^k$, for instance, is always true. This notation is known as Einstein's convention.

Let $S$ and $Q$ be manifolds with coordinate systems $\varphi: S \to \mathbb{R}^n$ and $\psi: Q \to \mathbb{R}^m$. A mapping $\lambda: S \to Q$ is said to be $C^\infty$ or smooth if $\psi \circ \lambda \circ \varphi^{-1}$ is a $C^\infty$ mapping from an open subset of $\mathbb{R}^n$ to $\mathbb{R}^m$. A necessary and sufficient condition for $\lambda$ to be $C^\infty$ is that $f \circ \lambda \in \mathcal{F}(S)$ for all $f \in \mathcal{F}(Q)$. If a $C^\infty$ mapping $\lambda$ is a bijection (i.e., one-to-one and $\lambda(S) = Q$) and the inverse $\lambda^{-1}$ is also $C^\infty$, then $\lambda$ is called a $C^\infty$ diffeomorphism from $S$ onto $Q$.

## 1.2 Tangent vectors and tangent spaces

The tangent space $T_p$ at a point $p \in S$ of a manifold $S$ is intuitively the vector space obtained by "locally linearizing" $S$ around $p$. Let $[\xi^i]$ be some coordinate system for $S$, and let $e_i$ denote the "tangent vector" which goes through point $p$ and is parallel to the $i^{\text{th}}$ coordinate curve (coordinate axis). By the $i^{\text{th}}$ coordinate curve we mean the curve which is obtained by fixing the values of all $\xi^j$ for $j \neq i$ and varying only the value of $\xi^i$. The $n$-dimensional space spanned by the $n$ tangent vectors $e_1, \cdots, e_n$ is the tangent space $T_p$ at point $p$ (Figure 1.3). Let $p'$ be a point "very close" to $p$, and let $[\xi^i]$ and $[\xi^i + \mathrm{d}\xi^i]$ (where $\mathrm{d}\xi^i$ is an infinitesimal) be the coordinates of $p$ and $p'$, respectively. Then the segment joining these two points may be described by $\overrightarrow{pp'} = \mathrm{d}\xi^i e_i$, an infinitesimal vector in $T_p$.

Let us make the above concepts more precise. To do so, we must first formally define what we mean by curves and the tangent vector of curves on a

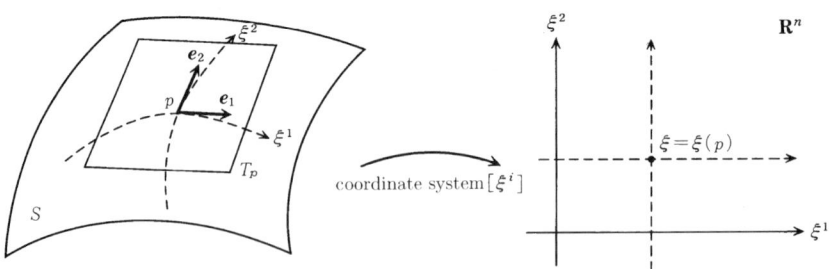

Figure 1.3: Tangent Space

manifold. Consider a one-to-one function $\gamma : I \to S$ from some interval $I \ (\subset \mathbb{R})$ to $S$. By defining $\gamma^i(t) \stackrel{\text{def}}{=} \xi^i(\gamma(t))$ we may express the point $\gamma(t)$ ($t \in I$) using coordinates as $\bar{\gamma}(t) = [\gamma^1(t), \cdots, \gamma^n(t)]$. If $\bar{\gamma}(t)$ is $C^\infty$ for $t \in I$, we call $\gamma$ a $\boldsymbol{C^\infty}$ **curve** on $S$. This definition is independent of coordinate system choice.

Now, given a curve $\gamma$ and a point $\gamma(a) = p$, let us consider what is meant by the "derivative" of $\gamma$ at $p$, or alternatively the "tangent vector" $\left(\frac{d\gamma}{dt}\right)_p = \dot{\gamma}(a)$. When $S$ is simply an open subset of $\mathbb{R}^n$, or can be embedded smoothly into $\mathbb{R}^\ell$ ($\ell \geq n$), the range of $\gamma$ is contained within a single linear space, and hence it suffices to consider the standard derivative

$$\dot{\gamma}(a) = \lim_{h \to 0} \frac{\gamma(a+h) - \gamma(a)}{h}. \tag{1.4}$$

In general, however, the equation above is not meaningful. On the other hand, if we take a $C^\infty$ function $f \in \mathcal{F}$ on $S$ and consider the value of $f(\gamma(t))$ on the curve, since this is a real-valued function, we may define the derivative $\frac{d}{dt} f(\gamma(t))$ in the usual way. Using coordinates, we have $f(\gamma(t)) = \bar{f}(\bar{\gamma}(t)) = \bar{f}(\gamma^1(t), \cdots, \gamma^n(t))$, and the derivatives may be rewritten as

$$\frac{d}{dt} f(\gamma(t)) = \left(\frac{\partial \bar{f}}{\partial \xi^i}\right)_{\bar{\gamma}(t)} \frac{d\gamma^i(t)}{dt} = \left(\frac{\partial f}{\partial \xi^i}\right)_{\gamma(t)} \frac{d\gamma^i(t)}{dt}. \tag{1.5}$$

We call this the directional derivative of $f$ along the curve $\gamma$. Let us consider this directional derivative as an expression of the tangent vector of $\gamma$. In other words, we take the operator : $\mathcal{F} \to \mathbb{R}$ which maps $f \in \mathcal{F}$ to $\frac{d}{dt} f(\gamma(t))|_{t=a}$, and simply define the tangent vector $\left(\frac{d\gamma}{dt}\right)_p = \dot{\gamma}(a)$ to be this operator. Then we may rewrite Equation (1.5) as

$$\dot{\gamma}(a) = \left(\frac{d\gamma}{dt}\right)_p = \dot{\gamma}^i(a) \left(\frac{\partial}{\partial \xi^i}\right)_p \tag{1.6}$$

($\dot{\gamma}^i(a) = \frac{d}{dt} \gamma^i(t)|_{t=a}$). Here $\left(\frac{\partial}{\partial \xi^i}\right)_p$ is an operator which maps $f \mapsto \left(\frac{\partial f}{\partial \xi^i}\right)_p$. It is possible to show that when the tangent vectors can be defined using Equation (1.4), there is a natural one-to-one correspondence between Equations (1.4)

## 1.3. VECTOR FIELDS AND TENSOR FIELDS

and (1.6). Hence the definition of tangent vectors as operators may be viewed as a generalization of Equation (1.4).

Since a partial derivative is simply a directional derivative along a coordinate axis, the operator $\left(\frac{\partial}{\partial \xi^i}\right)_p$ is the tangent vector at point $p$ of the $i^{\text{th}}$ coordinate curve. The $e_i$ mentioned previously corresponds to this $\left(\frac{\partial}{\partial \xi^i}\right)_p$. From Equation (1.3), we see that

$$\left(\frac{\partial}{\partial \rho^j}\right)_p = \left(\frac{\partial \xi^i}{\partial \rho^j}\right)_p \left(\frac{\partial}{\partial \xi^i}\right)_p \quad \text{and} \quad \left(\frac{\partial}{\partial \xi^i}\right)_p = \left(\frac{\partial \rho^j}{\partial \xi^i}\right)_p \left(\frac{\partial}{\partial \rho^j}\right)_p. \quad (1.7)$$

Consider all curves which pass through the point $p$. We denote the set of all tangent vectors corresponding to these curves by $T_p$, or $T_p(S)$. From Equation (1.6), we see that

$$T_p(S) = \left\{ c^i \left(\frac{\partial}{\partial \xi^i}\right)_p \middle| [c^1, \cdots, c^n] \in \mathbb{R}^n \right\}. \quad (1.8)$$

This forms a linear space, and since the operators $\left\{\left(\frac{\partial}{\partial \xi^i}\right)_p ; i = 1, \cdots, n\right\}$ are clearly linearly independent, the dimension of this space is $n$ $(= \dim S)$. We call $T_p(S)$ and its elements the **tangent space** and **tangent vectors**, of $S$ at the point $p$, respectively. In addition, we call $\left(\frac{\partial}{\partial \xi^i}\right)_p$ the **natural basis** of the coordinate system $[\xi^i]$.

Let $D \in T_p$ be some tangent vector. Then for all $f, g \in \mathcal{F}$ and all $a, b \in \mathbb{R}$, $D$ satisfies the following:

$$[\text{Linearity}] \quad D(af + bg) = aD(f) + bD(g). \quad (1.9)$$
$$[\text{Leibniz's rule}] \quad D(f \cdot g) = f(p)D(g) + g(p)D(f). \quad (1.10)$$

Conversely, it can be shown that any operator $D : \mathcal{F} \to \mathbb{R}$ satisfying these properties is an element of $T_p$. Hence, it is possible to define tangent vectors in terms of these properties.

Let $\lambda : S \to Q$ be a smooth mapping from a manifold $S$ to another manifold $Q$. Given a tangent vector $D \in T_p(S)$ of $S$, the mapping $D' : \mathcal{F}(Q) \to \mathbb{R}$ defined by $D'(f) = D(f \circ \lambda)$ satisfies Equations (1.9) (1.10) with $p$ replaced with $\lambda(p)$, and hence $D'$ belongs to $T_{\lambda(p)}(Q)$. Representing this correspondence as $D' = (\mathrm{d}\lambda)_p(D)$, we may define a linear mapping $(\mathrm{d}\lambda)_p : T_p(S) \to T_{\lambda(p)}(Q)$, which is called the **differential** of $\lambda$ at $p$. When $S$ and $Q$ are provided with coordinate systems $[\xi^i]$ and $[\rho^j]$ respectively, we have

$$(\mathrm{d}\lambda)_p\left(\left(\frac{\partial}{\partial \xi^i}\right)_p\right) = \left(\frac{\partial(\rho^j \circ \lambda)}{\partial \xi^i}\right)_p \left(\frac{\partial}{\partial \rho^j}\right)_{\lambda(p)}. \quad (1.11)$$

Moreover, for any curve $\gamma(t)$ on $S$ passing through the point $p$ it follows that

$$(\mathrm{d}\lambda)_p\left(\left(\frac{\mathrm{d}\gamma}{\mathrm{d}t}\right)_p\right) = \left(\frac{\mathrm{d}(\lambda \circ \gamma)}{\mathrm{d}t}\right)_{\lambda(p)}. \quad (1.12)$$

## 1.3 Vector fields and tensor fields

Let $X : p \mapsto X_p$ be a mapping which maps each point $p$ in the manifold $S$ to a tangent vector $X_p \in T_p(S)$. We call such a mapping a **vector field**. For example, if $[\xi^i]$ is a coordinate system, then we may define $n$ vector fields through the mappings $\frac{\partial}{\partial \xi^i} : p \mapsto \left(\frac{\partial}{\partial \xi^i}\right)_p$ $(i = 1, \cdots, n)$. These are the vector fields formed by the natural basis. Below, we shall write $\partial_i$ to mean $\frac{\partial}{\partial \xi^i}$. In general, given a vector field $X$, for each point $p$ there exists $n$ real numbers $\{X_p^1, \cdots, X_p^n\}$ which uniquely determine $X_p = X_p^i (\partial_i)_p$. Hence we may define the functions $X^i : p \mapsto X_p^i$ on $S$. We call the $n$ functions $\{X^1, \cdots, X^n\}$ the **components** of $X$ with respect to $[\xi^i]$. This allows us to write $X = X^i \partial_i$. If, in addition, we let $[\rho^j]$ be another coordinate system and $X = \tilde{X}^j \tilde{\partial}_j$ $\left(\tilde{\partial}_j \stackrel{\text{def}}{=} \frac{\partial}{\partial \rho^j}\right)$ be the component expression of $X$ with respect to $[\rho^j]$, then the following hold:

$$\tilde{X}^j = X^i \frac{\partial \rho^j}{\partial \xi^i} \quad \text{and} \quad X^i = \tilde{X}^j \frac{\partial \xi^i}{\partial \rho^j}. \tag{1.13}$$

If the components of a vector field are $C^\infty$ with respect to some coordinate system, then the components are $C^\infty$ with respect to any other. We call such a vector field a $C^\infty$ vector field. Since we consider only $C^\infty$ vector fields in this book, we shall refer to them as simply vector fields. We shall denote this family of vector fields by $\mathcal{T}(S)$, or simply $\mathcal{T}$. Clearly $\partial_i \in \mathcal{T}$ $(i = 1, \cdots, n)$.

Now for any $X, Y \in \mathcal{T}$ and any $c \in \mathbb{R}$, the mappings $X + Y : p \mapsto X_p + Y_p$ and $cX : p \mapsto cX_p$ are also members of $\mathcal{T}$. Hence $\mathcal{T}$ is a linear space. In addition, for any $f \in \mathcal{F}$, the mapping $fX : p \mapsto f(p)X_p$ is a member of $\mathcal{T}$.

We call $F : V_1 \times V_2 \times \cdots \times V_r \to W$, where $V_1, \cdots, V_r, W$ are linear spaces, a **multilinear mapping** if the following property holds. Let $\tilde{F}(v_i)$ denote a mapping of one variable equal to $F(v_1, \cdots, v_r)$ where some $v_i$ has been distinguished as the variable, and the other $v_j$ $(j \neq i)$ are held constant to some value $(\in V_j)$. Then $\tilde{F} : v_i \mapsto \tilde{F}(v_i)$ is a linear mapping from $V_i$ to $W$.

Now for each point $p \in S$, let $[T_p]_r^0$ denote the family of multilinear mappings of the form $\underbrace{T_p \times \cdots \times T_p}_{r \text{ direct products}} \to \mathbb{R}$, and let $[T_p]_r^1$ denote the family of the form $\underbrace{T_p \times \cdots \times T_p}_{r \text{ direct products}} \to T_p$. We call mappings $A : p \mapsto A_p$ which maps each point $p$ in $S$ to some element $A_p$ of $[T_p]_r^q$ $(q = 0, 1)$ a **tensor field of type $(q, r)$** on $S$. The types $(0, r)$ and $(1, r)$ are also respectively called **tensor fields of covariant degree $r$** and **tensor fields of contravariant degree 1 and covariant degree $r$**. Vector fields may be considered to be tensor fields of type $(1, 0)$. Although it is possible to define tensor fields of type $(q, r)$ for $q = 2, 3, \cdots$, they will not be used in this book. In addition, we shall occasionally refer to tensor fields as simply tensors.

Let $A$ be a tensor field of type $(q, r)$ and $X_1, \cdots, X_r$ be $r$ vector fields. Then

## 1.3. VECTOR FIELDS AND TENSOR FIELDS

we may consider a mapping with domain $S$ of the following form:
$$A(X_1,\cdots,X_r): p \mapsto A_p((X_1)_p,\cdots,(X_r)_p). \tag{1.14}$$

When $q=0$, $A_p((X_1)_p,\cdots,(X_r)_p) \in \mathbb{R}$ and hence this mapping is a real-valued function on $S$. When $q=1$, $A_p((X_1)_p,\cdots,(X_r)_p) \in T_p$, and hence this defines a vector field on $S$. Given $A$, if for all $C^\infty$ vector fields $X_1,\cdots,X_r \in \mathcal{T}$ the mapping $A(X_1,\cdots,X_r)$ is $C^\infty$ (i.e., when $q=0$ the mapping is in $\mathcal{F}$, and when $q=1$ it is in $\mathcal{T}$), we call $A$ a $C^\infty$ tensor field. Below, we consider only $C^\infty$ tensor fields, and shall simply call them tensor fields.

Consider the tensor field $A$ of type $(q,r)$ to be a mapping $(X_1,\cdots,X_r) \mapsto A(X_1,\cdots,X_r)$. Then when $q=0$ we have $A: \underbrace{\mathcal{T} \times \cdots \times \mathcal{T}}_{r \text{ direct products}} \to \mathcal{F}$, and when $q=1$ we have $A: \underbrace{\mathcal{T} \times \cdots \times \mathcal{T}}_{r \text{ direct products}} \to \mathcal{T}$. This, in addition to forming a multilinear mapping, has the following property: for all $f_1,\ldots,f_r \in \mathcal{F}$,
$$A(f_1 X_1,\cdots,f_r X_r) = f_1 \cdots f_r A(X_1,\cdots,X_r).$$

We call this the $\mathcal{F}$-**multilinearity** of $A$. Conversely, if the mapping $A: \mathcal{T} \times \cdots \times \mathcal{T} \to \mathcal{F}$, or alternatively $A: \mathcal{T} \times \cdots \times \mathcal{T} \to \mathcal{T}$ is $\mathcal{F}$-multilinear, then this determines a tensor field $p \mapsto A_p$ satisfying Equation (1.14).

The operation of a tensor field $A$ of type $(0,r)$ on the $r$ basis vector fields $\partial_{i_1},\cdots,\partial_{i_r}$ $\left(\partial_i \stackrel{\text{def}}{=} \frac{\partial}{\partial \xi^i}\right)$ defines a function. Let us denote this by
$$A(\partial_{i_1},\cdots,\partial_{i_r}) = A_{i_1 \cdots i_r}.$$

We call the $n^r$ functions $\{A_{i_1 \cdots i_r}\}$ obtained by changing the values of $i_1,\cdots,i_r$ the **components** of $A$ with respect to the coordinate system $[\xi^i]$. Let $X_1,\cdots,X_r$ be $r$ vector fields; these may be expressed component-wise as $X_j = X_j^i \partial_i$. Then from $\mathcal{F}$-multilinearity, we have
$$A(X_1,\cdots,X_r) = A_{i_1 \cdots i_r} X_1^{i_1} \cdots X_r^{i_r}$$

In the case of a tensor field $A$ of type $(1,r)$, $A(\partial_{i_1},\cdots,\partial_{i_r})$ is a vector field, and its component expression is given by
$$A(\partial_{i_1},\cdots,\partial_{i_r}) = A_{i_1 \cdots i_r}^k \partial_k.$$

The $n^{r+1}$ functions $\{A_{i_1 \cdots i_r}^k\}$ thus defined are called the components of $A$ with respect to $[\xi^i]$. As in the previous case, letting $X_j = X_j^i \partial_i$, the following holds:
$$A(X_1,\cdots,X_r) = (A_{i_1 \cdots i_r}^k X_1^{i_1} \cdots X_r^{i_r}) \partial_k.$$

Let $[\rho^j]$ be another coordinate system. Using $\tilde{\ }$ to denote components with respect to $[\rho^j]$, we have

$$\tilde{A}_{j_1 \cdots j_r} = A_{i_1 \cdots i_r} \left(\frac{\partial \xi^{i_1}}{\partial \rho^{j_1}}\right) \cdots \left(\frac{\partial \xi^{i_r}}{\partial \rho^{j_r}}\right) \quad \text{and} \tag{1.15}$$

$$\tilde{A}_{j_1 \cdots j_r}^\ell = A_{i_1 \cdots i_r}^k \left(\frac{\partial \xi^{i_1}}{\partial \rho^{j_1}}\right) \cdots \left(\frac{\partial \xi^{i_r}}{\partial \rho^{j_r}}\right) \left(\frac{\partial \rho^\ell}{\partial \xi^k}\right). \tag{1.16}$$

## 1.4 Submanifolds

Let $S$ and $M$ be manifolds, where $M$ is a subset of $S$. Let $[\xi^1, \cdots, \xi^n] = [\xi^i]$ and $[u^1, \cdots, u^m] = [u^a]$ be coordinate systems for $S$ and $M$, respectively, where $n = \dim S$ and $m = \dim M$. Below, we shall use the indices $i, j, k, \cdots$ over $\{1, \cdots, n\}$ for $S$ and $a, b, c, \cdots$ over $\{1, \cdots, m\}$ for $M$.

We call $M$ a **submanifold** of $S$ if the following conditions (i), (ii), and (iii) hold.

(i) The restriction $\xi^i|_M$ of each $\xi^i$ $(: S \to \mathbb{R})$ to $M$, is a $C^\infty$ function on $M$.

(ii) Let $B_a^i \stackrel{\text{def}}{=} \left(\frac{\partial \xi^i}{\partial u^a}\right)_p$ (more precisely, $\left(\frac{\partial \xi^i|_M}{\partial u^a}\right)_p$) and $B_a \stackrel{\text{def}}{=} [B_a^1, \cdots, B_a^n] \in \mathbb{R}^n$. Then for each point $p$ in $M$, $\{B_1, \cdots B_m\}$ are linearly independent (hence $m \leq n$).

(iii) For any open subset $W$ of $M$, there exists $U$, an open subset of $S$, such that $W = M \cap U$.

These conditions are independent of the choice of coordinate systems $[\xi^i]$ and $[u^a]$. Indeed, conditions (i) and (ii) mean that the embedding $\iota : M \to S$ defined by $\iota(p) = p, \forall p \in M$, is a $C^\infty$ mapping and that its differential $(\mathrm{d}\iota)_p$ is nondegenerate at each point $p$.

An open subset of $S$, as we noted in §1.1, forms an $n$-dimensional manifold; in addition, it is also a submanifold of $S$. We may construct an example of a submanifold of dimension $m$ $(< n)$ in the following way. Let $[\xi^i]$ be a coordinate system of $S$ and $\{c^{m+1}, \cdots, c^n\}$ be $n - m$ real numbers. Now define

$$M \stackrel{\text{def}}{=} \{p \in S | \xi^i(p) = c^i, m + 1 \leq i \leq n\}.$$

We assume that $M \neq \emptyset$ (the empty set). Then if we let $u^a \stackrel{\text{def}}{=} \xi^a|_M$ $(1 \leq a \leq m)$, $M$ is an $m$-dimensional manifold with coordinate system $[u^a]$, and hence it is a submanifold of $S$. The "reverse" of this is also true at least locally. In other words, if $M$ is an $m$-dimensional submanifold of $S$, with $[u^a]$ its coordinate system, and $\{c^{m+1}, \cdots, c^n\}$ is a set of $n - m$ real numbers, then it is possible to choose $U$, an open subset of $S$, and a coordinate system $[\xi^i]$, so that

$$M \cap U = \{p \in U | \xi^i(p) = c^i, m + 1 \leq i \leq n\},$$

and moreover, $u^a|_{M \cap U} = \xi^a|_{M \cap U}$ $(1 \leq a \leq m)$.

If $M$ is a submanifold of $S$ then a curve $\gamma : t \mapsto \gamma(t)$ in $M$ is also a curve in $S$. Hence letting $p$ be a point on $\gamma$, the tangent vector $\left(\frac{\mathrm{d}\gamma}{\mathrm{d}t}\right)_p$ of $\gamma$ may be considered both as an element of $T_p(M)$ and as one of $T_p(S)$. Using coordinate systems $[u^a]$ and $[\xi^i]$ for $M$ and $S$, respectively, and letting $\gamma^a \stackrel{\text{def}}{=} u^a \circ \gamma$ and $\gamma^i \stackrel{\text{def}}{=} \xi^i \circ \gamma$, these tangent vectors may be written as $\left(\frac{\mathrm{d}\gamma^a}{\mathrm{d}t}\right)_p (\partial_a)_p \in T_p(M)$

and $\left(\frac{\mathrm{d}\gamma^i}{\mathrm{d}t}\right)_p (\partial_i)_p \in T_p(S)$, where $\partial_a \stackrel{\text{def}}{=} \frac{\partial}{\partial u^a}$ and $\partial_i \stackrel{\text{def}}{=} \frac{\partial}{\partial \xi^i}$. Since

$$\left(\frac{\mathrm{d}\gamma^i}{\mathrm{d}t}\right)_p = \left(\frac{\partial \xi^i}{\partial u^a}\right)_p \left(\frac{\mathrm{d}\gamma^a}{\mathrm{d}t}\right)_p, \qquad (1.17)$$

from condition (ii) for submanifolds we see that there is a one-to-one correspondence between these tangent vectors. In other words, this correspondence is given by the differential $(\mathrm{d}\iota)_p$ of the embedding $\iota : M \to S$. By considering the corresponding pairs to be equivalent, we may view $T_p(M)$ as a linear subspace of $T_p(S)$. From Equation (1.17) we obtain

$$\left(\frac{\partial}{\partial u^a}\right)_p = \left(\frac{\partial \xi^i}{\partial u^a}\right)_p \left(\frac{\partial}{\partial \xi^i}\right)_p = B_a^i \partial_i. \qquad (1.18)$$

This shows that $B_a^i \partial_i$ is the natural basis vector $\partial_a$ of $M$ with respect to coordinate system $[u^a]$ seen as a vector in $T_p(S)$. In addition, this may be interpreted as the equality of the differential operators: for all $f \in \mathcal{F}(S)$, $\left(\frac{\partial f}{\partial u^a}\right)_p = \left(\frac{\partial \xi^i}{\partial u^a}\right)_p \left(\frac{\partial f}{\partial \xi^i}\right)_p$.

## 1.5 Riemannian metrics

Let $S$ be a manifold. For each point $p$ in $S$, let us assume that an inner product $\langle \, , \, \rangle_p$ has been defined on the tangent space $T_p(S)$. In other words, for any tangent vectors $D, D' \in T_p(S)$ we have $\langle D, D' \rangle_p \in \mathbb{R}$, and the following hold.

[Linearity] $\quad \langle aD + bD', D'' \rangle_p = a \langle D, D'' \rangle_p + b \langle D', D'' \rangle_p$
$\quad (\forall a, b \in \mathbb{R})$ (1.19)

[Symmetry] $\quad \langle D, D' \rangle_p = \langle D', D \rangle_p$ (1.20)

[Positive-definiteness] $\quad$ If $D \neq 0$ then $\langle D, D \rangle_p > 0$ (1.21)

Note that $\langle \, , \, \rangle_p \in [T_p(S)]_2^0$ since from Equations (1.19) and (1.20) we see that $\langle \, , \, \rangle_p$ is a bilinear form. Hence the mapping from points $p$ in $S$ to their inner product on $T_p(S)$, say $g : p \mapsto \langle \, , \, \rangle_p$, is a tensor field of covariant degree 2. We call this a ($C^\infty$) **Riemannian metric** on $S$. Such a metric, $g$, is not naturally determined by the structure of $S$ as a manifold; it is possible to consider an infinite number of Riemannian metrics on $S$. Given a Riemannian metric $g$ on $S$, we call $S$ (more precisely $(S, g)$) a **Riemannian manifold**.

Let $[\xi^i]$ be a coordinate system for $S$, and let $\partial_i \stackrel{\text{def}}{=} \frac{\partial}{\partial \xi^i}$. Then the components $\{g_{ij}; i, j = 1, \cdots, n\}$ ($n = \dim S$) of a Riemannian metric $g$ with respect to $[\xi^i]$ are determined by $g_{ij} = \langle \partial_i, \partial_j \rangle$. This is a $C^\infty$ function which maps each point $p$ in $S$ to $g_{ij}(p) = \langle (\partial_i)_p, (\partial_j)_p \rangle_p$. If we rewrite the tangent vectors $D, D' \in T_p$ in terms of their coordinates as $D = D^i (\partial_i)_p$ and $D' = D'^i (\partial_i)_p$, their inner product may then be written as:

$$\langle D, D' \rangle_p = g_{ij}(p) D^i D'^j.$$

Also, the length $\|D\|$ of the tangent vector $D$ is given by

$$\|D\|^2 = \langle D, D\rangle_p = g_{ij}(p)D^i D^j.$$

If we let $G(p) = [g_{ij}(p)]$ be an $n \times n$ matrix whose $(i,j)^{\text{th}}$ element is $g_{ij}(p)$, we see from Equations (1.20) and (1.21) that this is a positive definite symmetric matrix. Conversely, suppose we are given a coordinate system $[\xi^i]$ for an $n$-dimensional manifold $S$, and $n^2$ $C^\infty$ functions $\{g_{ij}\}$ ($\subset \mathcal{F}(S)$). Then if $G(p) = [g_{ij}(p)]$ is a positive definite symmetric matrix for every point $p \in S$, the corresponding Riemannian metric on $S$ which has $g_{ij}$ as its components with respect to $[\xi^i]$ is uniquely determined. The relationship between these components and the components $\tilde{g}_{k\ell} = \langle \tilde{\partial}_k, \tilde{\partial}_\ell \rangle$ ($\tilde{\partial}_k \stackrel{\text{def}}{=} \frac{\partial}{\partial \rho^k}$) with respect to a different coordinate system $[\rho^k]$ is given by the following transformations of covariant tensor fields of order 2 (refer to Equation (1.15)):

$$\tilde{g}_{k\ell} = g_{ij}\left(\frac{\partial \xi^i}{\partial \rho^k}\right)\left(\frac{\partial \xi^j}{\partial \rho^\ell}\right) \quad \text{and} \quad g_{ij} = \tilde{g}_{k\ell}\left(\frac{\partial \rho^k}{\partial \xi^i}\right)\left(\frac{\partial \rho^\ell}{\partial \xi^j}\right). \tag{1.22}$$

Let $g^{ij}(p)$ be the $(i,j)^{\text{th}}$ component of the inverse $G(p)^{-1}$ of $G(p) = [g_{ij}(p)]$ (this inverse is also positive definite symmetric). Now define the function $g^{ij} : p \mapsto g^{ij}(p)$ on $S$. Then

$$g_{ij}g^{jk} = \delta_i^k = \begin{cases} 1 & (k=i) \\ 0 & (k \neq i) \end{cases}, \tag{1.23}$$

and the relationship between this inverse and $\tilde{G}(p)^{-1} = [\tilde{g}^{k\ell}(p)]$, which is the inverse of $\tilde{G}(p) = [\tilde{g}_{k\ell}(p)]$, is given by the following.

$$\tilde{g}^{k\ell} = g^{ij}\left(\frac{\partial \rho^k}{\partial \xi^i}\right)\left(\frac{\partial \rho^\ell}{\partial \xi^j}\right) \quad \text{and} \quad g^{ij} = \tilde{g}^{k\ell}\left(\frac{\partial \xi^i}{\partial \rho^k}\right)\left(\frac{\partial \xi^j}{\partial \rho^\ell}\right). \tag{1.24}$$

Let $\gamma : [a,b] \to S$ be a curve in the Riemannian manifold $S$. We define its length $\|\gamma\|$ to be

$$\|\gamma\| \stackrel{\text{def}}{=} \int_a^b \left\|\frac{d\gamma}{dt}\right\| dt = \int_a^b \sqrt{g_{ij}\dot{\gamma}^i\dot{\gamma}^j}\, dt, \tag{1.25}$$

where $\dot{\gamma}^i$ is the derivative of $\gamma^i \stackrel{\text{def}}{=} \xi^i \circ \gamma$ (see Equation (1.6).)

Let $M$ be a submanifold of a Riemannian manifold $S$. As noted in §1.4, for each point $p \in M$, we may view $T_p(M)$ as a linear subspace of $T_p(S)$, and hence an inner product $g(p) = \langle \,,\, \rangle_p$ on $T_p(S)$ naturally defines an inner product on $T_p(M)$. Then, letting $g|_M(p)$ denote this inner product, $g|_M : p \mapsto g|_M(p)$ is a Riemannian metric on $M$. Given a coordinate system $[u^a]$ on $M$, we see from Equation (1.18) that the components of $g|_M$, $\{g_{ab}\}$ satisfy

$$g_{ab} = \left\langle \frac{\partial}{\partial u^a}, \frac{\partial}{\partial u^b} \right\rangle = g_{ij}\left(\frac{\partial \xi^i}{\partial u^a}\right)\left(\frac{\partial \xi^j}{\partial u^b}\right). \tag{1.26}$$

## 1.6 Affine connections and covariant derivatives

Let $S$ be an $n$-dimensional manifold. If $S$ is an open subset of $\mathbb{R}^n$, then by defining the tangent vector of a curve $\gamma$ according to Equation (1.4), the tangent space $T_p = T_p(S)$ at each point $p \in S$ may be considered equivalent to $\mathbb{R}^n$. This means that for $p$ and $q$ not equal, there is still a natural correspondence between $T_p$ and $T_q$. For a general manifold $S$, however, $T_p$ and $T_q$ are entirely different spaces when $p \neq q$. Hence, to consider relationships between $T_p$ and $T_q$, we must somehow augment the structure of $S$ as a manifold. Affine connections are such a structural augmentation.

Intuitively, defining an affine connection on a manifold $S$ means that for each point $p$ in $S$ and its "neighbor" $p'$, we define a linear one-to-one mapping between $T_p$ and $T_{p'}$. Here we call $p'$ a neighbor of $p$ if, given a coordinate system $[\xi^i]$ of $S$, the difference between the coordinates of $p$ and $p'$, $d\xi^i \overset{\text{def}}{=} \xi^i(p') - \xi^i(p)$, when construed as a first-order infinitesimal, is sufficiently small that we may ignore the second-order infinitesimals $(d\xi^i)(d\xi^j)$. Below we shall introduce the notion of affine connections in an intuitive manner using infinitesimals. (It is possible to formalize this discussion by using fiber bundles.)

As shown in Figure 1.4, in order to establish a linear mapping $\Pi_{p,p'}$ between $T_p$ and $T_{p'}$ we must specify, for each $j \in \{1, \cdots, n\}$, how to express $\Pi_{p,p'}((\partial_j)_p)$ in terms of a linear combination of $\{(\partial_1)_{p'}, \cdots, (\partial_n)_{p'}\}$ $\left(\partial_j \overset{\text{def}}{=} \frac{\partial}{\partial \xi^j}\right)$. Let us assume that the difference between $\Pi_{p,p'}((\partial_j)_p)$ and $(\partial_j)_{p'}$ is an infinitesimal, and that it may be expressed as a linear combination of $\{d\xi^1, \cdots, d\xi^n\}$. Then we have

$$\Pi_{p,p'}((\partial_j)_p) = (\partial_j)_{p'} - d\xi^i (\Gamma^k_{ij})_p (\partial_k)_{p'}, \qquad (1.27)$$

where $\{(\Gamma^k_{ij})_p; i,j,k = 1, \cdots, n\}$ are $n^3$ real numbers which depend on the point $p$.

If for each pair of neighboring points $p$ and $p'$ in $S$, there is defined a linear mapping $\Pi_{p,p'} : T_p \to T_{p'}$ of the form described in Equation (1.27), and if the $n^3$ functions $\Gamma^k_{ij} : p \mapsto (\Gamma^k_{ij})_p$ are all $C^\infty$, then we say that we have introduced an **affine connection** on $S$. In addition, we call $\{\Gamma^k_{ij}\}$ the **connection coefficients** of the affine connection with respect to the coordinate system $[\xi^i]$. Note that the only constraint on the connection coefficients are that they be $C^\infty$, and that therefore affine connections have this degree of freedom. Below, we often refer to affine connections as simply connections.

Let $[\rho^r] = [\rho^1, \cdots, \rho^n]$ be a coordinate system distinct from $[\xi^i]$, and let $\tilde{\partial}_r \overset{\text{def}}{=} \frac{\partial}{\partial \rho^r} = \frac{\partial \xi^i}{\partial \rho^r} \partial_i$. From Equation (1.27) and the linearity of $\Pi_{p,p'}$ we have

$$\Pi_{p,p'}((\tilde{\partial}_s)_p) = \left(\frac{\partial \xi^j}{\partial \rho^s}\right)_p \{(\partial_j)_{p'} - d\xi^i (\Gamma^k_{ij})_p (\partial_k)_{p'}\}.$$

By substituting into the right hand side of this equation

$$\left(\frac{\partial \xi^j}{\partial \rho^s}\right)_{p'} = \left(\frac{\partial \xi^j}{\partial \rho^s}\right)_p + \left(\frac{\partial^2 \xi^j}{\partial \rho^r \partial \rho^s}\right)_p d\rho^r \quad \text{and}$$

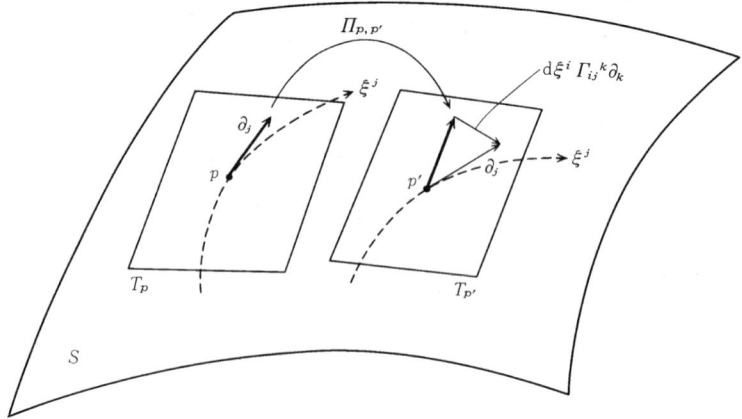

Figure 1.4: Affine connection (an infinitesimal translation)

$$d\xi^i = \left(\frac{\partial \xi^i}{\partial \rho^r}\right)_p d\rho^r \qquad \left(d\rho^r \stackrel{\text{def}}{=} \rho^r(p') - \rho^r(p)\right),$$

and ignoring second order infinitesimals, we obtain

$$\Pi_{p,p'}((\tilde{\partial}_s)_p) = (\tilde{\partial}_s)_{p'} - d\rho^r (\tilde{\Gamma}^t_{rs})_p (\tilde{\partial}_t)_{p'}, \tag{1.28}$$

where $(\tilde{\Gamma}^t_{rs})_p$ is the value of the function

$$\tilde{\Gamma}^t_{rs} = \left\{\Gamma^k_{ij} \frac{\partial \xi^i}{\partial \rho^r} \frac{\partial \xi^j}{\partial \rho^s} + \frac{\partial^2 \xi^k}{\partial \rho^r \partial \rho^s}\right\} \frac{\partial \rho^t}{\partial \xi^k} \tag{1.29}$$

at the point $p$. Note that Equations (1.27) and (1.28) are of the same form. Furthermore, if the functions $\Gamma^k_{ij}$ are $C^\infty$ for all $(i,j,k)$ then so are the functions $\tilde{\Gamma}^t_{rs}$ for all $(r,s,t)$. In other words, the notion of affine connections is independent of the choice of coordinate system. Their connection coefficients, however, are related according to Equation (1.29).

An affine connection determines, for neighboring points $p$ and $p'$, a correspondence between $T_p$ and $T_{p'}$. By connecting a sequence of such correspondences, we may find, for non-neighboring points $p$ and $q$, a correspondence between $T_p$ and $T_q$. This correspondence depends, however, on the curve $\gamma$ by which one connects $p$ and $q$. Let us define the notion of "translating tangent vectors along a curve" in the following way.

Let $\gamma : [a,b] \to S$, where $\gamma(a) = p$ and $\gamma(b) = q$, be a curve which connects points $p$ and $q$ in $S$. We call a mapping from each point $\gamma(t)$ to a tangent vector $X(t) \in T_{r(t)}$, say $X : t \mapsto X(t)$, a **vector field along** $\boldsymbol{\gamma}$. Given such a vector field $X$, if, for all $t \in [a,b]$ and the corresponding infinitesimal $dt$, the corresponding tangent vectors are linearly related as specified by the connection, i.e., if

$$X(t+dt) = \Pi_{\gamma(t),\gamma(t+dt)}(X(t)), \tag{1.30}$$

## 1.6. AFFINE CONNECTIONS AND COVARIANT DERIVATIVES

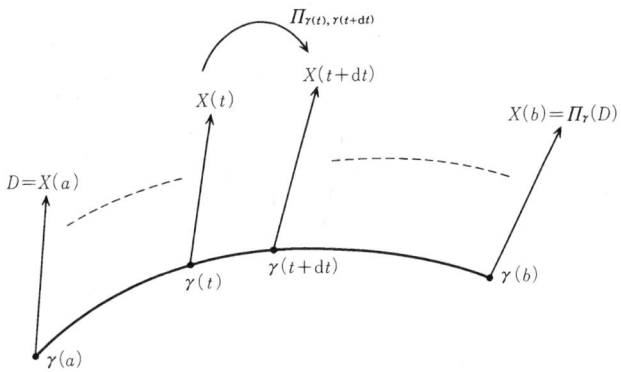

Figure 1.5: Translation of a tangent vector along a curve

then we say that $X$ is **parallel** along $\gamma$ (Figure 1.5).

Let us rewrite the equation above with respect to the coordinate system $[\xi^i]$. Letting $\partial_i = \frac{\partial}{\partial \xi^i}$, we have $X(t) = X^i(t)(\partial_i)_{\gamma(t)}$. From Equation (1.27) we have that

$$\Pi_{\gamma(t),\gamma(t+\mathrm{d}t)}(X(t)) = \left\{ X^k(t) - \mathrm{d}t\dot{\gamma}^i(t)X^j(t)(\Gamma^k_{ij})_{\gamma(t)} \right\}(\partial_k)_{\gamma(t+\mathrm{d}t)}, \quad (1.31)$$

where $\gamma^i \stackrel{\text{def}}{=} \xi^i \circ \gamma$, and $\dot{\gamma}^i(t)$ is its derivative with respect to $t$. Now since in addition, $X(t+\mathrm{d}t) = X^i(t+\mathrm{d}t)(\partial_i)_{\gamma(t+\mathrm{d}t)}$, substituting this into Equation (1.30) we obtain

$$\dot{X}^k(t) + \dot{\gamma}^i(t)X^j(t)(\Gamma^k_{ij})_{\gamma(t)} = 0, \quad (1.32)$$

where $\dot{X}^k(t) \stackrel{\text{def}}{=} \frac{\mathrm{d}X^k(t)}{\mathrm{d}t} = \frac{X^k(t+\mathrm{d}t)-X^k(t)}{\mathrm{d}t}$. Equation (1.32) is an ordinary linear differential equation on $X^1(t), \cdots, X^n(t)$, and hence given an initial (boundary) condition there exists a unique solution. From this, given $D \in T_{\gamma(a)} = T_p$, we see that there exists a unique parallel vector field along $\gamma$ such that $X(a) = D$. Then letting $\Pi_\gamma(D)$ denote the vector $X(b) \in T_{\gamma(b)} = T_q$ determined by $D$, we see that $\Pi_\gamma$ is a linear isomorphism from $T_p$ to $T_q$. We call $\Pi_\gamma$ the **parallel translation along $\gamma$**.

Let $\gamma : [a,b] \to S$ be a curve and $X$ be a vector field along $\gamma$. In general, $X(t)$ and $X(t+h)$ lie in different tangent spaces and hence it is not possible to consider the derivative $\frac{\mathrm{d}X(t)}{\mathrm{d}t} = \lim_{h \to 0} \frac{X(t+h)-X(t)}{h}$. However, if an affine connection is given on $S$, then the parallel translation of $X(t+h) \in T_{\gamma(t+h)}$ to the space $T_{\gamma(t)}$ along $\gamma$ gives us $X_t(t+h) = \Pi_{\gamma(t+h),\gamma(t)}(X(t+h))$, and using this we may consider within $T_{\gamma(t)}$ the quantity $\lim_{h \to 0} \frac{X_t(t+h)-X(t)}{h}$. We call this the **covariant derivative** of $X(t)$, and denote it by $\frac{\delta X(t)}{\mathrm{d}t}$. In other words, instead of $\mathrm{d}X(t) = X(t+\mathrm{d}t) - X(t)$, we use

$$\delta X(t) = \Pi_{\gamma(t+\mathrm{d}t),\gamma(t)}(X(t+\mathrm{d}t)) - X(t) \quad (1.33)$$

(see Figure 1.6).

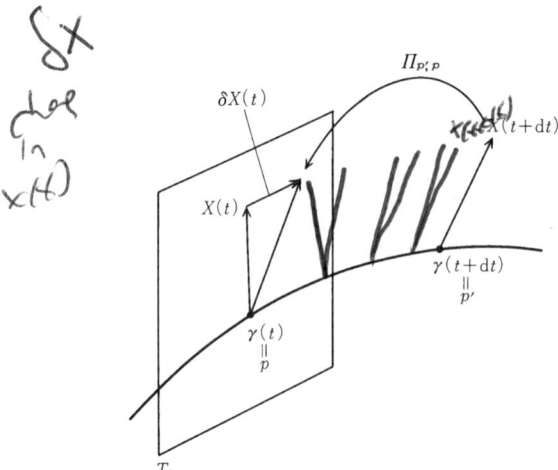

Figure 1.6: The covariant derivative along a curve.

Rewriting $X(t)$ as $X^j(t)(\partial_j)_{\gamma(t)}$, we have

$$\Pi_{\gamma(t+dt),\gamma(t)}(X(t+dt)) = \left\{ X^k(t+dt) + dt\,\dot{\gamma}^i(t) X^j(t) (\Gamma^k_{ij})_{\gamma(t)} \right\} (\partial_k)_{\gamma(t)}, \quad (1.34)$$

and substituting this into Equation (1.33), we obtain

$$\frac{\delta X(t)}{dt} = \left\{ \dot{X}^k(t) + \dot{\gamma}^i(t) X^j(t) (\Gamma^k_{ij})_{\gamma(t)} \right\} (\partial_k)_{\gamma(t)}. \quad (1.35)$$

This also forms a vector field along $\gamma$. In addition, we see that the parallel translation condition in Equation (1.32) may now be written simply as $\frac{\delta X}{dt} = 0$.

In this way, using an affine connection it is possible to define the infinitesimal $\delta X$ and the derivative $\frac{\delta X}{dt}$ of a vector field $X(t)$ along a curve. Extending this to "the directional derivative of a vector field $X = X^i \partial_i \in \mathcal{T}$ on $S$ along a tangent vector $D = D^i(\partial_i)_p \in T_p$" is straightforward as follows: consider a curve whose tangent vector at the point $p$ is $D$, and by taking the covariant derivative of $X$ along this curve we obtain

$$\nabla_D X = D^i \left\{ (\partial_i X^k)_p + X^j_p (\Gamma^k_{ij})_p \right\} (\partial_k)_p \in T_p(S). \quad (1.36)$$

In fact, letting $X_\gamma : t \mapsto X_{\gamma(t)}$ for an arbitrary curve $\gamma$, we have from Equations (1.35) and (1.36) that

$$\frac{\delta X_{\gamma(t)}}{dt} = \nabla_{\dot{\gamma}(t)} X. \quad (1.37)$$

We may also define for each $X, Y \in \mathcal{T}(S)$ the vector field $\nabla_X Y \in \mathcal{T}(S)$ by $(\nabla_X Y)_p = \nabla_{X_p} Y \in T_p(S)$. We call this the **covariant derivative** of $Y$ with respect to $X$. Given $X = X^i \partial_i$ and $Y = Y^i \partial_i$, we may write

$$\nabla_X Y = X^i \left\{ \partial_i Y^k + Y^j \Gamma^k_{ij} \right\} \partial_k. \quad (1.38)$$

## 1.7. FLATNESS

In particular, when $X = \partial_i$ and $Y = \partial_j$, we obtain the component expression of the covariant derivative

$$\nabla_{\partial_i}\partial_j = \Gamma_{ij}^k \partial_k. \tag{1.39}$$

This may be construed as the vector field which expresses the change in the basis vector $\partial_j$ as it is moved in the direction of $\partial_i$.

The operator $\nabla : \mathcal{T} \times \mathcal{T} \to \mathcal{T}$ which maps $(X,Y)$ to $\nabla_X Y$ satisfies the following properties: for arbitrary $X, Y, Z \in \mathcal{T}$ and $f \in \mathcal{F}$ (: the set of $C^\infty$ functions on $S$),

(i) $\nabla_{X+Y} Z = \nabla_X Z + \nabla_Y Z$.

(ii) $\nabla_X (Y + Z) = \nabla_X Y + \nabla_X Z$.

(iii) $\nabla_{fX} Y = f \nabla_X Y$.

(iv) $\nabla_X (fY) = f \nabla_X Y + (Xf)Y$.

Here, $Xf$ denotes the function $p \mapsto X_p f$ ($\in \mathcal{F}$). Note that $\nabla_X Y$ is $\mathcal{F}$-linear with respect to $X$, but not with respect to $Y$, and hence $\nabla$ is not a tensor field.

In fact, it is possible to consider the conditions (i)-(iv) as the defining properties of affine connections. In other words, we may define an affine connection on $S$ to be a mapping $\nabla : \mathcal{T}(S) \times \mathcal{T}(S) \to \mathcal{T}(S)$ which satisfies conditions (i)-(iv). In addition, we may define the connection coefficients $\{\Gamma_{ij}^k\}$ of $\nabla$ with respect to some coordinate system $[\xi^i]$ to be the $n^3$ functions determined by Equation (1.39). Then it is possible to prove Equations (1.38) and (1.29) from conditions (i)-(iv). It is also possible to reverse the derivation in Equations (1.32)-(1.37) to arrive at the definitions of $\frac{\delta X(t)}{dt}$ and $\Pi_\gamma$ from that of $\nabla$. This method would make the use of both infinitesimals and fiber bundles unnecessary. In this book, we shall often refer to the "connection $\nabla$".

Finally, we note that the totality of affine connections on a manifold forms an affine space. In other words, for any affine connections $\nabla$ and $\nabla'$ and for any real number $\alpha \in \mathbb{R}$, the affine combination $\alpha \nabla + (1-\alpha)\nabla'$ defines another affine connection. Note also that the difference of two affine connections is a tensor field of type $(1, 2)$.

## 1.7 Flatness

Let $X \in \mathcal{T}(S)$ be a vector field on $S$. If for any curve $\gamma$ on $S$, $X_\gamma : t \mapsto X_{\gamma(t)}$ is parallel along $\gamma$ (with respect to the connection $\nabla$), we say that $X$ is **parallel** on $S$ (with respect to $\nabla$.) In this case, for any curve $\gamma$ which connects points $p$ and $q$, $X_q = \Pi_\gamma(X_p)$ holds. A necessary and sufficient condition for an $X = X^i \partial_i$ to be parallel is that $\nabla_Y X = 0$ for all $Y \in \mathcal{T}(S)$, or equivalently that

$$\partial_i X^k + X^j \Gamma_{ij}^k = 0. \tag{1.40}$$

Note that nonzero parallel vector fields do not exist in general.

Let $[\xi^i]$ be a coordinate system of $S$, and suppose that with respect to this coordinate system the $n$ basis vector fields $\partial_i = \frac{\partial}{\partial \xi^i}$ ($i = 1, \cdots, n$) are all parallel on $S$. Then we call $[\xi^i]$ an **affine coordinate system** for $\nabla$. This condition is equivalent both to $\nabla_{\partial_i} \partial_j = 0$ and also to the condition that the connection coefficients $\{\Gamma_{ij}^k\}$ of $\nabla$ with respect to $[\xi^i]$ are all identically 0.

Given some connection, a corresponding affine coordinate system does not in general exist. If an affine coordinate system exists for connection $\nabla$, we say that $\nabla$ is **flat**, or alternatively that $S$ is flat with respect to $\nabla$. Let $[\xi^i]$ be an affine coordinate system. Then with respect to a different coordinate system $[\rho^r]$, we see from Equation (1.29) that the connection coefficients $\{\tilde{\Gamma}_{rs}^t\}$ may be written as $\tilde{\Gamma}_{rs}^t = \frac{\partial^2 \xi^k}{\partial \rho^r \partial \rho^s} \frac{\partial \rho^t}{\partial \xi^k}$. Hence a necessary and sufficient condition for $[\rho^r]$ to be another affine coordinate system is that $\frac{\partial^2 \xi^k}{\partial \rho^r \partial \rho^s} = 0$. This is equivalent to the condition that there exist an $n \times n$ matrix $A$ and an $n$-dimensional vector $B$ such that

$$\xi(p) = A\rho(p) + B \qquad (\forall p \in S) \tag{1.41}$$

($\xi(p) = [\xi^i(p)]$ and $\rho(p) = [\rho^r(p)]$.) We call a transformation of the form described in Equation (1.41) an **affine transformation** (when $B = 0$, this is simply a linear transformation). In addition, we see that this transformation is regular, i.e., one-to-one, and that $A$ is a regular matrix. The collection of such regular affine transformations form a group, and affine coordinate systems have this degree of freedom.

Let $\nabla$ be a connection on $S$. Then for vector fields $X, Y, Z \in \mathcal{T}$, if we define

$$R(X,Y)Z \stackrel{\text{def}}{=} \nabla_X(\nabla_Y Z) - \nabla_Y(\nabla_X Z) - \nabla_{[X,Y]}Z \quad \text{and} \tag{1.42}$$

$$T(X,Y) \stackrel{\text{def}}{=} \nabla_X Y - \nabla_Y X - [X,Y], \tag{1.43}$$

then these are also vector fields ($\in \mathcal{T}$). Here, letting $X = X^i \partial_i$ and $Y = Y^i \partial_i$, we have defined $[X,Y]$ to be the vector field

$$[X,Y] = (X^j \partial_j Y^i - Y^j \partial_j X^i) \partial_i$$

(this does not depend on the choice of coordinate system). The mappings $R : \mathcal{T} \times \mathcal{T} \times \mathcal{T} \to \mathcal{T}$ and $T : \mathcal{T} \times \mathcal{T} \to \mathcal{T}$ as defined above are both $\mathcal{F}$-multilinear. Hence $R$ and $T$ are respectively tensor fields of types $(1,3)$ and $(1,2)$. We call $R$ the **Riemann-Christoffel curvature tensor (field)** of $\nabla$, or more simply the **curvature tensor (field)**, and $T$ the **torsion tensor (field)** of $\nabla$. The component expressions of $R$ and $T$ with respect to coordinate system $[\xi^i]$ are given by

$$R(\partial_i, \partial_j)\partial_k = R_{ijk}^\ell \partial_\ell \quad \text{and} \quad T(\partial_i, \partial_j) = T_{ij}^k \partial_k \tag{1.44}$$

$\left(\partial_i \stackrel{\text{def}}{=} \frac{\partial}{\partial \xi^i}\right)$, and these may be computed in the following way:

$$\begin{aligned} R_{ijk}^\ell &= \partial_i \Gamma_{jk}^\ell - \partial_j \Gamma_{ik}^\ell + \Gamma_{ih}^\ell \Gamma_{jk}^h - \Gamma_{jh}^\ell \Gamma_{ik}^h \quad \text{and} \\ T_{ij}^k &= \Gamma_{ij}^k - \Gamma_{ji}^k. \end{aligned} \tag{1.45}$$
$$\tag{1.46}$$

If $[\xi^i]$ is an affine coordinate system for $\nabla$, then clearly $R^\ell_{ijk} = 0$ and $T^k_{ij} = 0$. In fact, in this case the components of $R$ and $T$, since they are tensors, are always all 0 with respect to any coordinate system. In other words, if $\nabla$ is flat, then $R = 0$ and $T = 0$. Conversely, if $R = 0$ and $T = 0$, it is known that $\nabla$ is locally flat in the following sense: for each point $p \in S$, there exists a neighborhood $U$ of $p$ such that $\nabla$ is flat on $U$. A proof will be found in standard textbooks of differential geometry.

In general, when $T = 0$ (i.e., $\Gamma^k_{ij} = \Gamma^k_{ji}$) holds, $\nabla$ is called a **symmetric connection** or **torsion-free connection**. The connections having appeared so far in information geometry are mostly symmetric connections. However, the incorporation of torsion into the framework of information geometry, which would relate it to such fields as quantum mechanics (noncommutative probability theory) and systems theory, is an interesting topic for the future. We will make an attempt in this direction in §7.3.

If a connection is flat, then parallel translation does not depend on the curve selected to connect the two points. In particular, the $n$ basis vector fields $\partial_i = \frac{\partial}{\partial \xi^i}$ ($i = 1, \cdots, n$) of an affine coordinate system $[\xi^i]$ are parallel vector fields, and hence $\Pi_\gamma((\partial_i)_p) = (\partial_i)_q$ regardless of the curve $\gamma$ used to connect the points $p$ and $q$. In addition, if the components $X^i$ of a vector field $X = X^i \partial_i$ are all constant on $S$, then $X$ is parallel, and $\Pi_\gamma(X_p) = X_q$.

In general, if parallel translation does not depend on curve choice, or in other words if there are $n$ linearly independent parallel vector fields on $S$ then $R = 0$, and in addition, when $S$ is simply connected (i.e., when arbitrary closed loops may be continuously contracted to a single point) it is known that the converse also holds.[2] There exist, however, connections for which $R = 0$ and $T \neq 0$. When this is the case, although parallel translation does not depend on the curve selected, there does not exist an affine coordinate system. Such spaces, called spaces of distant parallelism, were introduced by Einstein within the context of unified field theory, and also serve a major role within the theory of non-Riemannian plasticity. Another example will be shown in §7.3.

From Equations (1.45) and (1.46) we see that in general $R^\ell_{ijk} = -R^\ell_{jik}$ and $T^k_{ij} = -T^k_{ji}$. Hence, in the particular case when $S$ is 1-dimensional, $R = 0$ and $T = 0$ necessarily hold, and therefore $S$ is flat.

## 1.8  Autoparallel submanifolds

Let $S$ be an $n$-dimensional manifold and $M$ be an $m$-dimensional submanifold of $S$. Let $[\xi^i]$ and $[u^a]$ be coordinate systems for $S$ and $M$, respectively, and let $\partial_i = \frac{\partial}{\partial \xi^i}$ and $\partial_a = \frac{\partial}{\partial u^a}$. Suppose also that $\nabla$ is an affine connection on $S$ and that $\{\Gamma^k_{ij}\}$ are the connection coefficients of $\nabla$ with respect to $[\xi^i]$. Now letting $X = X^a \partial_a$ and $Y = Y^a \partial_a \in \mathcal{T}(M)$ be vector fields on $M$, we may consider $\nabla_{X_p} Y$, the "directional derivative of $Y$ along $X_p$", as we did in Equation (1.36). However, even though in general $\nabla_{X_p} Y$ is a tangent vector of $S$ ($\in T_p(S)$), it

---
[2] There are those who define "flat" to denote this case.

is not necessarily a tangent vector of $M$ ($\in T_p(M)$). If we let $\nabla_X Y$ denote the mapping from points $p$ in $M$ to $\nabla_{X_p} Y \in T_p(S)$, then using identities such as $\partial_a = (\partial_a \xi^i)\partial_i$ we have

$$\nabla_X Y = X^a(\partial_a Y^b)\partial_b + X^a Y^b \left\{ (\partial_a \xi^i)(\partial_b \xi^j)\Gamma_{ij}^k + \partial_a \partial_b \xi^k \right\} \partial_k. \qquad (1.47)$$

In particular, letting $X = \partial_a$ and $Y = \partial_b$, we obtain

$$\nabla_{\partial_a} \partial_b = \left\{ (\partial_a \xi^i)(\partial_b \xi^j)\Gamma_{ij}^k + \partial_a \partial_b \xi^k \right\} \partial_k. \qquad (1.48)$$

Note also that Equation (1.47) may be written as

$$\nabla_X Y = X^a(\partial_a Y^b)\partial_b + X^a Y^b \nabla_{\partial_a} \partial_b \qquad (1.49)$$

As we mentioned above, for $X, Y \in \mathcal{T}(M)$, $(\nabla_X Y)_p = \nabla_{X_p} Y$ is an element of $T_p(S)$, but not necessarily one of $T_p(M)$, i.e., in general, $\nabla_X Y \notin \mathcal{T}(M)$. If, however,

$$\nabla_X Y \in \mathcal{T}(M) \quad \text{for} \quad \forall X, Y \in \mathcal{T}(M), \qquad (1.50)$$

then $\nabla$ determines a covariant derivative on $M$. In fact, when this is the case the conditions (i)-(iv) from §1.6 hold for all $X, Y, Z \in \mathcal{T}(M)$ and all $f \in \mathcal{F}(M)$, and $\nabla$ is an affine connection on $M$. If we use this connection to define a parallel translation $\Pi_\gamma^M : T_{\gamma(a)}(M) \to T_{\gamma(b)}(M)$ on $M$ along the curves $\gamma : [a, b] \to M$, then this translation coincides exactly with the parallel translation $\Pi_\gamma : T_{\gamma(a)}(S) \to T_{\gamma(b)}(S)$ on $S$ restricted to the tangent spaces of $M$, using the original connection on $S$. In other words

$$\Pi_\gamma^M = \Pi_\gamma |_{T_{\gamma(a)}(M)}. \qquad (1.51)$$

If a submanifold $M$ of $S$ satisfies Equation (1.50), we say that $M$ is **autoparallel** with respect to $\nabla$. In particular, open subsets of $S$ are autoparallel. From Equation (1.49) we see that a necessary and sufficient condition for $M$ to be autoparallel is that $\nabla_{\partial_a} \partial_b \in \mathcal{T}(M)$ holds for all $a, b$. This, in turn, is equivalent to there existing $m^3$ functions $\{\Gamma_{ab}^c\}$ ($\in \mathcal{F}(M)$) which satisfy

$$\nabla_{\partial_a} \partial_b = \Gamma_{ab}^c \partial_c. \qquad (1.52)$$

These $\{\Gamma_{ab}^c\}$ form the connection coefficients of $\nabla$ with respect to $[u^a]$. Using Equation (1.48) we may rewrite Equation (1.52) in the following way:

$$\Gamma_{ab}^c \partial_c \xi^k = (\partial_a \xi^i)(\partial_b \xi^j)\Gamma_{ij}^k + \partial_a \partial_b \xi^k. \qquad (1.53)$$

We can also see that $M$ is autoparallel in $S$ if and only if $M$ is closed with respect to the parallel translation on $S$ in the following sense: for every curve $\gamma : [a, b] \to M$ in $M$ and for every tangent vector $D$ of $M$ at $\gamma(a)$, the result $\Pi_\gamma(D)$ of the parallel translation $\Pi_\gamma : T_{\gamma(a)}(S) \to T_{\gamma(b)}(S)$ belongs to the tangent space of $M$ at $\gamma(b)$.

## 1.8. AUTOPARALLEL SUBMANIFOLDS

1-dimensional autoparallel submanifolds are called **autoparallel curves** or **geodesics**. For a curve $\gamma : t \mapsto \gamma(t)$, the condition in Equation (1.52) may be rewritten using Equation (1.37) as

$$\frac{\delta}{dt}\frac{d\gamma}{dt} = \Gamma(t)\frac{d\gamma}{dt}, \qquad (1.54)$$

where $\Gamma : t \mapsto \Gamma(t)$ is a $C^\infty$ function. As we noted at the end of §1.7, connections on 1-dimensional manifolds are necessarily flat, and hence by substituting into Equation (1.54) a suitable one-to-one transformation (change of variable) of $t$, we may obtain $\Gamma(t) \equiv 0$. We call such a $t$ an **affine parameter** of $\gamma$. In this case Equation (1.54) reduces to

$$\frac{\delta}{dt}\frac{d\gamma}{dt} = 0, \qquad (1.55)$$

and implies that $\frac{d\gamma}{dt}$ is parallel along $\gamma$. It is possible to define geodesics using Equation (1.55). Rewriting Equation (1.55) using the coordinate system $[\xi^i]$ and the corresponding representation $\gamma^i = \xi^i \circ \gamma$, we obtain

$$\ddot{\gamma}(t) + \dot{\gamma}^i(t)\dot{\gamma}^j(t)(\Gamma_{ij}^k)_{\gamma(t)} = 0. \qquad (1.56)$$

Let $M$ be an autoparallel submanifold of $S$. If the torsion tensor of $S$ is 0, then the torsion tensor of $M$ is also 0. This is clear from Equations (1.46) and (1.53). The same holds for the curvature tensor. The latter fact may be derived using Equations (1.45) and (1.53), but it is in fact immediate from the analysis of parallel translation as follows: from Equation (1.51) we see that if the choice of curve does not affect parallel translation in $S$, then it similarly does not in $M$. Note that, in the case when parallel translation does not depend on curve choice, a necessary and sufficient condition for a submanifold $M$ to be autoparallel in $S$ is that there exist $m = (\dim M)$ linearly independent vector fields on $M$ which are parallel with respect to the connection on $S$.

Consider the case when $S$ is flat with respect to $\nabla$. Then by the argument above autoparallel submanifolds of $S$ are also flat. Hence without loss of generality we may assume that $[\xi^i]$ and $[u^a]$ are affine coordinate systems in Equation (1.53), the condition for a submanifold $M$ of $S$ to be autoparallel. Equation (1.53) then reduces to $\partial_a \partial_b \xi^k = 0$. This condition is equivalent to there existing an $n \times m$ matrix $A$ and an $n$-dimensional vector $B$ which satisfies

$$\xi(p) = Au(p) + B \qquad (\forall p \in M) \qquad (1.57)$$

($\xi(p) = [\xi^i(p)]$ and $u(p) = [u^a(p)]$.) In general, a subspace of $\mathbb{R}^n$ which may be expressed as $\{Au + B | u \in \mathbb{R}^m\}$ is called an **affine subspace** of $\mathbb{R}^n$ (; when $B = 0$ we have a linear subspace). We summarize the discussion above in the following theorem.

**Theorem 1.1** *If $S$ is flat, then a necessary and sufficient condition for a submanifold $M$ to be autoparallel is that $M$ is expressed as an affine subspace (or* an open subset of an affine subspace) of $S$ with respect to an affine coordinate system. In particular, geodesics may be expressed using linear equations (as a line or a segment) with respect to affine coordinate systems. In addition, if $M$ is autoparallel, then it is also flat.

## 1.9 Projection of connections and embedding curvature

If $M$ is a submanifold of $S$ which is not autoparallel with respect to $\nabla$ on $S$, then there is no natural connection on $M$ which may be derived from $\nabla$. However, if there is for each point $p$ a mapping $\pi_p$ from $T_p(S)$ to $T_p(M)$, then we may use this to define a connection on $M$. Assume that $\pi_p : T_p(S) \to T_p(M)$ is a linear mapping and that $\pi_p(D) = D$ for every $D \in T_p(M)$, and that the relation $p \mapsto \pi_p$ is $C^\infty$. Now suppose, for each $X, Y \in \mathcal{T}(M)$, we define $\nabla_X^{(\pi)} \in \mathcal{T}(M)$ in the following way:

$$(\nabla_X^{(\pi)} Y)_p = \pi_p((\nabla_X Y)_p) \qquad (\forall p \in M). \tag{1.58}$$

Then $\nabla^{(\pi)}$ is a connection on $M$. In particular, if a Riemannian metric $g = \langle \, , \, \rangle$ is given on $S$, we may take as $\pi_p$ the orthogonal projection with respect to $g$. This is defined to be that which satisfies, for all $D \in T_p(S)$ and all $D' \in T_p(M)$,

$$\langle \pi_p(D), D' \rangle_p = \langle D, D' \rangle_p. \tag{1.59}$$

We call such $\nabla^{(\pi)}$ the **projection of $\nabla$ onto $M$ with respect to $g$**.

If $S$ has a coordinate system $[\xi^i]$, then the connection coefficients $\{\Gamma_{ij}^k\}$ of $\nabla$ are determined by Equation (1.39). If $S$ also has a Riemannian metric $g$, then we may define $n^3$ additional functions $\{\Gamma_{ij,k}\}$ in the following way:

$$\Gamma_{ij,k} \overset{\text{def}}{=} \langle \nabla_{\partial_i} \partial_j, \partial_k \rangle = \Gamma_{ij}^h g_{hk}. \tag{1.60}$$

The quantities $\{\Gamma_{ij,k}\}$, like $\{\Gamma_{ij}^k\}$, may be considered as a different component expression of the same $\nabla$. With respect to a different coordinate system $[\rho^r]$ for $S$, these may be written as follows $\left(\tilde{\partial}_r \overset{\text{def}}{=} \frac{\partial}{\partial \rho^r}\right)$:

$$\tilde{\Gamma}_{rs,t} \overset{\text{def}}{=} \langle \nabla_{\tilde{\partial}_r} \tilde{\partial}_s, \tilde{\partial}_t \rangle = \left( \frac{\partial \xi^i}{\partial \rho^r} \frac{\partial \xi^j}{\partial \rho^s} \Gamma_{ij,k} + \frac{\partial^2 \xi^h}{\partial \rho^r \partial \rho^s} g_{hk} \right) \frac{\partial \xi^k}{\partial \rho^t}. \tag{1.61}$$

Similarly, for the projection $\nabla^{(\pi)}$ of $\nabla$ onto $M$, we may define, given a coordinate system $[u^a]$ for $M$, $\Gamma_{ab,c}^{(\pi)} \overset{\text{def}}{=} \langle \nabla_{\partial_a}^{(\pi)} \partial_b, \partial_c \rangle$ $\left( \partial_a \overset{\text{def}}{=} \frac{\partial}{\partial u^a} \right)$. Using Equations (1.58), (1.59) and (1.48) we may rewrite this as

$$\Gamma_{ab,c}^{(\pi)} = \langle \Gamma_{\partial_a} \partial_b, \partial_c \rangle = \left\{ (\partial_a \xi^i)(\partial_b \xi^j) \Gamma_{ij,k} + (\partial_a \partial_b \xi^j) g_{jk} \right\} (\partial_c \xi^k). \tag{1.62}$$

## 1.10. RIEMANNIAN CONNECTION

The connection coefficients of $\nabla^{(\pi)}$ are then given by $\Gamma^{(\pi)d}_{ab} = \Gamma^{(\pi)}_{ab,c} g^{cd}$. From this, we see that if $\nabla$ is symmetric, then so is $\nabla^{(\pi)}$.

Now let
$$H(X,Y) \stackrel{\text{def}}{=} \nabla_X Y - \nabla_X^{(\pi)} Y \tag{1.63}$$
for $X, Y \in \mathcal{T}(M)$. Then $(H(X,Y))_p = (\nabla_X Y)_p - \pi_p((\nabla_X Y)_p)$ is the orthogonal projection of $(\nabla_X Y)_p$ onto $[T_p(M)]^\perp$, the orthocomplement of $T_p(M)$. Given this, note that the autoparallel condition for $M$ in Equation (1.50) is equivalent to stating that $H(X,Y) = 0$ holds for all $X, Y \in \mathcal{T}(M)$, and that this, in turn, is equivalent to simply stating that $H = 0$. Intuitively, $H$ may be considered as measuring the degree to which $M$ is "not autoparallel" or "curved" in $S$. In addition, since $H(X,Y)$ is $\mathcal{F}(M)$-linear with respect to both $X$ and $Y$ (i.e., is $\mathcal{F}(M)$-bilinear), $H$ may be considered as "a kind of" tensor field, even though $H(X,Y)$ is not a vector field on $M$ in general. We call such an $H$ an **embedding curvature** of the submanifold $M$ ($\subset S$) with respect to $\nabla$.

Since $M$ has $\nabla^{(\pi)}$ as a connection, we may use this to compute its Riemann-Christoffel curvature $R^{(\pi)}$. This $R^{(\pi)}$ expresses the "inherent curvature" of $M$ itself, while the embedding curvature $H$ expresses the curvature of the arrangement of $M$ within $S$. As we noted in §1.8, if $R$, the Riemann-Christoffel curvature of $S$, is 0, and if, in addition, $H = 0$ (i.e., $M$ is autoparallel), then $R^{(\pi)} = 0$ also. However, $R^{(\pi)} = 0$ does not entail $H = 0$. For example, consider a cylinder surface $M$ embedded within a 3-dimensional Euclidean space. The 2-dimensional geometry on the surface of this cylinder is Euclidean, and $R^{(\pi)} = 0$. However, within the 3-dimensional space it is curved, and hence $H$ is not 0. It is important to distinguish these two notions of curvature.

For each point $p$ in $S$, let $\{(\partial_a)_p; 1 \leq a \leq m\}$ ($m = \dim M$) be a basis for $T_p(M)$, and let $\{(\partial_\kappa)_p; m+1 \leq \kappa \leq n\}$ ($n = \dim S$) be a basis for $[T_p(M)]^\perp$. Then we may define the $m^2(n-m)$ functions $\{H_{ab\kappa}\}$ in the following way:

$$H_{ab\kappa} \stackrel{\text{def}}{=} \langle H(\partial_a, \partial_b), \partial_\kappa \rangle = \langle \nabla_{\partial_a} \partial_b, \partial_\kappa \rangle. \tag{1.64}$$

It follows from the properties of tensors that $H = 0 \Leftrightarrow H_{ab\kappa} = 0$ ($\forall a, b, \kappa$).

## 1.10 Riemannian connection

Let $\nabla$ be an affine connection on a Riemannian manifold $(S, g = \langle\,,\,\rangle)$, and suppose $\nabla$ satisfies, for all vector fields $X, Y, Z \in \mathcal{T}(S)$,

$$Z \langle X, Y \rangle = \langle \nabla_Z X, Y \rangle + \langle X, \nabla_Z Y \rangle. \tag{1.65}$$

Then we say that $\nabla$ is a **metric connection** with respect to $g$. Using the coordinate expressions of $g$ and $\nabla$ we may rewrite this condition as follows:

$$\partial_k g_{ij} = \Gamma_{ki,j} + \Gamma_{kj,i}. \tag{1.66}$$

Let us show that, under a metric connection, the parallel translation of two vectors leaves their inner product unchanged. Consider a curve $\gamma : t \mapsto \gamma(t)$

on $S$ and two vector fields $X$ and $Y$ along $\gamma$. Letting $\frac{\delta X}{dt}$ and $\frac{\delta Y}{dt}$ respectively denote the covariant derivatives of $X$ and $Y$ with respect to $\nabla$, we see from Equation (1.65) that

$$\frac{d}{dt}\langle X(t), Y(t)\rangle = \left\langle \frac{\delta X(t)}{dt}, Y(t)\right\rangle + \left\langle X(t), \frac{\delta Y(t)}{dt}\right\rangle. \tag{1.67}$$

Now if $X$ and $Y$ are both parallel on $\gamma$ (i.e., $\frac{\delta X}{dt} = \frac{\delta Y}{dt} = 0$), then the right hand side of the equation above is 0, and hence $\langle X(t), Y(t)\rangle$ does not depend on $t$ and is constant. The parallel translation $\Pi_\gamma$ along $\gamma$, then, is a metric isomorphism which preserves inner products. In other words, letting $p$ and $q$ be the boundary points of $\gamma$, for any two tangent vectors $D_1, D_2 \in T_p$ the following holds:

$$\langle \Pi_\gamma(D_1), \Pi_\gamma(D_2)\rangle_q = \langle D_1, D_2\rangle_p. \tag{1.68}$$

We call a connection which is both metric and symmetric the **Riemannian connection** or the **Levi-Civita connection** with respect to $g$. For a given $g$, such a connection exists uniquely. In fact, combining Equation (1.66) with the requirement that $\Gamma_{ij,k} = \Gamma_{ji,k}$, we have

$$\Gamma_{ij,k} = \frac{1}{2}\left(\partial_i g_{jk} + \partial_j g_{ki} - \partial_k g_{ij}\right). \tag{1.69}$$

The geodesics with respect to the Riemannian connection $\nabla$ are known to (locally) coincide with the shortest curve joining two points (where we measure length according to Equation (1.25).) In addition, if we consider the case when $\nabla$ is flat and there exists an affine coordinate system $[\xi^i]$, we find that since $\partial_i = \frac{\partial}{\partial \xi^i}$ is parallel on $S$, $\langle \partial_i, \partial_j\rangle$ is constant on $S$. Since affine coordinate systems have a degree of freedom as given in Equation (1.41), we see in particular that there exists an affine coordinate system which satisfies

$$\langle \partial_i, \partial_j\rangle = \delta_{ij}. \tag{1.70}$$

A coordinate system which satisfies the equation above is called a **Euclidean coordinate system** (with respect to $g$). Hence the Riemannian connection is flat if and only if there exists a Euclidean coordinate system.

In most differential geometry textbooks, only Riemannian connections are introduced on Riemannian manifolds. Non-metric connections are not even discussed. However, when considering families of probability distributions as manifolds, we find that the natural connections which one would introduce are non-metric (see §2.3). As we shall discuss in Chapter 3, this leads us to the novel notion of dual connections.

# Chapter 2

# The geometric structure of statistical models

Information geometry began as the geometric study of statistical estimation. This involved viewing the set of probability distributions which constitute a statistical model as a manifold, and analyzing the relationship between the geometric structure of this manifold and statistical estimation using this model. In this chapter, we show that a Riemannian metric and a family of affine connections are naturally introduced on such a manifold, and analyze their fundamental properties. Investigation of deeper properties resulting from a duality which underlies these structures is postponed till the next chapter.

## 2.1 Statistical models

In this book, we shall represent **probability distributions** on a set $\mathcal{X}$ using functions defined in the following way. If $\mathcal{X}$ is a discrete set (with finite or countably infinite cardinality), then by a probability distribution on $\mathcal{X}$ we mean a function $p : \mathcal{X} \to \mathbb{R}$ which satisfies

$$p(x) \geq 0 \quad (\forall x \in \mathcal{X}) \quad \text{and} \quad \sum_{x \in \mathcal{X}} p(x) = 1 \tag{2.1}$$

($p$ is also referred to as a probability function). If $\mathcal{X} = \mathbb{R}^n$, then we mean a function $p : \mathcal{X} \to \mathbb{R}$ which satisfies

$$p(x) \geq 0 \quad (\forall x \in \mathcal{X}) \quad \text{and} \quad \int p(x) \mathrm{d}x = 1. \tag{2.2}$$

In other words, $p$ is a probability density function on $\mathcal{X}$. Here the domain of integration is the entire set $\mathcal{X}$, and when $n \geq 2$, $\int$ denotes a multiple integral. As a mathematical note, let us mention that in general, what we are considering may be viewed as the density function $p = \frac{\mathrm{d}P}{\mathrm{d}\nu} : \mathcal{X} \to \mathbb{R}$ (the Radon-Nikodym derivative), where $\nu$ is a $\sigma$-finite measure on a measurable space $(\mathcal{X}, \mathcal{B})$ with $\mathcal{B}$ being a completely additive class (i.e., Borel field) consisting of $\mathcal{X}$ and its subsets, and $P$ is a probability measure on $(\mathcal{X}, \mathcal{B})$ which is absolutely continuous with respect to $\nu$.

In the discussions below, we shall focus our attention on the cases most significant from the point of view of applications, namely the cases given in Equation (2.1) and (2.2). However, most of the analysis may be carried over to the general case of $(\mathcal{X}, \mathcal{B}, \nu)$. When we wish to consider the two cases in Equation (2.1) and (2.2) in a unified manner, we shall use the (integral) expression of Equation (2.2). The results for the discrete case may be obtained by simply replacing occurrences of the integral $\int \cdots \mathrm{d}x$ with the sum $\sum_{x \in \mathcal{X}} \cdots$.

Consider a family $S$ of probability distributions on $\mathcal{X}$. Suppose each element of $S$, a probability distribution, may be parameterized using $n$ real-valued variables $[\xi^1, \cdots, \xi^n]$ so that

$$S = \left\{ p_\xi = p(x; \xi) \,\middle|\, \xi = [\xi^1, \cdots, \xi^n] \in \Xi \right\}, \tag{2.3}$$

where $\Xi$ is a subset of $\mathbb{R}^n$ and the mapping $\xi \mapsto p_\xi$ is injective. We call such $S$ an $n$-dimensional **statistical model**, a **parametric model**, or simply a **model** on $\mathcal{X}$. We will often abbreviate Equation (2.3) as $S = \{p_\xi\}$, and also use expression such as $p_\xi(x) = p(x; \xi)$ and $S = \{p(x; \xi)\}$. When we say "a statistical model $S = \{p_\xi\}$," there shall be cases in which we refer simply to the set $S$, and other cases in which we refer in addition to the parameterization $\xi \mapsto p_\xi$. The intended meaning should be clear from the context.

Suppose we wish to estimate the underlying probability distribution which has produced the observations $x_1, \cdots, x_N$. Often what is first done is to establish a statistical model $S$ from which to select the possible candidates. In doing so, we assume that there exists some distribution $p^*$ which governs the generation of the data, and that we may consider the observed data to be the results of sampling a random variable governed by this distribution. We call this $p^*$ the **underlying distribution** or the **true distribution**. Although $p^*$ is unknown, it is often possible to use prior knowledge concerning the data to determine the "shape" of $p^*$. This "shape" typically contains several free parameters, and in order to specify a particular distribution, the values taken by these parameters must be given. Equation (2.3) is a general expression of this notion.

Let us now state several assumptions we make concerning statistical models $S = \{p_\xi \,|\, \xi \in \Xi\}$. So that we may freely differentiate with respect to the parameters, we assume that $\Xi$ is an open subset of $\mathbb{R}^n$ and that for each $x \in \mathcal{X}$, the function $\xi \mapsto p(x; \xi)$ $(\Xi \to \mathbb{R})$ is $C^\infty$. This allows such expressions as $\partial_i p(x; \xi)$ and $\partial_i \partial_j p(x; \xi)$ to be defined $\left( \partial_i \stackrel{\text{def}}{=} \frac{\partial}{\partial \xi^i} \right)$. In addition, we assume that the order of integration and differentiation may be freely rearranged. For example, we shall often use formulas such as

$$\int \partial_i p(x; \xi) \, \mathrm{d}x = \partial_i \int p(x; \xi) \, \mathrm{d}x = \partial_i 1 = 0. \tag{2.4}$$

For a probability distribution $p$ on $\mathcal{X}$, let $\mathrm{supp}(p) \stackrel{\text{def}}{=} \{x | p(x) > 0\}$ (the

## 2.1. STATISTICAL MODELS

support of $p$). The case when $\mathrm{supp}(p_\xi)$ varies with $\xi$ poses rather significant difficulties for analysis, and hence we shall only consider the case when $\mathrm{supp}(p_\xi)$ is constant with respect to $\xi$ below. Letting $\mathcal{X}$ be redefined as $\mathrm{supp}(p)$, this is equivalent to assuming that $p(x;\xi) > 0$ holds for all $\xi \in \Xi$ and all $x \in \mathcal{X}$. This means that the model $S$ is a subset of

$$\mathcal{P}(\mathcal{X}) \stackrel{\mathrm{def}}{=} \left\{ p : \mathcal{X} \to \mathbb{R} \,\bigg|\, p(x) > 0 (\forall x \in \mathcal{X}), \int p(x)\,\mathrm{d}x = 1 \right\}. \quad (2.5)$$

We give as examples several standard statistical models.

**Example 2.1 (Normal Distribution)**

$$\mathcal{X} = \mathbb{R}, \quad n = 2, \quad \xi = [\mu, \sigma], \quad \Xi = \{[\mu,\sigma] \mid -\infty < \mu < \infty, 0 < \sigma < \infty\}$$

$$p(x;\xi) = \frac{1}{\sqrt{2\pi}\sigma} \exp\left\{-\frac{(x-\mu)^2}{2\sigma^2}\right\}$$

**Example 2.2 (Multivariate Normal Distribution)**

$$\mathcal{X} = \mathbb{R}^k, \quad n = k + \frac{k(k+1)}{2}, \quad \xi = [\mu, \Sigma]$$

$$\Xi = \{[\mu, \Sigma] \mid \mu \in \mathbb{R}^k, \Sigma \in \mathbb{R}^{k \times k} : \text{positive definite}\}$$

$$p(x;\xi) = (2\pi)^{-k/2}(\det \Sigma)^{-1/2} \exp\left\{-\frac{1}{2}(x-\mu)^t \Sigma^{-1}(x-\mu)\right\}$$

**Example 2.3 (Poisson Distribution)**

$$\mathcal{X} = \{0, 1, 2, \cdots\}, \quad n = 1, \quad \Xi = \{\xi \mid \xi > 0\}$$

$$p(x;\xi) = e^{-\xi} \frac{\xi^x}{x!}$$

**Example 2.4 ($\mathcal{P}(\mathcal{X})$ for finite $\mathcal{X}$)**

$$\mathcal{X} = \{x_0, x_1, \cdots, x_n\}, \quad \Xi = \left\{ [\xi^1, \cdots, \xi^n] \,\bigg|\, \xi^i > 0\ (\forall i), \sum_{i=1}^n \xi^i < 1 \right\}$$

$$p(x_i;\xi) = \begin{cases} \xi^i & (1 \leq i \leq n) \\ 1 - \sum_{i=1}^n \xi^i & (i = 0) \end{cases}$$

Given a statistical model $S = \{p_\xi \mid \xi \in \Xi\}$, the mapping $\varphi : S \to \mathbb{R}^n$ defined by $\varphi(p_\xi) = \xi$ allows us to consider $\varphi = [\xi^i]$ as a coordinate system for $S$. In addition, suppose we have a $C^\infty$ diffeomorphism $\psi$ from $\Xi$ to $\psi(\Xi)$, the latter being an open subset of $\mathbb{R}^n$. In other words, suppose that $\psi$ is one-to-one, and that both $\psi$ and $\psi^{-1}$ are $C^\infty$. Then if we use $\rho = \psi(\xi)$ instead of $\xi$ as our parameters, we obtain $S = \{p_{\psi^{-1}(\rho)} \mid \rho \in \psi(\Xi)\}$. This expresses the same family of probability distributions as $S = \{p_\xi\}$.

If we consider parameterizations which are $C^\infty$ diffeomorphic to each other to be equivalent, then we may consider $S$ as a $C^\infty$ differentiable manifold (we may call $S$ a **statistical manifold**). In this case, a parameterization of $S$ is in fact also a coordinate system of $S$. Below, we shall often conflate the distribution $p_\xi$ and the coordinate $\xi$, and shall use phrases such as "the point $\xi$" and "the tangent space $T_\xi(S)$."

## 2.2 The Fisher metric

Let $S = \{p_\xi \mid \xi \in \Xi\}$ be an $n$-dimensional statistical model. Given a point $\xi$ ($\in \Xi$), the **Fisher information matrix** of $S$ at $\xi$ is the $n \times n$ matrix $G(\xi) = [g_{ij}(\xi)]$, where the $(i,j)^{\text{th}}$ element $g_{ij}(\xi)$ is defined by the equation below; in particular, when $n = 1$, we call this the **Fisher information**.

$$g_{ij}(\xi) \stackrel{\text{def}}{=} E_\xi[\partial_i \ell_\xi \partial_j \ell_\xi] = \int \partial_i \ell(x;\xi) \partial_j \ell(x;\xi) p(x;\xi) \mathrm{d}x, \qquad (2.6)$$

where $\partial_i \stackrel{\text{def}}{=} \frac{\partial}{\partial \xi^i}$,

$$\ell_\xi(x) = \ell(x;\xi) = \log p(x;\xi), \qquad (2.7)$$

(log denotes the natural logarithm), and $E_\xi$ denotes the expectation with respect to the distribution $p_\xi$, $E_\xi[f] \stackrel{\text{def}}{=} \int f(x) p(x;\xi) \mathrm{d}x$. Although there are some models for which the integral in Equation (2.6) diverges, we assume in our discussion below that $g_{ij}(\xi)$ is finite for all $\xi$ and all $i, j$, and that $g_{ij} : \Xi \to \mathbb{R}$ is $C^\infty$. We note that it is possible to write $g_{ij}$ as

$$g_{ij}(\xi) = -E_\xi[\partial_i \partial_j \ell_\xi]. \qquad (2.8)$$

This may be derived by rewriting Equation (2.4) as

$$E_\xi[\partial_i \ell_\xi] = 0 \qquad (2.9)$$

and applying $\partial_j$ to both sides. Another important representation is

$$g_{ij}(\xi) = 4 \int \partial_i \sqrt{p(x;\xi)} \, \partial_j \sqrt{p(x;\xi)} \, \mathrm{d}x, \qquad (2.10)$$

which will be revisited in §2.5.

The matrix $G(\xi)$ is symmetric ($g_{ij}(\xi) = g_{ji}(\xi)$), and since for any $n$-dimensional vector $c = [c^1, \cdots, c^n]^t$ ($^t$ denotes transpose),

$$c^t G(\xi) c = c^i c^j g_{ij}(\xi) = \int \left\{ \sum_{i=1}^n c^i \partial_i \ell(x;\xi) \right\}^2 p(x;\xi) \mathrm{d}x \geq 0, \qquad (2.11)$$

## 2.2. THE FISHER METRIC

it is also positive semidefinite. We assume further that $G$ is positive definite. From the equation above, we see that this is equivalent to stating that the elements of $\{\partial_1 \ell_\xi, \cdots, \partial_n \ell_\xi\}$ when viewed as functions on $\mathcal{X}$ are linearly independent, which, in turn, is equivalent to stating that the elements of $\{\partial_1 p_\xi, \cdots, \partial_n p_\xi\}$ are linearly independent.

As we noted in Example 2.4, when $\mathcal{X}$ is a finite set we may consider $\mathcal{P}(\mathcal{X})$ itself to be a statistical model which forms a $(|\mathcal{X}|-1)$-dimensional manifold ($|\mathcal{X}|$ denotes the cardinality of $\mathcal{X}$). In this case, the various assumptions we have been making concerning the model $S$ on $\mathcal{X}$ may be considered as formalizing the statement "$S$ is a submanifold of $\mathcal{P}(\mathcal{X})$." When $\mathcal{X}$ is an infinite set, however, this statement is not meaningful since it is not possible to view $\mathcal{P}(\mathcal{X})$ as a manifold; nevertheless, a similar intuition may be said to hold for this case also.

Now suppose that the assumptions above hold, and define the inner product of the natural basis of the coordinate system $[\xi^i]$ by $g_{ij} = \langle \partial_i, \partial_j \rangle$. This uniquely determines a Riemannian metric $g = \langle\;,\;\rangle$. We call this the **Fisher metric**, or alternatively, the **information metric**. Since $g_{ij}$ defined by Equation (2.6) behaves according to Equation (1.22) under coordinate transformations, we see that the Fisher metric is invariant over the choice of coordinate system. Indeed we may write $\langle X, Y \rangle_\xi = E_\xi [(X\ell)(Y\ell)]$ for all tangent vectors $X, Y \in T_\xi(S)$.

In order to grasp the nature of this metric, and to make preparations for later arguments as well, let us review some fundamental results in statistics concerning the Fisher information matrix.

Let $F : \mathcal{X} \to \mathcal{Y}$ be a mapping which transforms the value of random variable $X$ to $Y = F(X)$. Then, given the distribution $p(x; \xi)$ of $X$, this determines the distribution $q(y; \xi)$ governing $Y$. In addition, letting

$$r(x; \xi) = \frac{p(x; \xi)}{q(F(x); \xi)}, \qquad (2.12)$$

$$p(x \mid y; \xi) = r(x; \xi) \delta_{F(x)}(y) \quad \text{and} \qquad (2.13)$$

$$\Pr(A \mid y; \xi) = \int_A p(x \mid y; \xi) \, dx, \quad A \subset \mathcal{X}, \qquad (2.14)$$

where $\delta_{F(x)}$ is the delta function on $(\mathcal{Y}, dy)$ concentrated on the point $F(x)$, we have for any $B \subset \mathcal{Y}$

$$\int_{A \cap F^{-1}(B)} p(x; \xi) \, dx = \int_B \Pr(A \mid y; \xi) q(y; \xi) \, dy. \qquad (2.15)$$

This means that $\Pr(A \mid y; \xi)$ is the conditional probability of the event $\{X \in A\}$ given $Y = y$. If the value $\Pr(A \mid y; \xi)$ does not depend on $\xi$ for all $A$ and $y$, or equivalently, if $r(x; \xi)$ does not depend on $\xi$ for all $x$, then we say that $F$ is a **sufficient statistic** for the model $S$. One-to-one mappings are trivial examples of sufficient statistics. More important examples will be given later in connection with exponential families (§4.2).

If $F$ is a sufficient statistic, then we may rewrite Equation (2.12) as

$$p(x; \xi) = q(F(x); \xi) \, r(x). \qquad (2.16)$$

Then the portion of the distribution $p(x;\xi)$ which depends on $\xi$ is wholly contained within the distribution $q(y;\xi)$ of $Y = F(X)$, and hence in order to estimate the unknown parameter $\xi$ (the unknown distribution $p(x;\xi)$), it suffices to know the value of $Y$. (Indeed, given the value of $Y$, we can "simulate" $X$ by using the random number generator $p(x \mid y)$ which does not depend on $\xi$.) This is reason for its name. The following fact, known as the factorization theorem, gives a simple criterion for sufficiency: $F$ is a sufficient statistic if and only if there exist some functions $s : \mathcal{Y} \times \Xi \to \mathbb{R}$ and $t : \mathcal{X} \to \mathbb{R}$ such that for all $x$ and $\xi$

$$p(x;\xi) = s(F(x);\xi) \, t(x). \tag{2.17}$$

Now we have the following theorem.

**Theorem 2.1** *The Fisher information matrix $G_F(\xi) = [g_{ij}^F(\xi)]$ of the induced model $S_F \stackrel{\text{def}}{=} \{q(y;\xi)\}$ satisfies $G_F(\xi) \leq G(\xi)$, where $G(\xi) = [g_{ij}(\xi)]$ is the Fisher information matrix of the original model $S$, in the sense that $\Delta G(\xi) \stackrel{\text{def}}{=} G(\xi) - G_F(\xi)$ is positive semidefinite. A necessary and sufficient condition for the equality $G_F(\xi) = G(\xi)$ to identically hold is that $F$ is a sufficient statistic for $S$. In general, the information loss $\Delta G(\xi) = [\Delta g_{ij}(\xi)]$ caused by summarizing the data $x$ into $y = F(x)$ is given by*

$$\begin{aligned} \Delta g_{ij}(\xi) &= E_\xi[\partial_i \log r(X;\xi) \partial_j \log r(X;\xi)] \\ &= E_\xi\left[\mathrm{Cov}_\xi[\partial_i \ell(X;\xi), \partial_j \ell(X;\xi) \mid Y]\right], \end{aligned} \tag{2.18}$$

*where $E_\xi[\mathrm{Cov}_\xi[\cdot,\cdot \mid Y]] = \int \mathrm{Cov}_\xi[\cdot,\cdot \mid y] \, q(y;\xi) \, dy$, and $\mathrm{Cov}_\xi[\cdot,\cdot \mid y]$ for a fixed $y$ denotes the covariance with respect to the conditional distribution $p(x \mid y;\xi)$.*

**Proof:** For any $B \subset \mathcal{Y}$ we have

$$\begin{aligned} \int_B \partial_i \log q(y;\xi) \, q(y;\xi) \, dy &= \partial_i \int_B q(y;\xi) \, dy \\ &= \partial_i \int_{F^{-1}(B)} p(x;\xi) \, dx \\ &= \int_{F^{-1}(B)} \partial_i \ell(x;\xi) \, p(x;\xi) \, dx, \end{aligned}$$

and hence

$$\partial_i \log q(y;\xi) = E_\xi[\partial_i \ell(X;\xi) \mid y]. \tag{2.19}$$

On the other hand, Equation (2.12) leads to

$$\partial_i \ell(x;\xi) = \partial_i \log q(F(x);\xi) + \partial_i \log r(x;\xi). \tag{2.20}$$

From these equations, we have

$$E_\xi[\partial_i \log r(X;\xi) \mid F(x)] = 0.$$

## 2.2. THE FISHER METRIC

This implies that $\partial_i \log r(x; \xi)$ as a function of $x$ is orthogonal to any function of the form $\varphi(F(x))$, where $\varphi$ is an arbitrary function on $\mathcal{Y}$, and is orthogonal to $\partial_j \log q(F(x); \xi)$ in particular, with respect to the inner product

$$\langle\!\langle \Phi, \Psi \rangle\!\rangle_\xi = E_\xi[\Phi(X)\Psi(X)]. \tag{2.21}$$

It is then easy to see that Equation (2.18) follows from Equations (2.20), (2.19) and the definition of the conditional covariance:

$$\operatorname{Cov}_\xi[\partial_i \ell_\xi, \partial_j \ell_\xi \,|\, y] =$$
$$E_\xi\Big[\big\{\partial_i \ell_\xi - E_\xi[\partial_i \ell_\xi \,|\, y]\big\}\big\{\partial_j \ell_\xi - E_\xi[\partial_j \ell_\xi \,|\, y]\big\} \,\Big|\, y\Big].$$

The nonnegativity of $\Delta G(\xi)$ is now obvious, and the condition for $\Delta G(\xi)$ to identically vanish is that $\partial_i \log r(x; \xi) = 0$ for all $\xi, i, x$, which is equivalent to the sufficiency of $F$. ∎

The properties $G_F(\xi) \leq G(\xi)$ and $G(\xi) = G_F(\xi) + \Delta G(\xi)$ with Equation (2.18) are often referred to as the **monotonicity** and the **chain rule**, respectively, of the Fisher metric. Let us make some remarks on these notions. First, it is easy to see that these are extended to $G_\kappa(\xi) \leq G(\xi)$ and $G(\xi) = G_\kappa(\xi) + \Delta G(\xi)$ for any transition (or conditional) probability distribution (i.e. Markov kernel) $\kappa = \{\kappa(y \,|\, x) \geq 0 \,;\, x \in \mathcal{X}, y \in \mathcal{Y}\}$ such that $\int \kappa(y \,|\, x)\, dy = 1, \forall x$, where $G_\kappa(\xi)$ is the Fisher information matrix of the induced model: $q(y; \xi) = \int \kappa(y \,|\, x)\, p(x; \xi)\, dx$. The previous case for a deterministic mapping $F$ corresponds to $\kappa(y \,|\, x) = \delta_{F(x)}(y)$. Secondly, as a special case of the chain rule, the **additivity**

$$G_{12}(\xi) = G_1(\xi) + G_2(\xi) \tag{2.22}$$

holds for a product model: $p_{12}(x_1, x_2; \xi) = p_1(x_1; \xi)\, p_2(x_2; \xi)$. Thirdly, given two models $\{p_1(x; \xi)\}$ and $\{p_2(x; \xi)\}$ having common sample space $\mathcal{X}$ and parameter space $\Xi$, the following convexity follows from the monotonicity (cf. Equation (3.23) in §3.2 and its proof):

$$G_\lambda(\xi) \leq \lambda\, G_1(\xi) + (1-\lambda)\, G_2(\xi), \quad 0 \leq \forall \lambda \leq 1, \tag{2.23}$$

where $G_1(\xi), G_2(\xi)$ and $G_\lambda(\xi)$ are the Fisher information matrices of $\{p_1(x; \xi)\}$, $\{p_2(x; \xi)\}$ and $\{\lambda p_1(x; \xi) + (1-\lambda) p_2(x; \xi)\}$, respectively.

Next, we shall review the well-known **Cramér-Rao inequality**. Suppose that a data $x$ is randomly generated subject to a probability distribution which is unknown but is assumed to be in a prescribed $n$-dimensional model $S = \{p_\xi \,|\, \xi = [\xi^1, \cdots, \xi^n] \in \Xi\}$. We consider the problem of estimating the unknown parameter $\xi$ by a function $\hat{\xi}(x)$ of the data $x$. A mapping $\hat{\xi} = [\hat{\xi}^1, \ldots, \hat{\xi}^n] : \mathcal{X} \to \mathbb{R}^n$ introduced for this purpose is called an estimator. We say that $\hat{\xi}$ is an **unbiased estimator** if

$$E_\xi\big[\hat{\xi}(X)\big] = \xi \quad \text{for} \quad \forall \xi \in \Xi. \tag{2.24}$$

The mean squared error of an unbiased estimator $\hat{\xi}$ may be expressed as the variance-covariance matrix $V_\xi[\hat{\xi}] = [v_\xi^{ij}]$ where

$$v_\xi^{ij} \stackrel{\text{def}}{=} E_\xi\left[\left(\hat{\xi}^i(X) - \xi^i\right)\left(\hat{\xi}^j(X) - \xi^j\right)\right].$$

**Theorem 2.2 (Cramér-Rao inequality)** *The variance-covariance matrix $V_\xi[\hat{\xi}]$ of an unbiased estimator $\hat{\xi}$ satisfies $V_\xi[\hat{\xi}] \geq G(\xi)^{-1}$ in the sense that $V_\xi[\hat{\xi}] - G(\xi)^{-1}$ is positive semidefinite.*

A proof of the theorem will be given in §2.5, which is essentially the same as a standard proof, but is rather tinged with a geometrical flavor.

An unbiased estimator $\hat{\xi}$ achieving the equality $V_\xi[\hat{\xi}] = G(\xi)^{-1}$ for all $\xi$ is called an **efficient estimator**. It is obvious that an efficient estimator is the best unbiased estimator in the sense that its variance is the minimum among those of all the unbiased estimators. The converse, however, is not generally true. In other words, the best unbiased estimator does not always achieve the Cramér-Rao bound. Furthermore, there is a case where a biased estimator has a smaller mean square error than the efficient estimator. See textbooks of statistics for further details.

It should be noted that there does not generally exist an efficient estimator on a model $S = \{p_\xi\}$. A necessary and sufficient condition for the existence of efficient estimator will be given in §3.5, which imposes restrictions both on the shape of $S$ and on the parametrization $\xi \mapsto p_\xi$. However, it is worth emphasizing that there always exists a sequence of estimators $\left\{\hat{\xi}_N = \hat{\xi}_N(x_1, \cdots x_N)\right\}_{N=1}^{\infty}$ which asymptotically achieves the equality in the CraméR-Rao inequality as the number $N$ of independent observations goes to $\infty$. Such a sequence of estimators is called an **asymptotically efficient estimator** or a **first-order efficient estimator**; see §4.1 and §4.4. Thus the matrix $G(\xi)^{-1}$ represents the degree to which an asymptotically optimal estimator fluctuates around the true value of parameter $\xi$. In other words, a smaller $G(\xi)^{-1}$ (i.e., a larger $G(\xi)$) indicates a more accurate estimator. To be able to accurately estimate the parameter $\xi$ means that as the value of $\xi$ is changed, the "character" (i.e., $p_\xi$) of the data changes dramatically. The Fisher metric may be considered to be a geometric expression of the size of this change.

## 2.3 The $\alpha$-connection

Let $S = \{p_\xi\}$ be an $n$-dimensional model, and consider the function $\Gamma_{ij,k}^{(\alpha)}$ which maps each point $\xi$ to the following value:

$$\left(\Gamma_{ij,k}^{(\alpha)}\right)_\xi \stackrel{\text{def}}{=} E_\xi\left[\left(\partial_i\partial_j\ell_\xi + \frac{1-\alpha}{2}\partial_i\ell_\xi\partial_j\ell_\xi\right)(\partial_k\ell_\xi)\right], \quad (2.25)$$

where $\alpha$ is some arbitrary real number. The $n^3$ functions $\Gamma_{ij,k}^{(\alpha)}$ thus defined behave according to Equation (1.61) under coordinate transformations, and

## 2.3. THE $\alpha$-CONNECTION

hence we have an affine connection $\nabla^{(\alpha)}$ on $S$ defined by

$$\left\langle \nabla^{(\alpha)}_{\partial_i} \partial_j, \partial_k \right\rangle = \Gamma^{(\alpha)}_{ij,k}, \tag{2.26}$$

where $g = \langle\ ,\ \rangle$ is the Fisher metric. We call this $\nabla^{(\alpha)}$ the **$\alpha$-connection** .[1]

The $\alpha$-connection is clearly a symmetric connection. The relationship between the $\alpha$-connection and the $\beta$-connection is given by

$$\Gamma^{(\beta)}_{ij,k} = \Gamma^{(\alpha)}_{ij,k} + \frac{\alpha - \beta}{2} T_{ijk}, \tag{2.27}$$

where $T_{ijk}$ is a covariant symmetric tensor of degree 3 defined by

$$(T_{ijk})_\xi \stackrel{\text{def}}{=} E_\xi \left[ \partial_i \ell_\xi \partial_j \ell_\xi \partial_k \ell_\xi \right]. \tag{2.28}$$

(Note that this is in no way related to the torsion tensor $T^k_{ij} = \Gamma^k_{ij} - \Gamma^k_{ji}$.) We also have

$$\begin{aligned}
\nabla^{(\alpha)} &= (1-\alpha)\nabla^{(0)} + \alpha \nabla^{(1)} \\
&= \frac{1+\alpha}{2} \nabla^{(1)} + \frac{1-\alpha}{2} \nabla^{(-1)}.
\end{aligned} \tag{2.29}$$

(See the last paragraph of § 1.6.) In addition, for a submanifold $M$ of $S$, the $\alpha$-connection on $M$ is simply the projection with respect to $g$ of the $\alpha$-connection on $S$ (see §1.9).

Let us investigate some fundamental properties of $\nabla^{(\alpha)}$ for several particular values of $\alpha$. First, taking the partial derivative of the definition of $g_{ij}$ in Equation (2.6) with respect to $\xi^k$, we obtain

$$\begin{aligned}
\partial_k g_{ij} &= E_\xi[(\partial_k \partial_i \ell_\xi)(\partial_j \ell_\xi)] + E_\xi[(\partial_i \ell_\xi)(\partial_k \partial_j \ell_\xi)] + E_\xi[(\partial_i \ell_\xi)(\partial_j \ell_\xi)(\partial_k \ell_\xi)] \\
&= \Gamma^{(0)}_{ki,j} + \Gamma^{(0)}_{kj,i},
\end{aligned} \tag{2.30}$$

which leads to:

**Theorem 2.3** *The 0-connection is the Riemannian connection with respect to the Fisher metric.*

In general, when $\alpha \neq 0$, $\nabla^{(\alpha)}$ is not metric.

Let us introduce now the notion of exponential family, which will be shown to have close relation to $\nabla^{(1)}$. In general, if an $n$-dimensional model $S = \{p_\theta \mid \theta \in \Theta\}$ can be expressed in terms of functions $\{C, F_1, \cdots, F_n\}$ on $\mathcal{X}$ and a function $\psi$ on $\Theta$ as

$$p(x;\theta) = \exp\left[ C(x) + \sum_{i=1}^{n} \theta^i F_i(x) - \psi(\theta) \right], \tag{2.31}$$

---

[1] In the sequel, the terms "flat", "affine", "parallel", "autoparallel", etc. with respect to the $\alpha$-connection are referred to as **$\alpha$-flat**, **$\alpha$-affine**, **$\alpha$-parallel**, **$\alpha$-autoparallel**, etc.

then we say that $S$ is an **exponential family**, and that the $[\theta^i]$ are its **natural** or its **canonical parameters**. From the normalization condition $\int p(x;\theta)\mathrm{d}x = 1$ we obtain

$$\psi(\theta) = \log \int \exp\left[C(x) + \sum_{i=1}^{n} \theta^i F_i(x)\right] \mathrm{d}x. \tag{2.32}$$

It is easy to see that the parametrization $\theta \mapsto p_\theta$ is one-to-one if and only if the $n+1$ functions $\{F_1, \cdots, F_n, 1\}$ are linearly independent, where 1 denotes the constant function which identically takes the value 1. From now on, this linear independence will always be assumed when an exponential family is considered.

Many practically important models are shown to be exponential families, including all of the examples given in §2.1.

**Example 2.5** *(Example 2.1: Normal Distribution)*

$$C(x) = 0, \quad F_1(x) = x, \quad F_2(x) = x^2, \quad \theta^1 = \frac{\mu}{\sigma^2} \quad \theta^2 = -\frac{1}{2\sigma^2}$$

$$\psi(\theta) = \frac{\mu^2}{2\sigma^2} + \log(\sqrt{2\pi}\sigma) = -\frac{(\theta^1)^2}{4\theta^2} + \frac{1}{2}\log\left(-\frac{\pi}{\theta^2}\right)$$

**Example 2.6** *(Example 2.2: Multivariate Normal Distribution)*
Letting
$$C(x) = 0, \quad F_i(x) = x_i, \quad F_{ij}(x) = x_i x_j \quad (i \leq j),$$

$$\theta^i = \sum_j (\Sigma^{-1})^{ij} \mu_j, \quad \theta^{ii} = -\frac{1}{2}(\Sigma^{-1})^{ii}, \quad \theta^{ij} = -(\Sigma^{-1})^{ij} \quad (i < j),$$

and
$$F_A(x) = x, \quad F_B(x) = xx^t, \quad \theta^A = \Sigma^{-1}\mu, \quad \theta^B = -\frac{1}{2}\Sigma^{-1},$$

we have
$$\begin{aligned}p(x;\theta) &= \exp\left[\sum_{1 \leq i \leq k} \theta^i F_i(x) + \sum_{1 \leq i \leq j \leq k} \theta^{ij} F_{ij}(x) - \psi(\theta)\right] \\ &= \exp\left[(\theta^A)^t F_A(x) + \mathrm{tr}(\theta^B F_B(x)) - \psi(\theta)\right]\end{aligned}$$

where
$$\begin{aligned}\psi(\theta) &= \frac{1}{2}\mu^t \Sigma^{-1} \mu + \frac{1}{2} \log \det(2\pi\Sigma) \\ &= -\frac{1}{4}(\theta^A)^t (\theta^B)^{-1} \theta^A + \frac{1}{2} \log \det(-\pi (\theta^B)^{-1}).\end{aligned}$$

**Example 2.7** *(Example 2.3: Poisson Distribution)*

$$C(x) = -\log x!, \quad F(x) = x, \quad \theta = \log \xi$$

$$\psi(\theta) = \xi = \exp\theta$$

## 2.3. THE α-CONNECTION

**Example 2.8** *(Example 2.4: $\mathcal{P}(\mathcal{X})$ for finite $\mathcal{X}$)*

$$C(x) = 0, \quad F_i(x) = \begin{cases} 1 & (x = x_i) \\ 0 & (x \neq x_i) \end{cases}$$

$$\theta^i = \log \frac{p(x_i)}{p(x_0)} = \log \frac{\xi^i}{1 - \sum_{j=1}^n \xi^j} \quad (i = 1, \cdots, n)$$

$$\psi(\theta) = -\log p(\theta) = -\log\left(1 - \sum_{i=1}^n \xi^i\right) = \log\left(1 + \sum_{i=1}^n \exp\theta^i\right)$$

From the definition of an exponential family given in Equation (2.31), and letting $\partial_i = \frac{\partial}{\partial \theta^i}$, we may obtain

$$\partial_i \ell(x;\theta) = F_i(x) - \partial_i \psi(\theta) \quad \text{and} \tag{2.33}$$
$$\partial_i \partial_j \ell(x;\theta) = -\partial_i \partial_j \psi(\theta). \tag{2.34}$$

Hence we have $\Gamma^{(1)}_{ij,k} = -\partial_i \partial_j \psi(\theta) E_\theta[\partial_k \ell_\theta]$, which is 0 from Equation (2.9). In other words, we see that $[\theta^i]$ is a 1-affine coordinate system, and $S$ is 1-flat. We therefore call $\nabla^{(1)}$ the **exponential connection**, or the **e-connection**, and shall write $\nabla^{(1)} = \nabla^{(e)}$.

Next, let us consider the case when an $n$-dimensional model $S = \{p_\theta\}$ can be expressed in terms of functions $\{C, F_1, \cdots, F_n\}$ on $\mathcal{X}$ as

$$p(x;\theta) = C(x) + \sum_{i=1}^n \theta^i F_i(x), \tag{2.35}$$

or in other words, when $S$ forms an affine subspace of $\mathcal{P}(\mathcal{X})$. In this case we say that $S$ is a **mixture family** with **mixture parameters** $[\theta^i]$. In particular, $\mathcal{P}(\mathcal{X})$ itself is a mixture family when $\mathcal{X}$ is finite. A representative form of a mixture family is given by the mixture of $n+1$ probability distributions $\{p_0, p_1, \cdots, p_n\}$:

$$\begin{aligned} p(x;\theta) &= \sum_{i=1}^n \theta^i p_i(x) + \left(1 - \sum_{i=1}^n \theta^i\right) p_0(x) \\ &= p_0(x) + \sum_{i=1}^n \theta^i \{p_i(x) - p_0(x)\} \end{aligned} \tag{2.36}$$

where $[\theta^i]$ are subject to $\theta^i > 0$ and $\sum_i \theta^i < 1$. The distribution family $\mathcal{P}(\{x_0, \cdots, x_n\})$ may be expressed in this form by letting the distributions $\{p_0, \cdots, p_n\}$ be defined by $p_i(x_j) = \delta_{ij}$. In general, for a mixture family we have

$$\partial_i \ell(x;\theta) = \frac{F_i(x)}{p(x;\theta)} \quad \text{and} \tag{2.37}$$

$$\partial_i\partial_j \ell(x;\theta) = -\frac{F_i(x)F_j(x)}{p(x;\theta)^2}, \qquad (2.38)$$

from which we see that $\partial_i\partial_j\ell + \partial_i\ell\partial_j\ell = 0$, and hence $\Gamma^{(-1)}_{ij,k} = 0$. Therefore $[\theta^i]$ is a $(-1)$-affine coordinate system, and $S$ is $(-1)$-flat. We call $\nabla^{(-1)}$ the **mixture connection** or the **m-connection**, and we write $\nabla^{(-1)} = \nabla^{(m)}$.

The discussion above is summarized in the next theorem.

**Theorem 2.4** *An exponential family (a mixture family, respectively) is e-flat (m-flat) and its natural parameters (mixture parameters) form an e-affine (m-affine) coordinate system.*

In the next chapter (§ 3.3) we shall see that an exponential family (a mixture family, respectively) turns out to be also m-flat (e-flat, resp.) as a consequence of the duality between the e-connection and the m-connection.

**Theorem 2.5** *Let $S$ be an exponential family (a mixture family, respectively) and $M$ be a submanifold of $S$. Then $M$ is an exponential family (a mixture family) if and only if $M$ is e-autoparallel (m-autoparallel) in $S$.*

**Proof:** We only give here the proof for the part that if $S$ and $M$ are exponential families then $M$ is e-autoparallel in $S$, because the rest is straightforward from Theorem 1.1. Let $S = \{p(x;\theta)\}$ and $M = \{q(x;u)\}$ be given by

$$p(x;\theta) = \exp\left[C(x) + \sum_{i=1}^{n} \theta^i F_i(x) - \psi(\theta)\right],$$

$$q(x;u) = p(x;\theta(u)) = \exp\left[D(x) + \sum_{a=1}^{m} u^a G_a(x) - \varphi(u)\right].$$

Then we have

$$\begin{aligned} G_a(x) - \partial_a\varphi(u) &= \partial_a \log q(x;u) \\ &= (\partial_a\theta^i)_u \, \partial_i \log p(x;\theta(u)) \\ &= (\partial_a\theta^i)_u \{F_i(x) - \partial_i\psi(\theta(u))\}, \end{aligned}$$

and hence

$$(\partial_a\theta^i)_u F_i(x) + \lambda_a(u) = G_a(x),$$

where $\lambda_a(u)$ is constant with respect to $x$. Since $G_a(x)$ does not depend on $u$ and since $\{F_1,\cdots,F_n,1\}$ are assumed to be linearly independent, we see that $(\partial_a\theta^i)_u$ is constant with respect to $u$ for all $i$ and all $a$. This, combined with Theorem 1.1, implies that $M$ is e-autoparallel in $S$. ∎

In the theorem above the part for exponential families is of practical significance. A submanifold $M$ of an exponential family $S$ is called a curved exponential family and will play the leading role in chapter 4. In general, the degree to which a curved exponential family $M$ differs from an exponential family can be measured by the embedding curvature with respect to the e-connection (see §4.5).

We conclude this section by demonstrating an instructive example. Suppose that we are given a smooth probability density function $q$ on $\mathbb{R}$, and let $q^{(k)}$ be the $k$th i.i.d. extension; i.e., for $y = (y_1, \ldots, y_k)^t$, $q^{(k)}(y) = q(y_1) \cdots q(y_k)$. For a regular matrix $A \in \mathbb{R}^{k \times k}$ and a vector $\mu = (\mu_1, \ldots, \mu_k)^t \in \mathbb{R}^k$, define the probability density function $p_{A,\mu}$ on $\mathbb{R}^k$ by

$$p(x; A, \mu) = q^{(k)}(A^{-1}(x - \mu))/|\det A|,$$

which gives the probability distribution for the random variable $AY + \mu$ when $Y$ is supposed to distribute according to $q^{(k)}(y)$. For instance, letting $q$ be the standard normal distribution, we obtain

$$p(x; A, \mu) = (2\pi)^{-k/2}(\det \Sigma)^{-1/2} \exp\left\{-\frac{1}{2}(x-\mu)^t \Sigma^{-1}(x-\mu)\right\},$$

where $\Sigma \stackrel{\text{def}}{=} AA^t$. Now let $q$ and $A$ be arbitrarily fixed and consider the statistical model $S \stackrel{\text{def}}{=} \{p_{A,\mu} \mid \mu \in \mathbb{R}^k\}$. This model is not in general an exponential family nor a mixture family, but is always $\alpha$-flat for all $\alpha$ and, in particular, is a Euclidean space with respect to the Fisher metric. This may be regarded as a consequence of the fact that $S$ is essentially the direct product, including its $\alpha$-connections and Fisher metric, of $k$ copies of the 1-dimensional statistical model $\{q(y - \nu) \mid \nu \in \mathbb{R}\}$ on which affine connections are always flat (see the last paragraph of §1.7). Moreover, when $q(y)$ (or $q(y - \nu)$ for some $\nu$) is an even function as in the case of normal distributions, the tensor $T$ defined by Equation (2.28) vanishes, which means that all the $\alpha$-connections are identical, and the vector parameter $\mu$, as well as its regular affine transformations, forms an affine coordinate system for the $\alpha$-connections.

## 2.4 Chentsov's theorem and some historical remarks

As we noted in §1.5 and §1.6, it is possible to define an infinite number of distinct Riemannian metrics and affine connections on a manifold. This is, of course, also true for the manifolds which are formed by statistical models. Hence the Fisher metric and the $\alpha$-connections defined above are simply instances among the infinite number of possible metrics and connections. This may lead one to ask, however, whether there is anything which distinguishes the Fisher metric and $\alpha$-connections from the others; the answer is that indeed there is. A statistical model $S$, in addition to its structure as a manifold, has the property that "each point denotes a probability distribution." Taking this property into consideration, we find that there are natural structural conditions which are uniquely met by the Fisher metric and the $\alpha$-connections. We formalize this below.

Let $S = \{p(x;\xi)\}$ be a model on $\mathcal{X}$ and $F : \mathcal{X} \to \mathcal{Y}$ be some mapping, which induces a model $S_F = \{q(y;\xi)\}$ on $\mathcal{Y}$. If $F$ is a sufficient statistic for $S$, then $\partial_i \log p(x;\xi) = \partial_i \log q(F(x);\xi)$ from Equation (2.16), and hence $g_{ij}$ and $\Gamma^{(\alpha)}_{ij,k}$ are the same on both $S$ and $S_F$. We refer to this as "the **invariance** of the Fisher metric and the $\alpha$-connection with respect to $F$." Denoting the Fisher metric and the $\alpha$-connection on $S$ by $g = \langle\,,\,\rangle$ and $\nabla^{(\alpha)}$, while denoting those on $S_F$ by $g' = \langle\,,\,\rangle'$ and $\nabla'^{(\alpha)}$, the invariance properties are represented as

$$\langle X, Y \rangle_p = \langle \lambda_*(X), \lambda_*(Y) \rangle'_{\lambda(p)} \quad \text{and}$$
$$\lambda_*\left(\nabla^{(\alpha)}_X Y\right) = \nabla'^{(\alpha)}_{\lambda_*(X)} \lambda_*(Y), \quad \forall X, Y, Z \in \mathcal{T}(S),$$

where $\lambda$ is the diffeomorphism from $S$ onto $S_F$ defined by $\lambda(p_\xi) = q_\xi$, and the mapping $\lambda_* : \mathcal{T}(S) \to \mathcal{T}(S_F)$ is defined by $(\lambda_*(X))_{\lambda(p)} = (d\lambda)_p(X_p)$. In general, this invariance plays a crucial role when analyzing the relationship between statistics/probability theory and the structure formed by introducing a metric and a connection on a statistical model. The Fisher metric and the $\alpha$-connections are uniquely characterized by this invariance in the sense described below.

First let us consider a manifold which has, as its points, distributions on some finite set. Let $\mathcal{X}_n \stackrel{\text{def}}{=} \{0, 1, \cdots, n\}$ and $\mathcal{P}_n \stackrel{\text{def}}{=} \mathcal{P}(\mathcal{X}_n)$ for the natural numbers $n = 1, 2, \cdots$. Suppose that we have a sequence $\{(g_n, \nabla_n)\}_{n=1}^\infty$, where the $g_n$ and the $\nabla_n$ are an arbitrary Riemannian metric and an arbitrary affine connection, respectively, on $\mathcal{P}_n$ for each $n$. Then letting $S$ be a model on $\mathcal{X}_n$ and $F : \mathcal{X}_n \to \mathcal{X}_m$ ($n \geq m$) be a surjective mapping, $(g_n, \nabla_n)$ and $(g_m, \nabla_m)$ induce metrics and connections on $S$ ($\subset \mathcal{P}_n$) and $S_F$ ($\subset \mathcal{P}_m$) by restriction and projection (see §1.5, §1.9). Now, we have the following theorem, which restates a result by N. N. Chentsov (Čencov) [65] from a slightly different point of view. (We omit the proof.)

**Theorem 2.6** *Assume that $\{(g_n, \nabla_n)\}_{n=1}^\infty$ is invariant with respect to sufficient statistics; i.e., for all $n$, $m$, $S \subset \mathcal{P}_n$, and $F : \mathcal{X}_n \to \mathcal{X}_m$ such that $F$ is a sufficient statistic for $S$, the induced metrics and connections on $S$ and $S_F$ are assumed to be invariant. Then there exist a positive real number $c$ and a real number $\alpha$ such that, for all $n$, $g_n$ coincides with the Fisher metric on $\mathcal{P}_n$ scaled by a factor of $c$, and $\nabla_n$ coincides with the $\alpha$-connection on $\mathcal{P}_n$.*

In this way, we see that for models on finite sets the Fisher metric (up to a constant factor) and the $\alpha$-connections are characterized by the invariance with respect to sufficient statistics. It is natural to expect that the characterization will also apply to models on infinite sets so that the Fisher metric and the $\alpha$-connections can always be considered as the only metric and connections which meet the invariance requirement. However, it is not so easy to extend Chentsov's theorem to the case when the underlying set $\mathcal{X}$ may be infinite, because both the formulation and the proof of the theorem essentially rely upon the finiteness

## 2.4. CHENTSOV'S THEOREM AND SOME HISTORICAL REMARKS

of $\mathcal{X}$. Here, we shall only observe that Chentsov's theorem leads to the Fisher metric and the $\alpha$-connections for a model $S = \{p(x;\xi)\}$ on an infinite set $\mathcal{X}$ if a kind of limiting procedure is permitted.

First, let us finitely partition $\mathcal{X}$ into the regions $\delta_1, \delta_2, \cdots, \delta_n$. In other words, each $\delta_i$ is a subset of $\mathcal{X}$, $\delta_i \cap \delta_j = \emptyset$ ($i \neq j$), and $\bigcup_{i=1}^n \delta_i = \mathcal{X}$. Now fix a particular partition $\Delta = \{\delta_1, \cdots, \delta_n\}$ and let

$$p_\Delta(\delta_i; \xi) \stackrel{\text{def}}{=} \int_{\delta_i} p(x;\xi)\, dx.$$

Then $S_\Delta \stackrel{\text{def}}{=} \{p_\Delta(\delta_i;\xi)\}$ forms a model on $\Delta$. Since $\Delta$ is a finite set, from Chentsov's theorem we know that the Fisher metric and the $\alpha$-connections are introduced on $S_\Delta$ by the invariance requirement. Now we may consider $S$ to be the limit of $S_\Delta$ as $\Delta$ becomes finer and finer. Hence, if we require that the desired metrics and connections on models should be "continuous" with respect to such a limit, it is concluded that the metric and the connections on $S$ should be given by the limit of the Fisher metric and the $\alpha$-connections on $S_\Delta$, and under some regularity condition they coincide with the Fisher metric and the $\alpha$-connections on $S$.

Let us now mention some history. The actual idea of considering statistical models as a manifold and analyzing them from the point of view of differential geometry is quite old. C.R. Rao had already pointed out in the 1945 paper [190] that the Fisher information matrix determines a Riemannian metric, and had written on the importance of analyzing model structures from the perspective of Riemannian geometry. After this, there were a variety of efforts to pursue this line of research, but unfortunately very few of these produced results which related directly back to statistical problems.

In 1975, B. Efron [83] introduced the notion of "statistical curvature" for 1-parameter models, and found that it serves an important role in the asymptotic theory of statistical estimation. Building on this result, A.P. Dawid [76] introduced a connection on the space $\mathcal{P}$ of all the positive probability distributions and showed that the statistical curvature could be expressed as the embedding curvature with respect to this connection. This connection is the e-connection introduced above. In addition, Dawid pointed out that it was possible to introduce a variety of connections on $\mathcal{P}$, and as examples gave the Riemannian connection with respect to the Fisher metric (the 0-connection) and the m-connection. However, in general $\mathcal{P}$ is a difficult space to treat rigorously since it is of infinite dimension, and cannot even be viewed as an infinite-dimensional manifold in the usual sense (; see the next section). Amari [5, 6], in 1980, reformulated Dawid's discussion by "projecting" them onto finite-dimensional models, and introduced the natural generalization to $\alpha$-connections. In addition, the importance of considering the e-connection and the m-connection as a pair was clearly shown for the first time. This result leads to the topic of the next chapter, that of understanding the structure of dual connections. The notion of the duality of connections was introduced by Nagaoka and Amari [161] in 1982, where the general theory of dually flat spaces was developed to provide a mathematical foundation to the theory of $\alpha$-connections, and since then it has come to serve a central role in information geometry. On the other hand, the notion of $\alpha$-connections itself had in fact been introduced from a different perspective (that of its invariance) in 1972 by the Russian mathematician N.N. Chentsov [65]. However, because its relation to curvature and statistical estimation was not pursued, and because the paper was written in Russian, this result was not widely known within the international statistical community.

## 2.5 The geometry of $\mathcal{P}(\mathcal{X})$

Let $\mathcal{X}$ be a finite set. As mentioned previously, an arbitrary model $S$ on $\mathcal{X}$ is a submanifold of $\mathcal{P} = \mathcal{P}(\mathcal{X})$, and the Fisher metric and the $\alpha$-connection on $S$ are the projections of those on $\mathcal{P}$ onto $S$. The geometry of $\mathcal{P}$ is most fundamental in this respect. In this section we shall focus our attention upon the Fisher metric and the e, m-connections on $\mathcal{P}$, and postpone the study of general $\alpha$-connections until the next section for the reason that the e, m-connections are particularly important in most applications and, in addition, have much simpler structures than the other $\alpha$-connections. It is also noted that some of the important properties of the e, m-connections are closely related to the notion of dual connections and will be discussed in §3.5. As for the the case when $\mathcal{X}$ is infinite, we shall briefly mention it at the end of this section.

In the first place, the model $\mathcal{P}$ is a subset of $\mathbb{R}^{\mathcal{X}} \stackrel{\text{def}}{=} \{A \mid A : \mathcal{X} \to \mathbb{R}\}$: the totality of $\mathbb{R}$-valued functions on $\mathcal{X}$. More specifically $\mathcal{P}$ is an open set of the affine subspace $\mathcal{A}_1 \stackrel{\text{def}}{=} \{A \mid \sum_x A(x) = 1\}$ of $\mathbb{R}^{\mathcal{X}}$ and hence the tangent space $T_p(\mathcal{P})$ can naturally be identified with the linear subspace $\mathcal{A}_0 \stackrel{\text{def}}{=} \{A \mid \sum_x A(x) = 0\}$. When a tangent vector $X \in T_p(\mathcal{P})$ is considered as an element of $\mathcal{A}_0$, we denote it by $X^{(\text{m})}$ and call it the **mixture representation** or the **m-representation** of $X$, and write as

$$T_p^{(\text{m})}(\mathcal{P}) \stackrel{\text{def}}{=} \left\{ X^{(\text{m})} \,\middle|\, X \in T_p(\mathcal{P}) \right\} = \mathcal{A}_0.$$

For the natural basis $\partial_i$ of a coordinate system $\xi = [\xi^i]$, we have $(\partial_i)_\xi^{(\text{m})} = \partial_i p_\xi$. As observed in §2.3, $\mathcal{P}$ is a mixture family which is m-flat, and its mixture parameters form an m-affine coordinate system. This means that the m-connection on $\mathcal{P}$ is nothing but the natural connection induced from the affine structure of $\mathcal{A}_1$. Obviously, the parallel translation $\Pi_{p,q}^{(\text{m})} : T_p(\mathcal{P}) \to T_q(\mathcal{P})$ with respect to the m-connection is given by

$$\Pi_{p,q}^{(\text{m})}(X) = X' \iff X'^{(\text{m})} = X^{(\text{m})}. \tag{2.39}$$

We have thus seen that the natural embedding of $\mathcal{P}$ into $\mathbb{R}^{\mathcal{X}}$ makes the meaning of the m-connection clear.

Next, let us take another embedding $p \mapsto \log p$, and identify $\mathcal{P}$ with the subset $\{\log p \mid p \in \mathcal{P}\}$ of $\mathbb{R}^{\mathcal{X}}$. A tangent vector $X \in T_p(\mathcal{P})$ is then represented by the result of operating $X$ to $p \mapsto \log p$, which we denote by $X^{(\text{e})}$ and call the

## 2.5. THE GEOMETRY OF $\mathcal{P}(\mathcal{X})$

**exponential representation** or the **e-representation** of $X$ (although some people might prefer to call it the *logarithmic representation*.) In particular we have $(\partial_i)_\xi^{(e)} = \partial_i \log p_\xi$. It is obvious that

$$X^{(e)}(x) = X^{(m)}(x)/p(x) \tag{2.40}$$

and that (cf. Equation (2.9))

$$T_p^{(e)}(\mathcal{P}) \stackrel{\text{def}}{=} \left\{ X^{(e)} \,\middle|\, X \in T_p(\mathcal{P}) \right\} = \left\{ A \in \mathbb{R}^\mathcal{X} \,\middle|\, E_p[A] = 0 \right\}, \tag{2.41}$$

where $E_p[A] = \sum_x p(x) A(x)$. Note that the definition of the Fisher metric is expressed in terms of the present notation as

$$\langle X, Y \rangle_p = E_p\left[ X^{(e)} Y^{(e)} \right]. \tag{2.42}$$

Unlike $T_p^{(m)}(\mathcal{P})$, the space $T_p^{(e)}(\mathcal{P})$ depends on $p$, and an element $A$ of $T_p^{(e)}(\mathcal{P})$ does not generally belong to $T_q^{(e)}(\mathcal{P})$ when $p \neq q$. Nevertheless, the shifted function $A' = A - E_q[A]$ always belongs to $T_q^{(e)}(\mathcal{P})$, and the correspondence $A \mapsto A'$ establishes a linear isomorphism between $T_p^{(e)}(\mathcal{P})$ and $T_q^{(e)}(\mathcal{P})$. Note that this is different from the correspondence between the e-representations of $X$ and $X' = \Pi_{p,q}^{(m)}(X)$, which is represented as $A' = Ap/q$. In fact, we have

$$\Pi_{p,q}^{(e)}(X) = X' \iff X'^{(e)} = X^{(e)} - E_q\left[ X^{(e)} \right], \tag{2.43}$$

where $\Pi_{p,q}^{(e)}$ denotes the parallel translation from $T_p(\mathcal{P})$ to $T_q(\mathcal{P})$ with respect to the e-connection. This may be seen as follows. Let $X : p \mapsto X_p$ be an arbitrary vector field on $\mathcal{P}$, and $\{\partial_i\}$ the natural basis of a coordinate system $[\xi^i]$. From the definition of the e-connection (Equation (2.25) for $\alpha = 1$) we have

$$\left\langle \nabla_{\partial_i}^{(e)} X, \partial_k \right\rangle_p = E_p\left[ (\partial_i X^{(e)})_p (\partial_k)_p^{(e)} \right], \tag{2.44}$$

where $(\partial_i X^{(e)})_p$ denotes the $i^{\text{th}}$ partial derivative of the mapping $X^{(e)} : p \mapsto X_p^{(e)}$ ($\mathcal{P} \to \mathbb{R}^\mathcal{X}$) at a point $p$. Now suppose that $X^{(e)}$ is represented as $X_p^{(e)} = F - E_p[F], \forall p \in \mathcal{P}$ by a function (random variable) $F$ on $\mathcal{X}$. Then $(\partial_i X^{(e)})_p = -\partial_i E_p[F]$ is constant as a function on $\mathcal{X}$, and hence $\left\langle \nabla_{\partial_i}^{(e)} X, \partial_k \right\rangle_p = 0$ at each point $p$ by Equation (2.44). This implies that $X$ is e-parallel and proves Equation (2.43).

Let us investigate some properties of the Fisher metric through the e-representation. We begin with mathematical preliminaries concerning the **cotangent space** $T_p^*(\mathcal{P}) = [T_p(\mathcal{P})]_1^0$ (the totality of $\mathbb{R}$-valued linear functions on $T_p(\mathcal{P})$; see §1.3) of a Riemannian manifold $(S, g = \langle \,,\, \rangle)$. First, for each tangent vector $X \in T_p(\mathcal{P})$ let $\omega_X \in T_p^*(S)$ be a cotangent vector such that $\omega_X : Y \mapsto \langle X, Y \rangle_p$. Then the correspondence $X \leftrightarrow \omega_X$ establishes a linear isomorphism between

$T_p(S)$ and $T_p^*(S)$, and the inner product and the norm (length) for cotangent vectors are naturally defined by $\langle \omega_X, \omega_Y \rangle_p = \langle X, Y \rangle_p$ and $\|\omega_X\|_p = \|X\|_p$. Next, for an arbitrary smooth function $f \in \mathcal{F}(S)$ let $(\mathrm{d}f)_p \in T_p^*(S)$, the **differential** of $f$ at $p$, be defined by $(\mathrm{d}f)_p : X \mapsto X(f)$, and $(\mathrm{grad}\, f)_p \in T_p(S)$, the **gradient** of $f$ at $p$, be the tangent vector corresponding to $(\mathrm{d}f)_p$:

$$\langle (\mathrm{grad}\, f)_p, X \rangle_p = (\mathrm{d}f)_p(X) = X(f), \quad \forall X \in T_p(S). \tag{2.45}$$

(Recall that a tangent vector is a mapping : $\mathcal{F}(S) \to \mathbb{R}$). We have the following coordinate expressions:

$$(\mathrm{grad}\, f)_p = (\partial_i f)_p g^{ij}(p) (\partial_j)_p, \tag{2.46}$$

$$\|(\mathrm{d}f)_p\|_p^2 = \|(\mathrm{grad}\, f)_p\|_p^2 = (\partial_i f)_p (\partial_j f)_p g^{ij}(p), \tag{2.47}$$

where $[g^{ij}(p)]$ denotes the inverse of the matrix $[g_{ij}(p)] = [\langle \partial_i, \partial_j \rangle_p]$.

The following theorem states that the variance $V_p[A] = E_p[(A - E_p[A])^2]$ of a random variable $A$ is measured by the sensitivity of the expectation $E_p[A]$ to perturbation of $p$.

**Theorem 2.7** *Suppose that an $A : \mathcal{X} \to \mathbb{R}$ is given and let $E[A]$ denote the function $p \mapsto E_p[A]$ on $\mathcal{P}$. Then we have*

$$V_p[A] = \|(\mathrm{d}E[A])_p\|_p^2, \tag{2.48}$$

*where the norm is the one induced from the Fisher metric.*

**Proof:** For every $X \in T_p(\mathcal{P})$ we have

$$X(E[A]) = \sum_x X^{(\mathrm{m})}(x) A(x) = E_p\!\left[ X^{(\mathrm{e})} A \right]$$
$$= E_p\!\left[ X^{(\mathrm{e})}(A - E_p[A]) \right], \tag{2.49}$$

where the last equality follows from $E_p[X^{(\mathrm{e})}] = 0$. Since $A - E_p[A] \in T_p^{(\mathrm{e})}(\mathcal{P})$ by Equation (2.41), there exists a vector $Y_p \in T_p(\mathcal{P})$ satisfying $Y_p^{(\mathrm{e})} = A - E_p[A]$, which turns out to be $(\mathrm{grad}\, E[A])_p$ from Equations (2.45) and (2.49). Hence we obtain

$$\|(\mathrm{d}E[A])_p\|_p^2 = \|Y_p\|_p^2 = E_p\!\left[ (Y_p^{(\mathrm{e})})^2 \right] = V_p[A]. $$

∎

When the domain of the function $E[A]$ is restricted to a submanifold $S \subset \mathcal{P}$, the previous theorem is modified as follows.

**Theorem 2.8** *We have*

$$V_p[A] \geq \|(\mathrm{d}E[A]|_S)_p\|_p^2, \tag{2.50}$$

## 2.5. THE GEOMETRY OF $\mathcal{P}(\mathcal{X})$

where the equality holds if and only if

$$A - E_p[A] \in T_p^{(e)}(S) \stackrel{\text{def}}{=} \left\{ X^{(e)} \,\middle|\, X \in T_p(S) \right\}. \tag{2.51}$$

**Proof:** Obvious from the previous theorem and the fact that for every $f \in \mathcal{F}(\mathcal{P})$ the gradient $(\text{grad } f|_S)_p$ of the restricted function $f|_S$ is the orthogonal projection of $(\text{grad } f)_p$ onto $T_p(S)$. ∎

Now, Theorem 2.2 (the Cramér-Rao inequality) for a finite $\mathcal{X}$ is straightforward from the theorem above. Indeed, letting $A \stackrel{\text{def}}{=} c_i \hat{\xi}^i$ for an unbiased estimator $\hat{\xi}$ and an arbitrary column vector $c = [c_i] \in \mathbb{R}^n$, and using Equation (2.47), we can easily verify that Equation (2.50) yields $c^t V_\xi[\hat{\xi}]c \geq c^t G(\xi)^{-1}c$. Moreover we see that a necessary and sufficient condition for the equality $V_\xi[\hat{\xi}] = G(\xi)^{-1}$ to hold at $p$ is that $c_i(\hat{\xi}^i - \xi^i(p))$ belongs to $T_p^{(e)}(S)$ for all $[c_i] \in \mathbb{R}^n$, or equivalently that $\hat{\xi}^i - \xi^i(p)$ belongs to $T_p^{(e)}(S)$ for all $i$. Therefore, if $\hat{\xi}$ is an efficient estimator, there exist $n$ vector fields $\{X^1, \cdots, X^n\}$ on $S$ such that $(X_p^i)^{(e)} = \hat{\xi}^i - \xi^i(p)$, $\forall i, \forall p$, which turn out to be parallel with respect to the e-connection on $\mathcal{P}$ due to Equation (2.43). This means that the existence of an efficient estimator for $S$ implies that $S$ is e-autoparallel in $S$ (see §1.8), or in other words that $S$ is an exponential family (Theorem 2.5). We shall revisit this subject in §3.5, where a necessary and sufficient condition for $(S, [\xi^i])$ to have an efficient estimator will be given (Theorem 3.12).

**Note:** An estimator $\hat{\xi}$ is said to be **locally unbiased** at a point $\xi$ when the unbiasedness approximately holds around $\xi$ in the following sense:

$$E_{\xi + \Delta \xi}[\hat{\xi}] = \xi + \Delta \xi + o(\Delta \xi), \tag{2.52}$$

which means that

$$E_\xi[\hat{\xi}] = \xi \quad \text{and} \quad \partial_i E_\xi[\hat{\xi}^j] = \delta_i^j. \tag{2.53}$$

Obviously, $\hat{\xi}$ is unbiased if and only if it is locally unbiased at every $\xi$. We see from the above discussion that the Cramér-Rao inequality $V_\xi[\hat{\xi}] \geq G(\xi)^{-1}$ at a point $\xi$ holds if $\hat{\xi}$ is locally unbiased at $\xi$. Moreover, the equality can be always attained for each $\xi$ by the locally unbiased estimator

$$\hat{\xi}^i(x) = \xi^i + g^{ij}(\xi) \partial_i \log p(x; \xi). \tag{2.54}$$

As we have seen above, the e-representation is particularly useful for connecting the Fisher metric with some statistical notions like expectation and variance. However, if one is concerned only with purely geometrical aspects of the Fisher metric, it might be better to use another representation. Let us consider the third way of embedding $\mathcal{P}$ into $\mathbb{R}^{\mathcal{X}} : p \mapsto 2\sqrt{p}$, whereby $\mathcal{P}$ is identified with the set $\{2\sqrt{p} \mid p \in \mathcal{P}\}$ which forms a part of the sphere of radius 2 : $\{a \in \mathbb{R}^{\mathcal{X}} \mid \sum_x \{a(x)\}^2 = 4\}$. A tangent vector $X \in T_p(\mathcal{P})$ is then represented by $X^{(0)} \stackrel{\text{def}}{=} X^{(m)}/\sqrt{p} = \sqrt{p} X^{(e)}$, which is called the **0-representation** of $X$ (see the next section for the general $\alpha$-representation), and the Fisher metric takes the form

$$\langle X, Y \rangle_p = \sum_x X^{(0)}(x) Y^{(0)}(x), \qquad (2.55)$$

which is equivalent to Equation (2.10). This indicates that the Fisher metric is nothing but the Riemannian metric induced on the sphere from the natural Euclidean metric of $\mathbb{R}^{\mathcal{X}}$.

Here, we discuss Jeffreys' prior distribution to illustrate an application of the observation made above. Let $S = \{p_\xi \mid \xi = [\xi^1, \cdots, \xi^n] \in \Xi\}$ be a statistical model, and let $G(\xi)$ denote the Fisher information matrix at point $\xi$. Now suppose that the volume $V \stackrel{\text{def}}{=} \int_\Xi \sqrt{\det G(\xi)} d\xi$ of $S$ with respect to the Fisher metric is finite (where the integral is implicitly $n$-fold.) Then $Q(\xi) \stackrel{\text{def}}{=} \frac{1}{V} \sqrt{\det G(\xi)}$ defines a probability density function on $\Xi$. Since this is invariant over the choice of coordinate system $[\xi^i]$, we may consider it as a probability distribution on the model $S$. This distribution is called **Jeffreys' prior** within the field of Bayesian statistics, and has recently been found to play an important role in universal data compression (see Clarke and Barron [67]). Now consider the case when $S = \mathcal{P}(\mathcal{X})$ and $\mathcal{X} = \{0, 1, \cdots, n\}$. Then we can see that $V$ equals to the surface area of the $n$ dimensional sphere of radius 2 divided by $2^{n+1}$, and letting $\Gamma$ be the gamma function we have

$$V = \frac{\pi^{(n+1)/2}}{\Gamma(\frac{n+1}{2})}.$$

In addition, $Q(\xi)d\xi$ is then the uniform distribution on the sphere.

Finally, let us give a brief look at what happens when $\mathcal{X}$ is infinite. For a finite $\mathcal{X}$, we have seen that $\mathcal{P}$ is an open set of the affine space $\mathcal{A}_1$ and therefore is a manifold with the tangent space $T_p(\mathcal{P})$ isomorphic to $\mathcal{A}_0$ at each point $p$. Intuitively speaking, the essence of this fact is that the constraint $p(x) > 0$ ($\forall x \in \mathcal{X}$) imposed on the elements of $\mathcal{P}$ is loose enough for $\mathcal{P}$ to have the same dimension as $\mathcal{A}_1$. According as the cardinality $|\mathcal{X}|$ gets larger, however, the constraint becomes relatively more restrictive, or in other words, the portion of $\mathcal{A}_1$ occupied by $\mathcal{P}$ becomes relatively smaller. Eventually when $\mathcal{X}$ gets infinite, $\mathcal{P}$ cannot have the same dimension as $\mathcal{A}_1$, nor even be regarded as an infinite-dimensional manifold in the usual sense, and the geometrical arguments on $\mathcal{P}$ in the present section lose their mathematical basis. Nevertheless, this does not mean that the arguments become totally meaningless. For instance, the Cramér-Rao inequality for a model on an infinite $\mathcal{X}$ can be proved almost in parallel with the proof given above for a finite $\mathcal{X}$, even though the differential geometrical framework does not completely work any more. It should be noted that the geometrical proof for a finite $\mathcal{X}$ is essentially based on the nature of

$\mathcal{P}$ as an exponential family. We may regard $\mathcal{P}$ for an infinite $\mathcal{X}$ as an infinite dimensional version of exponential family in some sense, but should be careful making mathematical reasoning based on such an analogy in general. It is interesting and important to try to construct the geometry of $\mathcal{P}$, or of an appropriate variant of $\mathcal{P}$ if necessary, by rigorous functional analytic methods, so that many significant statistical arguments including those of the present section and of chapter 4 can be treated within the scope of such a geometry. Several attempts have already been made in this direction; see von Friedrich [215], Kambayashi [115], Lafferty [136], Pistone and Sempi [189], Gibilisco and Pistone [97] and Pistone and Rogantin [188].

## 2.6  $\alpha$-affine manifolds and $\alpha$-families

The present section is aimed at extending some of the results on the e- and m-connections obtained in §2.3 to the general $\alpha$-connections, and has little relation to the other parts of this book except for §3.6 and §8.4. Although we are mainly motivated by a purely geometrical interest here, it is worth mentioning that the results of this section and §3.6 are closely related to the theory of Tsallis entropy [207, 74, 208] developed in statistical physics.

We begin with loosening a fundamental assumption made so far. Previously when we considered a statistical model $S = \{p_\xi \mid \xi \in \Xi\}$, each element $p_\xi$ was a probability distribution ($\in \mathcal{P}(\mathcal{X})$). In this section we relax this condition and consider cases in which $\int p(x;\xi)\,dx$ is allowed to take arbitrary finite values other than 1. In other words, we suppose that $S = \{p_\xi\}$ is a subset of the following set:

$$\tilde{\mathcal{P}}(\mathcal{X}) \stackrel{\text{def}}{=} \left\{ p : \mathcal{X} \to \mathbb{R} \,\bigg|\, p(x) > 0 \ (\forall x \in \mathcal{X}), \int p(x)dx < \infty \right\}. \tag{2.56}$$

This extension allows a more natural understanding of the properties of the $\alpha$-connection. Aside from this, we retain the assumptions made in §2.1 and §2.2. In this case we again find that $S$ is a manifold, and that the Fisher metric $g$ and the $\alpha$-connection $\nabla^{(\alpha)}$ may be defined using Equations (2.6) and (2.25). There are, however, previous equations such as Equations (2.8) and (2.9) which no longer hold in general.

Now for each $\alpha \in \mathbb{R}$ define the following:

$$L^{(\alpha)}(u) \stackrel{\text{def}}{=} \begin{cases} \dfrac{2}{1-\alpha} u^{\frac{1-\alpha}{2}} & (\alpha \neq 1) \\ \log u & (\alpha = 1) \end{cases} \tag{2.57}$$

$$\ell^{(\alpha)}(x;\xi) \stackrel{\text{def}}{=} L^{(\alpha)}(p(x;\xi)) \tag{2.58}$$

Note in particular that $\ell^{(1)}(x;\xi) = \ell(x;\xi)$ and that $\ell^{(-1)}(x;\xi) = p(x;\xi)$. For a tangent vector $X \in T_\xi(S)$, we call

$$X^{(\alpha)}(x) \stackrel{\text{def}}{=} X\ell^{(\alpha)}(x;\xi) \tag{2.59}$$

as a function of $x$ the **$\alpha$-representation** of $X$. The e-, m- and 0-representations introduced in the previous section are special cases corresponding to $\alpha = 1, -1, 0$, respectively. Now since

$$\partial_i \ell^{(\alpha)} = p^{(1-\alpha)/2} \partial_i \ell \quad \text{and} \quad \partial_i \partial_j \ell^{(\alpha)} = p^{(1-\alpha)/2} \left( \partial_i \partial_j \ell + \frac{1-\alpha}{2} \partial_i \ell \partial_j \ell \right),$$

we may rewrite Equations (2.6) and (2.25) as

$$g_{ij}(\xi) = \int \partial_i \ell^{(\alpha)}(x;\xi) \partial_j \ell^{(-\alpha)}(x;\xi) dx, \quad \text{and} \quad (2.60)$$

$$\Gamma_{ij,k}^{(\alpha)}(\xi) = \int \partial_i \partial_j \ell^{(\alpha)}(x;\xi) \partial_k \ell^{(-\alpha)}(x;\xi) dx. \quad (2.61)$$

We can read these equations to mean that the $\alpha$-connection on $S$ is induced from the affine structure of the space $\mathbb{R}^{\mathcal{X}}$ of functions on $\mathcal{X}$ through the embedding $\xi \mapsto \ell^{(\alpha)}(x;\xi)$. In addition, they lead to the duality of $\pm \alpha$-connections as seen in the next chapter. Note also that if $S = \{p_\xi\}$ consists of probability distributions then, corresponding to Equations (2.8) and (2.9), we have

$$\int p(x;\xi)^{\frac{1+\alpha}{2}} \partial_i \ell^{(\alpha)}(x;\xi) \, dx = 0, \quad \text{and} \quad (2.62)$$

$$\frac{1+\alpha}{2} g_{ij}(\xi) = - \int p(x;\xi)^{\frac{1+\alpha}{2}} \partial_i \partial_j \ell^{(\alpha)}(x;\xi) \, dx. \quad (2.63)$$

Let us fix $\alpha$ to a particular value. If for some coordinate system $[\theta^i]$

$$\partial_i \partial_j \ell^{(\alpha)}(x;\theta) = 0 \quad (2.64)$$

$\left( \partial_i \stackrel{\text{def}}{=} \frac{\partial}{\partial \theta^i} \right)$, then from Equation (2.61) we see that $[\theta^i]$ is an $\alpha$-affine coordinate system, and that $S = \{p_\theta\}$ is $\alpha$-flat. We call such an $S$ an **$\alpha$-affine manifold**. The condition in Equation (2.64) is equivalent to the existence of the functions $\{C, F_1, \cdots, F_n\}$ ($n = \dim S$) on $\mathcal{X}$ such that

$$\ell^{(\alpha)}(x;\theta) = C(x) + \sum_{i=1}^{n} \theta^i F_i(x). \quad (2.65)$$

In addition, from Theorem 1.1 (§1.8) we see that a necessary and sufficient condition for a submanifold $M$ of $S$ to be $\alpha$-affine is for $M$ to be $\alpha$-autoparallel in $S$.

**Example 2.9** *A mixture family is a $(-1)$-affine manifold, while an exponential family is not a 1- affine manifold.*

**Example 2.10 ($\tilde{\mathcal{P}}(\mathcal{X})$ for finite $\mathcal{X}$)** *Consider the case when $\mathcal{X}$ is a finite set $\{x_1, \cdots, x_n\}$. Let $F_i : \mathcal{X} \to \mathbb{R}$ be the functions defined by $F_i(x_j) = \delta_{ij}$ for $i, j = 1, \cdots, n$. Then for each $p \in \tilde{\mathcal{P}}(\mathcal{X})$, using the independent parameters $\theta^1, \cdots, \theta^n$, we have $L^{(\alpha)}(p(x)) = \theta^i F_i(x)$ (here $\theta^i = L^{(\alpha)}(p(x_i))$.) Therefore $\tilde{\mathcal{P}}(\mathcal{X})$ is an $\alpha$-affine manifold for every $\alpha \in \mathbb{R}$.*

## 2.6. $\alpha$-AFFINE MANIFOLDS AND $\alpha$-FAMILIES

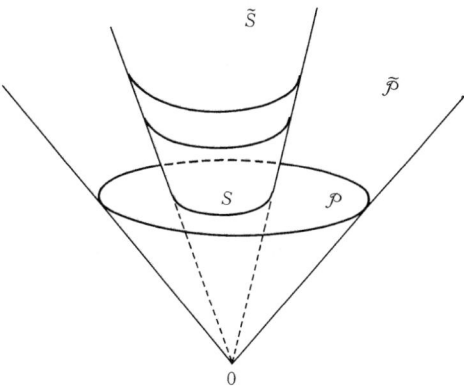

Figure 2.1: The denormalization of a statistical model

When $\mathcal{X}$ is infinite, on the other hand, $\tilde{\mathcal{P}}(\mathcal{X})$ may not be viewed as a manifold as in the case of $\mathcal{P}(\mathcal{X})$ mentioned in the last section, but in an informal sense its structure may still be considered similar to that of an $\alpha$-affine manifold.

Let $S = \{p_\xi \mid \xi \in \Xi\}$ $(\subset \mathcal{P}(\mathcal{X}))$ be a statistical model. If we let

$$\tilde{S} \stackrel{\text{def}}{=} \{\tau p_\xi \mid \xi \in \Xi,\ \tau > 0\} \quad (\subset \tilde{\mathcal{P}}(\mathcal{X})), \tag{2.66}$$

where $\tau p_\xi$ denotes the function on $\mathcal{X}$ which maps $x$ to $\tau p(x;\xi)$, then $\tilde{S}$ is a manifold which contains $S$ as a submanifold with $\dim \tilde{S} = \dim S + 1$. We call $\tilde{S}$ the **denormalization** of $S$ (Figure 2.1). As a coordinate system of $\tilde{S}$ we adopt $[\xi, \tau] = [\xi^1, \cdots, \xi^n, \tau]$ and denote its natural basis by $\widetilde{\partial}_i = \frac{\partial}{\partial \xi^i}$ and $\widetilde{\partial}_\tau = \frac{\partial}{\partial \tau}$, while the natural basis of the coordinate system $[\xi^i]$ of $S$ is denoted by $\partial_i$. In general, for a vector field $X = X^i \partial_i$ $(\in \mathcal{T}(S))$ on $S$, we define $\tilde{X} = X^i \widetilde{\partial}_i$ $(\in \mathcal{T}(\tilde{S}))$ as the vector field on $\tilde{S}$ which maps $\tau p_\xi$ to $(X^i)_\xi (\widetilde{\partial}_i)_{(\xi,\tau)}$.

Letting $\tilde{\ell}^{(\alpha)} = \ell^{(\alpha)}(x;\xi,\tau) = L^{(\alpha)}(\tau p(x;\xi))$ and $\ell^{(\alpha)} = \ell^{(\alpha)}(x;\xi)$, we have

$$\widetilde{\partial}_i \tilde{\ell}^{(\alpha)} = \tau^{\frac{1-\alpha}{2}} \partial_i \ell^{(\alpha)}, \quad \widetilde{\partial}_\tau \tilde{\ell}^{(\alpha)} = \tau^{-\frac{1+\alpha}{2}} p^{\frac{1-\alpha}{2}},$$

$$\widetilde{\partial}_i \widetilde{\partial}_j \tilde{\ell}^{(\alpha)} = \tau^{\frac{1-\alpha}{2}} \partial_i \partial_j \ell^{(\alpha)}, \quad \widetilde{\partial}_\tau \widetilde{\partial}_\tau \tilde{\ell}^{(\alpha)} = -\frac{1+\alpha}{2} \tau^{-\frac{3+\alpha}{2}} p^{\frac{1-\alpha}{2}},$$

$$\widetilde{\partial}_\tau \widetilde{\partial}_i \tilde{\ell}^{(\alpha)} = \widetilde{\partial}_i \widetilde{\partial}_\tau \tilde{\ell}^{(\alpha)} = \frac{1-\alpha}{2} \tau^{-\frac{1+\alpha}{2}} \partial_i \ell^{(\alpha)}.$$

Using these relations and Equations (2.60) (2.61), we obtain

$$\tilde{g}_{ij} = \tau g_{ij}, \quad \tilde{g}_{i\tau} = 0, \quad \tilde{g}_{\tau\tau} = \tau^{-1}, \tag{2.67}$$

$$\widetilde{\Gamma}^{(\alpha)}_{ij,k} = \tau \Gamma^{(\alpha)}_{ij,k}, \quad \widetilde{\Gamma}^{(\alpha)}_{ij,\tau} = -\frac{1+\alpha}{2} g_{ij}, \tag{2.68}$$

$$\widetilde{\Gamma}^{(\alpha)}_{i\tau,k} = \widetilde{\Gamma}^{(\alpha)}_{\tau i,k} = \frac{1-\alpha}{2} g_{ik}, \quad \widetilde{\Gamma}^{(\alpha)}_{i\tau,\tau} = \widetilde{\Gamma}^{(\alpha)}_{\tau i,\tau} = 0, \tag{2.69}$$

$$\widetilde{\Gamma}^{(\alpha)}_{\tau\tau,k} = 0, \quad \widetilde{\Gamma}^{(\alpha)}_{\tau\tau,\tau} = -\frac{1+\alpha}{2}\tau^{-2}. \tag{2.70}$$

Here $\tilde{g}$ and $\widetilde{\Gamma}^{(\alpha)}$ are the components of the Fisher metric and the $\alpha$-connection on $\tilde{S}$, while $g$ and $\Gamma^{(\alpha)}$ are those on $S$. These equations enable us to verify that the following relations hold for the covariant derivatives of the $\alpha$-connections on $\tilde{S}$ and $S$, which we denote by $\widetilde{\nabla}^{(\alpha)}$ and $\nabla^{(\alpha)}$, respectively:

$$\widetilde{\nabla}^{(\alpha)}_{\tilde{X}}\tilde{Y} = (\nabla^{(\alpha)}_X Y)^\sim - \frac{1+\alpha}{2}\left\langle \tilde{X}, \tilde{Y} \right\rangle \tilde{\partial}_\tau, \tag{2.71}$$

$$\widetilde{\nabla}^{(\alpha)}_{\tilde{X}}\tilde{\partial}_\tau = \widetilde{\nabla}^{(\alpha)}_{\tilde{\partial}_\tau}\tilde{X} = \frac{1-\alpha}{2}\tau^{-1}\tilde{X} \quad \text{and} \tag{2.72}$$

$$\widetilde{\nabla}^{(\alpha)}_{\tilde{\partial}_\tau}\tilde{\partial}_\tau = -\frac{1+\alpha}{2}\tau^{-1}\tilde{\partial}_\tau, \tag{2.73}$$

where $X$ and $Y$ are arbitrary vector fields on $S$, and $(\cdots)^\sim$ denotes $\widetilde{(\cdots)}$.

Now we are ready to show two theorems about the relation between $S$ and $\tilde{S}$.

**Theorem 2.9** $S$ is $(-1)$-autoparallel in $\tilde{S}$.

**Proof:** Obvious from Equation (2.71) or from the definition of $(-1)$-connection. ∎

**Theorem 2.10** Let $M$ be a submanifold of $S$ and $\tilde{M}$ be its denormalization. For every $\alpha \in \mathbb{R}$, the following conditions (i) and (ii) are equivalent.

(i) $M$ is $\alpha$-autoparallel in $S$.

(ii) $\tilde{M}$ is $\alpha$-autoparallel in $\tilde{S}$.

**Proof:** Let $\widetilde{\nabla}^{(\alpha)}$ and $\nabla^{(\alpha)}$ be the $\alpha$-connections on $\tilde{S}$ and $S$ as above. Noting that every vector field on $\tilde{M}$ can be represented as $f^i \tilde{X}_i + f^0 \tilde{\partial}_\tau|_{\tilde{M}}$ by vector fields $\{X_i\}$ on $M$ and functions $\{f^i\}$ and $f^0$ on $\tilde{M}$, we see from Equations (2.72) and (2.73) that condition (ii) is equivalent to stating that $\widetilde{\nabla}^{(\alpha)}_{\tilde{X}}\tilde{Y} \in \mathcal{T}(\tilde{M})$ for all $X, Y \in \mathcal{T}(M)$. On the other hand, we have for all $X, Y \in \mathcal{T}(M)$,

$$\widetilde{\nabla}^{(\alpha)}_{\tilde{X}}\tilde{Y} \in \mathcal{T}(\tilde{M}) \iff (\nabla^{(\alpha)}_X Y)^\sim \in \mathcal{T}(\tilde{M})$$
$$\iff \nabla^{(\alpha)}_X Y \in \mathcal{T}(M),$$

where the first equivalence follows from Equation (2.71) and the second one is obvious. It is thus concluded that (i) and (ii) are equivalent. ∎

We call a statistical model $S = \{p_\xi\}$ whose denormalization $\tilde{S}$ is an $\alpha$-affine manifold an **$\alpha$-family**.

## 2.6. $\alpha$-AFFINE MANIFOLDS AND $\alpha$-FAMILIES

**Example 2.11 ($\mathcal{P}(\mathcal{X})$ for finite $\mathcal{X}$)** *When $\mathcal{X}$ is finite, $\mathcal{P}(\mathcal{X})$ is an $\alpha$-family for every $\alpha \in \mathbb{R}$.*

For an $\alpha$-family $S$, its denormalization $\tilde{S}$ may be written in the form of Equation (2.65). We rewrite this as

$$L^{(\alpha)}(\tau p(x;\xi)) = C(x) + \sum_{\lambda=0}^{n} \theta^\lambda F_\lambda(x), \qquad (2.74)$$

where a one-to-one correspondence between $[\xi^1, \cdots, \xi^n, \tau]$ and $[\theta^0, \theta^1, \cdots, \theta^n]$ is implicitly assumed. When $\alpha \neq 1$, considering either the asymptote of $\tau \to 0$ ($\alpha < 1$) or $\tau \to \infty$ ($\alpha > 1$), we see that the affine space spanned by $L^{(\alpha)}(\tau p(x;\xi))$ must contain the origin and hence be a linear space. This means that $C(x)$ in the equation above vanishes and that the elements of the model $S$ are represented as

$$\ell^{(\alpha)}(x;\xi) = \sum_{\lambda=0}^{n} \theta^\lambda(\xi) F_\lambda(x), \qquad (2.75)$$

or, rescaling $\theta^\lambda$ by a constant factor, we have

$$p(x;\xi) = \left\{ \sum_{\lambda=0}^{n} \theta^\lambda(\xi) F_\lambda(x) \right\}^{2/(1-\alpha)}. \qquad (2.76)$$

**Example 2.12 ($\alpha = -1$: mixture family)** *When $\alpha = -1$ this becomes a mixture family. Indeed, if $F_i(x) - (a_i/a_o)F_0(x)$ for $1 \leq i \leq n$ and $F_0(x)/a_0$, where $a_\lambda \stackrel{\text{def}}{=} \int F_\lambda(x)\mathrm{d}x$, are respectively renamed $F_i(x)$ and $C(x)$, then Equation (2.76) is rewritten as Equation (2.35). The fact that a $(-1)$-family is a mixture family (and vice versa) is also viewed as a consequence of Theorem 2.9.*

**Example 2.13 ($\alpha = 1$: exponential family)** *A 1-family is an exponential family (and vice versa). Indeed, when $\alpha = 1$, observing that $L^{(1)}(\tau p(x;\xi)) = \log p(x;\tau) + \log \tau$, we see that the basis functions $\{F_0, \cdots, F_n\}$ in Equation (2.74) can be chosen so that $F_0 = 1$, and then, the elements of $S$ are written as Equation (2.31) with $\theta^0 = \psi(\theta^1, \ldots, \theta^n)$.*

From Theorem 2.10 we see that the following theorem holds for any $\alpha$, which is a generalization of Theorem 2.5.

**Theorem 2.11** *Let $S$ be an $\alpha$-family. Then a necessary and sufficient condition for a submanifold $M$ of $S$ to be an $\alpha$-family is for $M$ to be $\alpha$-autoparallel in $S$.*

In particular, when $\mathcal{X}$ is finite, the $\alpha$-families on $\mathcal{X}$ are characterized as the $\alpha$-autoparallel submanifolds of $\mathcal{P}(\mathcal{X})$.

It should be noted that, unlike Theorem 2.5, Theorem 2.4 has no $\alpha$-version because $\alpha$-family is not $\alpha$-flat in general. What then makes the cases $\alpha = \pm 1$ special? It is easy to answer the question for the case $\alpha = -1$. Indeed, this simply means that a $(-1)$-family $S$ inherits the $(-1)$-flatness from $\tilde{S}$ due to Theorem 2.9. The speciality of the case $\alpha = 1$ can also be understood through Theorem 2.9 with the aid of the duality between $(\pm 1)$-connections. We will revisit this in §3.5.

# Chapter 3

# Dual connections

When investigating the properties of the Fisher metric $g$ and the $\alpha$-connection $\nabla^{(\alpha)}$, and also when applying them to particular problems, it is important to consider them not individually, but rather as the triple $(g, \nabla^{(\alpha)}, \nabla^{(-\alpha)})$. The reason for this is that, through $g$, there exists a kind of duality between $\nabla^{(\alpha)}$ and $\nabla^{(-\alpha)}$ which is of fundamental significance. This notion of duality emerges not only when considering statistical models, but in a variety of problems relevant to information geometry. Through duality, it is possible to analyze these problems from a unified perspective; to do so is one of the principal motivations underlying this book. The present chapter is devoted to both the general theory of dual connections and its applications to the geometric structure of statistical models.

## 3.1 Duality of connections

Let $S$ be a manifold on which there is given a Riemannian metric $g = \langle \, , \, \rangle$ and two affine connections $\nabla$ and $\nabla^*$. If for all vector fields $X, Y, Z \in \mathcal{T}(S)$

$$Z \langle X, Y \rangle = \langle \nabla_Z X, Y \rangle + \langle X, \nabla_Z^* Y \rangle \qquad (3.1)$$

holds, then we say that $\nabla$ and $\nabla^*$ are **duals** of each other with respect to $g$, and call one either the **dual connection** or the **conjugate connection** of the other. In addition, we call such a triple $(g, \nabla, \nabla^*)$ a **dualistic structure** on $S$. By respectively using the coordinate expressions $g_{ij}$, $\Gamma_{ij,k}$, and $\Gamma_{ij,k}^*$, of $g$, $\nabla$, and $\nabla^*$ with respect to a coordinate system $[\xi^i]$, we may rewrite the duality constraint in Equation (3.1) as

$$\partial_k g_{ij} = \Gamma_{ki,j} + \Gamma_{kj,i}^*. \qquad (3.2)$$

In general, given a metric $g$ and a connection $\nabla$ on $S$, there exists a unique dual connection $\nabla^*$ of $\nabla$ with respect to $g$. In addition, $(\nabla^*)^* = \nabla$ holds. We also see that $(\nabla + \nabla^*)/2$ becomes a metric connection. And conversely, if a connection $\nabla'$ has the same torsion as $\nabla^*$ and if $(\nabla + \nabla')/2$ is metric, then $\nabla' = \nabla^*$.

The following result may be verified in a manner similar to the derivation of Equation (2.30), or is immediate from Equations (2.60) and (2.61).

**Theorem 3.1** *For any statistical model, or more generally, for any manifold of finite measures treated in §2.6, the $\alpha$-connection and the $(-\alpha)$-connection are dual with respect to the Fisher metric.*

In particular, the duality of the e-connection and the m-connection are of practical importance.

Suppose we have a dualistic structure $(g, \nabla, \nabla^*)$ on $S$. Letting $M$ be a submanifold of $S$, consider $\nabla_M$ and $\nabla^*_M$, which are respectively the projections of $\nabla$ and $\nabla^*$ onto $M$ with respect to $g$. These are dual with respect to $g_M$ (the metric on $M$ determined by $g$). We call $(g_M, \nabla_M, \nabla^*_M)$ the dualistic structure on $M$ **induced** by $(g, \nabla, \nabla^*)$, or the **induced dualistic structure** on $M$.

The condition for a connection $\nabla$ to be metric in Equation (1.65) is equivalent to requiring self-duality, $\nabla = \nabla^*$. Hence we see that the duality of connections may be considered as a generalization of the notion of metric connection. This becomes all the more clear if we analyze the meaning of duality using the parallel translation of vectors. Let $\gamma : t \mapsto \gamma(t)$ be a curve in $S$ and let $X$ and $Y$ be vector fields along $\gamma$. In addition, let $\frac{\delta X}{dt}$ and $\frac{\delta^* Y}{dt}$ respectively denote the covariant derivatives of $X$ with respect to $\nabla$ and $Y$ with respect to $\nabla^*$. Then from Equation (3.1) we see that

$$\frac{d}{dt}\langle X(t), Y(t)\rangle = \left\langle \frac{\delta X(t)}{dt}, Y(t) \right\rangle + \left\langle X(t), \frac{\delta^* Y(t)}{dt} \right\rangle. \tag{3.3}$$

Now suppose that $X$ is parallel with respect to $\nabla$, and that $Y$ is parallel with respect to $\nabla^*$. I.e., suppose that $\frac{\delta X}{dt} = \frac{\delta^* Y}{dt} = 0$. Then the right hand side of Equation (3.3) is 0, and hence the inner product $\langle X(t), Y(t)\rangle$ is constant on $\gamma$. Therefore, we obtain the following theorem.

**Theorem 3.2** *Letting $\Pi_\gamma$ and $\Pi^*_\gamma$ ($: T_p(S) \to T_q(S)$, where $p$ and $q$ are the boundary points of $\gamma$) respectively denote the parallel translation along $\gamma$ with respect to $\nabla$ and $\nabla^*$, then for all $X, Y \in T_p(S)$ we have*

$$\langle \Pi_\gamma(X), \Pi^*_\gamma(Y)\rangle_q = \langle X, Y\rangle_p. \tag{3.4}$$

This is a generalization of "the invariance of the inner product under parallel translation through metric connections" discussed in §1.10.

The relationship between $\Pi_\gamma$ and $\Pi^*_\gamma$ is completely determined by Equation (3.4). In other words, if either one of $\Pi_\gamma$ or $\Pi^*_\gamma$ is known, then the other can be obtained using Equation (3.4). Hence if $\Pi_\gamma$ is independent of the actual curve joining $p$ and $q$, and hence may be written as $\Pi_\gamma = \Pi_{p,q}$, then this is true of $\Pi^*_\gamma$ also. This means that the following theorem holds.

**Theorem 3.3** *Letting the curvature tensors of $\nabla$ and $\nabla^*$ be denoted by $R$ and $R^*$, respectively, we have*

$$R = 0 \iff R^* = 0. \tag{3.5}$$

Actually, what follows is that for any vector fields $X, Y, Z, W \in \mathcal{T}(S)$

$$\langle R(X,Y)Z, W \rangle = -\langle R^*(X,Y)W, Z \rangle, \tag{3.6}$$

from which Equation (3.5) is immediate. On the other hand, the corresponding property does not hold for the torsion tensors $T$ of $\nabla$ and $T^*$ of $\nabla^*$.

## 3.2 Divergences: general contrast functions

Let $S$ be a manifold and suppose that we are given a smooth function $D = D(\cdot \| \cdot) : S \times S \to \mathbb{R}$ satisfying for any $p, q \in S$

$$D(p \| q) \geq 0, \quad \text{and} \quad D(p \| q) = 0 \quad \text{iff} \quad p = q. \tag{3.7}$$

In other words, $D$ is a distance-like measure of the separation between two points. However, it does not in general satisfy the axioms of distance (symmetry and the triangle inequality). The aim of the present section is to show that such a $D$ is closely related to the duality of connections.

Let us introduce some notations. Given an arbitrary coordinate system $[\xi^i]$ of $S$, let us represent a pair of points $(p, p') \in S \times S$ by a pair of coordinates $([\xi^i], [\xi'^i])$ and denote the partial derivatives of $D(p \| p')$ with respect to $p$ and $p'$ by

$$D((\partial_i)_p \| p') \stackrel{\text{def}}{=} \partial_i D(p \| p'),$$
$$D((\partial_i)_p \| (\partial_j)_{p'}) \stackrel{\text{def}}{=} \partial_i \partial_j' D(p \| p'),$$
$$D((\partial_i \partial_j)_p \| (\partial_k)_{p'}) \stackrel{\text{def}}{=} \partial_i \partial_j \partial_k' D(p \| p'), \text{ etc.,}$$

where in the right-hand sides $\partial_i = \frac{\partial}{\partial \xi^i}$ is applied to the first variable of $D$ at $p$ while $\partial_i' = \frac{\partial}{\partial \xi'^i}$ to the second variable at $p'$. These definitions are naturally extended to those of

$$D((X_1 \cdots X_l)_p \| p'), \quad D(p \| (Y_1 \cdots Y_m)_{p'})$$
$$\text{and} \quad D((X_1 \cdots X_l)_p \| (Y_1 \cdots Y_m)_{p'})$$

for any vector fields $X_1, \cdots, X_l, Y_1, \cdots, Y_m \in \mathcal{T}(S)$. Now we consider their restrictions onto the diagonal $\{(p,p) \,|\, p \in S\} \subset S \times S$ and denote the functions induced on $S$ by

$$D[X_1 \cdots X_l \| \cdot] \; : \; p \mapsto D((X_1 \cdots X_l)_p \| p),$$
$$D[\cdot \| Y_1 \cdots Y_m] \; : \; p \mapsto D(p \| (Y_1 \cdots Y_m)_p),$$
$$D[X_1 \cdots X_l \| Y_1 \cdots Y_m] \; : \; p \mapsto D((X_1 \cdots X_l)_p \| (Y_1 \cdots Y_m)_p).$$

It then follows from Equation (3.7) that

$$D[\partial_i \| \cdot] = D[\cdot \| \partial_i] = 0, \tag{3.8}$$

$$D[\partial_i \partial_j \| \cdot] = D[\cdot \| \partial_i \partial_j] = -D[\partial_i \| \partial_j] \ (\stackrel{\text{def}}{=} g_{ij}^{(D)}), \tag{3.9}$$

and that the matrix $[g_{ij}^{(D)}]$ is positive semidefinite. When $[g_{ij}^{(D)}]$ is strictly positive definite everywhere on $S$, we say that $D$ is a **divergence**[1] or a **contrast function** on $S$. For a divergence $D$, a unique Riemannian metric $g^{(D)} = \langle \ , \ \rangle^{(D)}$ on $S$ is defined by $\langle \partial_i, \partial_j \rangle^{(D)} = g_{ij}^{(D)}$, or equivalently by

$$\langle X, Y \rangle^{(D)} = -D[X \| Y]. \tag{3.10}$$

This metric gives the second order approximation of $D$ as

$$D(p \| q) = \frac{1}{2} g_{ij}^{(D)}(q) \Delta \xi^i \Delta \xi^j + o(\|\Delta \xi\|^2), \tag{3.11}$$

where $\Delta \xi^i \stackrel{\text{def}}{=} \xi^i(p) - \xi^i(q)$ and $o(\|\Delta \xi\|^2)$ is a term vanishing faster than $\|\Delta \xi\|^2$ as $p$ tends to $q$.

Given a divergence $D$, we can also define an affine connection $\nabla^{(D)}$ with coefficients $\Gamma_{ij,k}^{(D)}$ by

$$\Gamma_{ij,k}^{(D)} = -D[\partial_i \partial_j \| \partial_k], \tag{3.12}$$

or equivalently by

$$\left\langle \nabla_X^{(D)} Y, Z \right\rangle^{(D)} = -D[XY \| Z], \tag{3.13}$$

because $\{\Gamma_{ij,k}^{(D)}\}$ obeys the same transformation rule as Equation (1.61). Note that $\nabla^{(D)}$ is necessarily symmetric ($\Gamma_{ij,k}^{(D)} = \Gamma_{ji,k}^{(D)}$). Combined with the metric $g^{(D)}$, the connection $\nabla^{(D)}$ gives the third order approximation of the divergence $D$:

$$D(p \| q) = \frac{1}{2} g_{ij}^{(D)}(q) \Delta \xi^i \Delta \xi^j + \frac{1}{6} h_{ijk}^{(D)}(q) \Delta \xi^i \Delta \xi^j \Delta \xi^k + o(\|\Delta \xi\|^3) \tag{3.14}$$

where

$$h_{ijk}^{(D)} \stackrel{\text{def}}{=} D[\partial_i \partial_j \partial_k \| \cdot]. \tag{3.15}$$

Indeed, the coefficients $\{h_{ijk}^{(D)}\}$ are determined from $\{g_{ij}^{(D)}\}$ and $\{\Gamma_{ij,k}^{(D)}\}$ by

$$h_{ijk}^{(D)} = \partial_i g_{jk}^{(D)} + \Gamma_{jk,i}^{(D)}. \tag{3.16}$$

Conversely, we see that $g^{(D)}$ and $\nabla^{(D)}$ are determined by the expansion (3.14) through Equation (3.16).

---

[1] In some literature the term *divergence* is used to mean a narrower notion such as the canonical divergence (as in the original Japanese edition of this book) introduced in §3.4 or the Kullback divergence.

## 3.2. DIVERGENCES: GENERAL CONTRAST FUNCTIONS

Let us replace the divergence $D(p \| q)$ with its **dual** $D^*(p \| q) = D(q \| p)$. Then we obtain $g^{(D^*)} = g^{(D)}$ and

$$\Gamma^{(D^*)}_{ij,k} = -D[\partial_k \| \partial_i \partial_j]. \tag{3.17}$$

Now it is easy to see the following theorem.

**Theorem 3.4** $\nabla^{(D)}$ and $\nabla^{(D^*)}$ are dual with respect to $g^{(D)}$.

This general construction of the dualistic structure $(g^{(D)}, \nabla^{(D)}, \nabla^{(D^*)})$ from a divergence $D$, together with the convenient notation of $D[X_1 \cdots X_l \| Y_1 \cdots Y_m]$, is due to Eguchi [84, 85, 87]. Corresponding to Equations (3.14) through (3.16), we have

$$D(p \| q) = D^*(q \| p) \tag{3.18}$$
$$= \frac{1}{2} g^{(D)}_{ij}(p) \Delta \xi^i \Delta \xi^j - \frac{1}{6} h^{(D^*)}_{ijk}(p) \Delta \xi^i \Delta \xi^j \Delta \xi^k + o(\|\Delta \xi\|^3), \tag{3.19}$$

where

$$h^{(D^*)}_{ijk} \stackrel{\text{def}}{=} D[\cdot \| \partial_i \partial_j \partial_k] = \partial_i g^{(D)}_{jk} + \Gamma^{(D^*)}_{jk,i}. \tag{3.20}$$

We thus see that any divergence induces a torsion-free dualistic structure. Conversely, any triple $(g, \nabla, \nabla^*)$ of a metric and mutually dual symmetric connections are induced from a divergence. Given such a triple together with a coordinate system $[\xi^i]$, let

$$D(p \| q) \stackrel{\text{def}}{=} \frac{1}{2} g_{ij}(q) \Delta \xi^i \Delta \xi^j + \frac{1}{6} h_{ijk}(q) \Delta \xi^i \Delta \xi^j \Delta \xi^k,$$

where $\Delta \xi^i = \xi^i(p) - \xi^i(q)$ and

$$h_{ijk} \stackrel{\text{def}}{=} \partial_i g_{jk} + \Gamma_{jk,i} = \Gamma_{ij,k} + \Gamma^*_{ik,j} + \Gamma_{jk,i}.$$

Then noting that $h_{ijk}$ is symmetric under all the permutations on $\{i, j, k\}$, we have $D[\partial_i \partial_j \| \cdot] = g_{ij}$ and $D[\partial_i \partial_j \partial_k \| \cdot] = h_{ijk}$, which immediately implies that $(g, \nabla, \nabla^*) = (g^{(D)}, \nabla^{(D)}, \nabla^{(D^*)})$. Note that the positivity (3.7) for sufficiently near $p, q$ follows from the positive-definiteness of $[g_{ij}]$. Also, an alternative choice of $D$ is given by

$$D(p \| q) \stackrel{\text{def}}{=} \frac{1}{2} g_{ij}(p) \Delta \xi^i \Delta \xi^j - \frac{1}{6} h^*_{ijk}(p) \Delta \xi^i \Delta \xi^j \Delta \xi^k,$$

where

$$h^*_{ijk} \stackrel{\text{def}}{=} \partial_i g_{jk} + \Gamma^*_{jk,i} = \Gamma^*_{ij,k} + \Gamma_{ik,j} + \Gamma^*_{jk,i}.$$

While these constructions are local, T. Matumoto [150] has proved that every torsion-free dualistic structure is induced from a globally defined divergence.

Now let us introduce an important class of divergences on statistical models. Let $f(u)$ be a convex function on $u > 0$. For each probability distributions $p, q$, we define

$$D_f(p \| q) \stackrel{\text{def}}{=} \int p(x) f\left(\frac{q(x)}{p(x)}\right) dx$$

and call it the **$f$-divergence** following I. Csiszàr. Let us review some fundamental properties of the $f$-divergence [70, 71, 214]. First, using Jensen's inequality we have

$$D_f(p \| q) \geq f\left(\int p(x) \frac{q(x)}{p(x)} dx\right) = f(1), \qquad (3.21)$$

where the equality holds if $p = q$ and, conversely, the equality implies $p = q$ when $f(u)$ is strictly convex at $u = 1$. Secondly, $D_f$ is kept invariant when $f(u)$ is replaced with $f(u) + c(u-1)$ for any $c \in \mathbb{R}$. Thirdly, $D_f^* = D_{f^*}$, where $f^*(u) = uf(1/u)$. Fourthly, and most importantly, $D_f$ has the following property similar to the monotonicity of the Fisher metric mentioned in § 2.2. Let $\kappa = \{\kappa(y \,|\, x) \geq 0 \;;\; x \in \mathcal{X}, y \in \mathcal{Y}\}$ be an arbitrary transition probability distribution satisfying $\int \kappa(y \,|\, x) dy = 1, \forall x$, whereby the value of $x$ is randomly transformed to $y$ according to the probability $\kappa(y \,|\, x)$. Note that the deterministic transformation $y = F(x)$ by a mapping $F : \mathcal{X} \to \mathcal{Y}$ corresponds to the case when $\kappa(y \,|\, x) = \delta_{F(x)}(y)$. Denoting the distributions of $y$ derived from $p(x)$ and $q(x)$ by $p_\kappa(y)$ and $q_\kappa(y)$ respectively, we have

$$D_f(p \| q) \geq D_f(p_\kappa \| q_\kappa). \qquad (3.22)$$

This inequality is referred to as the **monotonicity** of $D_f$ and is proved by Jensen's inequality using the conditional distribution $p_\kappa(x \,|\, y)$ as follows:

$$\begin{aligned} D_f(p \| q) &= \int\int p(x) \kappa(y \,|\, x) f\left(\frac{q(x)}{p(x)}\right) dx dy \\ &= \int\int p_\kappa(y) p_\kappa(x \,|\, y) f\left(\frac{q(x)}{p(x)}\right) dx dy \\ &\geq \int p_\kappa(y) f\left(\int p_\kappa(x \,|\, y) \frac{q(x)}{p(x)} dx\right) dy \\ &= D_f(p_\kappa \| q_\kappa). \end{aligned}$$

The equality in Equation (3.22) holds if $\kappa$ is induced from a sufficient statistic with respect to $\{p, q\}$, which means that $p_\kappa(x \,|\, y) = q_\kappa(x \,|\, y)$ for all $x$ and $y$, and the converse statement is also true when $f$ is a strictly convex function. Note that the **joint convexity**

$$\begin{aligned} D_f(\lambda p_1 + (1-\lambda) p_2 \| \lambda q_1 + (1-\lambda) q_2) \\ \leq \lambda D_f(p_1 \| q_1) + (1-\lambda) D_f(p_2 \| q_2), \quad 0 \leq \lambda \leq 1 \end{aligned} \qquad (3.23)$$

follows from the monotonicity by setting $\lambda_1 = \lambda$, $\lambda_2 = 1 - \lambda$, $p(x, i) = \lambda_i p_i(x)$, $q(x, i) = \lambda_i q_i(x)$ and $F(x, i) = x$ for $i = 1, 2$. In addition, we may now consider Equation (3.21) as the monotonicity for a constant mapping $F$.

## 3.2. DIVERGENCES: GENERAL CONTRAST FUNCTIONS

Let us assume that $f$ is a strictly convex and smooth function satisfying $f(1) = 0$. Then $D_f$ becomes a divergence in our sense on each statistical model, and induces the metric $g^{(D_f)} = g^{(f)}$ and the connection $\nabla^{(D_f)} = \nabla^{(f)}$ in the aforementioned manner. Since $D_f$ is invariant with respect to sufficient statistics, so are $g^{(f)}$ and $\nabla^{(f)}$. According to Theorem 2.6, this implies that they are represented as $g^{(f)} = cg$ and $\nabla^{(f)} = \nabla^{(\alpha)}$ by some $c > 0$ and $\alpha$, where $g$ is the Fisher metric and $\nabla^{(\alpha)}$ is the $\alpha$-connection. Indeed, a direct calculation shows that this is true for $c = f''(1)$ and $\alpha = 3 + 2f'''(1)/f''(1)$.

Important examples of smooth $f$-divergences are given by the **$\alpha$-divergence** $D^{(\alpha)} = D_{f^{(\alpha)}}$ for a real number $\alpha$, which is defined by

$$f^{(\alpha)}(u) = \begin{cases} \frac{4}{1-\alpha^2}\{1 - u^{(1+\alpha)/2}\} & (\alpha \neq \pm 1) \\ u \log u & (\alpha = 1) \\ -\log u & (\alpha = -1). \end{cases} \quad (3.24)$$

We have for $\alpha \neq \pm 1$

$$D^{(\alpha)}(p \parallel q) = \frac{4}{1-\alpha^2}\left\{1 - \int p(x)^{\frac{1-\alpha}{2}} q(x)^{\frac{1+\alpha}{2}} dx\right\} \quad (3.25)$$

and for $\alpha = \pm 1$

$$D^{(-1)}(p \parallel q) = D^{(1)}(q \parallel p) = \int p(x) \log \frac{p(x)}{q(x)} dx. \quad (3.26)$$

We can immediately see that the $\alpha$-divergence $D^{(\alpha)}$ induces $(g^{(f^{(\alpha)})}, \nabla^{(f^{(\alpha)})}) = (g, \nabla^{(\alpha)})$, whereas a deeper relation between $D^{(\alpha)}$ and $(g, \nabla^{(\alpha)})$ will be elucidated in § 3.6. Note that $D^{(\alpha)}(p \parallel q) = D^{(-\alpha)}(q \parallel p)$ generally holds. In particular, $D^{(0)}(p \parallel q)$ is symmetric, and moreover $\sqrt{D^{(0)}(p \parallel q)}$ satisfies the axioms of distance, which follows since

$$D^{(0)}(p \parallel q) = 2 \int (\sqrt{p(x)} - \sqrt{q(x)})^2 dx.$$

$\sqrt{D^{(0)}(p\parallel q)}$ (or $D^{(0)}(p\parallel q)$ itself) is called the **Hellinger distance**. The $\alpha$-divergence is closely related to the $\alpha$-entropy of Rényi [194], the Chernoff distance [66] and the Tsallis entropy [207, 74, 208].

The $\pm 1$-divergence is called by several different names, among which are the **Kullback divergence**, the **Kullback-Leibler information**, the **relative entropy** or simply the **divergence**. It is particularly important and has many applications in fields related to probability and information. Following the convention, we refer to $D^{(-1)}$ as the Kullback divergence and $D^{(1)}$ as its dual in this book. Unlike other $f$-divergences, the Kullback divergence satisfies the following **chain rule**:

$$\begin{aligned} D^{(-1)}(p \parallel q) &= D^{(-1)}(p_\kappa \parallel q_\kappa) \\ &+ \int D^{(-1)}(p_\kappa(\cdot \mid y) \parallel q_\kappa(\cdot \mid y)) p_\kappa(y) dy, \end{aligned} \quad (3.27)$$

which gives another proof of the monotonicity. In particular, it satisfies the **additivity**:

$$D^{(\pm 1)}(p_{12} \| q_{12}) = D^{(\pm 1)}(p_1 \| q_1) + D^{(\pm 1)}(p_2 \| q_2) \qquad (3.28)$$

for product distributions $p_{12}(x_1, x_2) = p_1(x_1)p_2(x_2)$ and $q_{12}(x_1, x_2) = q_1(x_1)q_2(x_2)$. See § 2.2 for the chain rule and the additivity of the Fisher metric.

## 3.3 Dually flat spaces

Let $(g, \nabla, \nabla^*)$ be a dualistic structure on a manifold $S$. If the connections $\nabla$ and $\nabla^*$ are both symmetric ($T = T^* = 0$), then from Theorem 3.3 we see that $\nabla$-flatness and $\nabla^*$-flatness are equivalent. For example, since the $\alpha$-connections are always symmetric, we have for any statistical model (or more generally for any manifold consisting of finite measures) $S$ and for any real number $\alpha$ that

$$S \text{ is } \alpha\text{-flat} \iff S \text{ is } (-\alpha)\text{-flat}. \qquad (3.29)$$

In particular, recalling that an exponential family is 1-flat and that a mixture family is $(-1)$-flat (§2.3), we now see in addition that they are both $(\pm 1)$-flat.

In general, we call $(S, g, \nabla, \nabla^*)$ a **dually flat space** if both duals $\nabla$ and $\nabla^*$ are flat.

**Theorem 3.5** *Let $(S, g, \nabla, \nabla^*)$ be a dually flat space. If a submanifold $M$ of $S$ is autoparallel with respect to either $\nabla$ or $\nabla^*$, then $M$ is a dually flat space with respect to the dualistic structure $(g_M, \nabla_M, \nabla_M^*)$ induced on $M$ by $(g, \nabla, \nabla^*)$.*

**Proof:** Suppose $M$ is $\nabla$-autoparallel. Then from Theorem 1.1 (§1.8) we know that $\nabla_M$ is flat. Hence by Equation (3.5) the curvature tensor of $\nabla_M^*$ is 0. On the other hand, since $\nabla^*$ is flat, it is a symmetric connection, and hence its projection $\nabla_M^*$ is symmetric also. From the above we see that $\nabla_M^*$ is flat, and that $M$ is a dually flat space. The argument for the case when $M$ is $\nabla^*$-autoparallel is similar. ∎

For instance, an m-autoparallel submanifold of an exponential family and an e-autoparallel submanifold of a mixture family are both $(\pm 1)$-flat, even though they are no longer exponential nor mixture families in general.

Now let us investigate the general structure of a dually flat space $(S, g, \nabla, \nabla^*)$. First, from the definition it follows that there exist for $S$ a $\nabla$-affine coordinate system $[\theta^i]$ and a $\nabla^*$-affine coordinate system $[\eta_j]$,[2] for which we let $\partial_i \stackrel{\text{def}}{=} \frac{\partial}{\partial \theta^i}$ and $\partial^j \stackrel{\text{def}}{=} \frac{\partial}{\partial \eta_j}$. Since $\partial_i$ is a $\nabla$-flat vector field and $\partial^j$ is a $\nabla^*$-flat vector field, we see from Theorem 3.2 that $\langle \partial_i, \partial^j \rangle$ is constant on $S$. From Equation (1.41) which describes the degree of freedom in an affine coordinate system under

---
[2] By superscripting one of the indices and subscripting the other, we obtain forms which are naturally suited to the use of Einstein's convention

## 3.3. DUALLY FLAT SPACES

regular affine transformations, we see that for a particular $\nabla$-affine coordinate system $[\theta^i]$, one may choose a corresponding $\nabla^*$-affine coordinate system $[\eta_j]$ such that

$$\langle \partial_i, \partial^j \rangle = \delta_i^j. \tag{3.30}$$

In general, if two coordinate systems $[\theta^i]$ and $[\eta_j]$ for a Riemannian manifold $(S, g)$ satisfy the condition above, we call the coordinate systems **mutually dual** (with respect to $g$), and call one the **dual coordinate system** of the other. We see then that the Euclidean coordinate system defined in Equation (1.70) is self-dual. In general, there do not exist dual coordinate systems for a Riemannian manifold $(S, g)$. However, if $(S, g, \nabla, \nabla^*)$ is a dually flat space, then such a pair of coordinate systems exist. Conversely, if for a Riemannian manifold $(S, g)$ there exists such coordinate systems $[\theta^i]$ and $[\eta_j]$, then the connections $\nabla$ and $\nabla^*$ for which they are affine are determined, and $(S, g, \nabla, \nabla^*)$ is a dually flat space.

Let the components of $g$ with respect to $[\theta^i]$ and $[\eta_j]$ be defined by

$$g_{ij} \stackrel{\text{def}}{=} \langle \partial_i, \partial_j \rangle \quad \text{and} \quad g^{ij} \stackrel{\text{def}}{=} \langle \partial^i, \partial^j \rangle. \tag{3.31}$$

By considering the coordinate transformation between $[\theta^i]$ and $[\eta_j]$, we have

$$\partial^j = (\partial^j \theta^i) \partial_i \quad \text{and} \quad \partial_i = (\partial_i \eta_j) \partial^j.$$

From this we see that Equation (3.30) is equivalent to

$$\frac{\partial \eta_j}{\partial \theta^i} = g_{ij} \quad \text{and} \quad \frac{\partial \theta^i}{\partial \eta_j} = g^{ij}, \tag{3.32}$$

and therefore $g_{ij} g^{jk} = \delta_i^k$, which is consistent with Equation (1.23).

Now suppose that we are given mutually dual coordinate systems $[\theta^i]$ and $[\eta_j]$, and consider the following partial differential equation for a function $\psi : S \to \mathbb{R}$:

$$\partial_i \psi = \eta_i. \tag{3.33}$$

We may rewrite this as $d\psi = \eta_i d\theta^i$, and a solution exists if and only if $\partial_i \eta_j = \partial_j \eta_i$. Since from Equation (3.32) we see that $\partial_i \eta_j = g_{ij} = \partial_j \eta_i$, in the context of our discussion a solution $\psi$ always exists. From Equations (3.33) and (3.32) we see that

$$\partial_i \partial_j \psi = g_{ij}. \tag{3.34}$$

Hence the second derivatives of $\psi$ form a positive definite matrix, and therefore $\psi$ is a strictly convex function of $[\theta^1, \cdots, \theta^n]$. Similarly, a solution $\varphi$ to

$$\partial^i \varphi = \theta^i \tag{3.35}$$

exists. In particular, using a solution $\psi$ to Equation (3.33), let

$$\varphi = \theta^i \eta_i - \psi. \tag{3.36}$$

Then we have
$$d\varphi = \theta^i d\eta_i + \eta_i d\theta^i - d\psi.$$

Substituting $d\psi = \eta_i d\theta^i$ into this equation, we obtain $d\varphi = \theta^i d\eta_i$, which is equivalent to Equation (3.35). From Equations (3.35) and (3.32) we see that $\varphi$ satisfies

$$\partial^i \partial^j \varphi = g^{ij}, \tag{3.37}$$

and hence it is a strictly convex function of $[\eta_1, \cdots, \eta_n]$. Furthermore, it follows from the convexity of $\psi$ and Equations (3.33) and (3.36) that for every $q \in S$

$$\varphi(q) = \max_{p \in S} \left\{ \theta^i(p)\eta_i(q) - \psi(p) \right\}. \tag{3.38}$$

Similarly, for every $p \in S$ we have

$$\psi(p) = \max_{q \in S} \left\{ \theta^i(p)\eta_i(q) - \varphi(q) \right\}. \tag{3.39}$$

Sometimes it is more natural to view these relations as

$$\varphi(\eta) = \max_{\theta \in \Theta} \left\{ \theta^i \eta_i - \psi(\theta) \right\} \tag{3.40}$$

$$\psi(\theta) = \max_{\eta \in \mathrm{H}} \left\{ \theta^i \eta_i - \varphi(\eta) \right\}, \tag{3.41}$$

where $\psi$ and $\varphi$ are simply convex functions defined on convex regions $\Theta$ and H in $\mathbb{R}^n$.

In general, those coordinate transformations $[\theta^i] \leftrightarrow [\eta_i]$ which may be expressed in the form given in Equations (3.33) through (3.39) are called **Legendre transformations**, and $\psi$ and $\varphi$ are called their **potentials**. Note also that

$$\Gamma^*_{ij,k} \stackrel{\text{def}}{=} \langle \nabla^*_{\partial_i} \partial_j, \partial_k \rangle = \partial_i \partial_j \partial_k \psi \quad \text{and} \tag{3.42}$$

$$\Gamma^{ij,k} \stackrel{\text{def}}{=} \langle \nabla_{\partial^i} \partial^j, \partial^k \rangle = \partial^i \partial^j \partial^k \varphi, \tag{3.43}$$

which are derived from Equation (3.2) combined with $\Gamma_{ij,k} = \Gamma^{*ij,k} = 0$.

We summarize the discussion above in the following theorem.

**Theorem 3.6** *Let $[\theta^i]$ be a $\nabla$-affine coordinate system on a dually flat space $(S, g, \nabla, \nabla^*)$. Then with respect to $g$ there exists a dual coordinate system $[\eta_i]$ of $[\theta^i]$, where $[\eta_i]$ turns out to be a $\nabla^*$-affine coordinate system. These two coordinate systems are related by the Legendre transformation given using potentials $\psi$ and $\varphi$ in Equations (3.33) through (3.39). In addition, the components of the metric $g$ with respect to these coordinate systems are given by the second derivatives of the potentials as given in Equations (3.34) and (3.37).*

## 3.4 Canonical divergence

In §3.2, we observed that an arbitrary divergence induces a torsion-free dualistic structure and that the converse statement is also true. However, it should be noted that the correspondence between divergences and dualistic structures is not one-to-one in that infinitely many divergences correspond to one dualistic structure. In this section, we show that a kind of canonical divergence is uniquely defined on a dually flat space.

Let $(S, g, \nabla, \nabla^*)$ be a dually flat space, on which we are given mutually dual affine coordinate systems $\{[\theta^i], [\eta_i]\}$ and their potentials $\{\psi, \varphi\}$. Given two points $p, q \in S$, let

$$D(p \,\|\, q) \stackrel{\text{def}}{=} \psi(p) + \varphi(q) - \theta^i(p)\eta_i(q). \tag{3.44}$$

Then from Equations (3.38) and (3.39) we see that $D(p \,\|\, q) \geq 0$ and $D(p \,\|\, q) = 0 \Leftrightarrow p = q$. Moreover, it is easy to verify the equations

$$D((\partial_i \partial_j)_p \,\|\, q) = g_{ij}(p) \quad \text{and} \quad D(p \,\|\, (\partial^i \partial^j)_q) = g^{ij}(q) \tag{3.45}$$

which immediately implies that $D$ is a divergence and induces $g$. We can also conclude from these equations that $\nabla = \nabla^{(D)}$ and $\nabla^* = \nabla^{(D^*)}$ since we have $\Gamma_{ij,k} = \Gamma^{*ij,k} = 0$ due to the $\nabla$-affinity of $[\theta^i]$ and the $\nabla^*$-affinity of $[\eta_i]$.

On a dually flat space $(S, g, \nabla, \nabla^*)$, the degrees of freedom of the dual affine coordinate systems $\{[\theta^i], [\eta_i]\}$ and the potentials $\{\psi, \varphi\}$ are expressed by

$$\tilde{\theta}^j = A_i^j \theta^i + B^j, \qquad \tilde{\eta}_j = C_j^i \eta_i + D_j,$$
$$\tilde{\psi} = \psi + D_j \tilde{\theta}^j + c, \quad \text{and} \quad \tilde{\varphi} = \varphi + B^j \tilde{\eta}_j - B^j D_j - c.$$

Here $[A_i^j]$ is a regular matrix, $[C_j^i]$ is its inverse, $[B^j]$ and $[D_j]$ are real-valued vectors, and $c$ is a real number. These degrees of freedom completely cancel each other in Equation (3.44) so that $D$ is uniquely determined from $(S, g, \nabla, \nabla^*)$. We call this $D$ the **canonical divergence of $(S, g, \nabla, \nabla^*)$** or the **$(g, \nabla)$-divergence** on $S$ for short.[3]

By interchanging the roles of $\nabla$ and $\nabla^*$ on $S$, the roles of $[\theta^i]$ and $[\eta_i]$, and also those of $\psi$ and $\varphi$ are exchanged, and we find the $(g, \nabla^*)$-divergence $D^*$ to be

$$D^*(p \,\|\, q) = D(q \,\|\, p). \tag{3.46}$$

In addition, if we let $M$ be a submanifold which is autoparallel with respect to either $\nabla$ or $\nabla^*$, and consider the induced dually flat structure $(g_M, \nabla_M, \nabla_M^*)$ (see Theorem 3.5), then it is possible to prove that the $(g_M, \nabla_M)$-divergence $D_M$ is given by the restriction $D_M = D|_{M \times M}$, or in other words

$$D_M(p \,\|\, q) = D(p \,\|\, q) \qquad (\forall p, q \in M). \tag{3.47}$$

---

[3] In some literature (including the original Japanese edition of this book), the canonical divergence is simply called the divergence.

A similar argument can be made for $D_M^*$.

When $\nabla$ is a Riemannian connection ($\nabla = \nabla^*$), the condition for "dually flat" reduces to $\nabla$ being flat, and hence in this case there exists a Euclidean coordinate system $[\theta^i]$. This coordinate system is self-dual ($\theta^i = \eta_i$), and its potential is given by $\psi = \varphi = \frac{1}{2}\sum_i (\theta^i)^2$. Substituting this into Equation (3.44), we obtain

$$\begin{aligned} D(p \| q) &= \frac{1}{2}\sum_i \left\{ (\theta^i(p))^2 + (\theta^i(q))^2 - 2\theta^i(p)\theta^i(q) \right\} \\ &= \frac{1}{2}\{d(p,q)\}^2, \end{aligned} \qquad (3.48)$$

where $d$ is the Euclidean distance $d(p,q) \stackrel{\text{def}}{=} \sqrt{\sum_i \{\theta^i(p) - \theta^i(q)\}^2}$. In general, the canonical divergence $D(p \| q)$ on a dually flat space is only approximately equal to $\frac{1}{2}(d(p,q))^2$ in the sense of Equation (3.11).

Let us give an important characterization of the canonical divergence.

**Theorem 3.7** *Let $\{[\theta^i], [\eta_i]\}$ be mutually dual affine coordinate systems of a dually flat space $(S, g, \nabla, \nabla^*)$, and let $D$ be a divergence on $S$. Then a necessary and sufficient condition for $D$ to be the $(g, \nabla)$-divergence is that for all $p, q, r \in S$ the following* **triangular relation** *holds:*

$$\begin{aligned} &D(p \| q) + D(q \| r) - D(p \| r) \\ &= \{\theta^i(p) - \theta^i(q)\}\{\eta_i(r) - \eta_i(q)\}. \end{aligned} \qquad (3.49)$$

**Proof:** The necessity of Equation (3.49) is straightforward from Equations (3.36) and (3.44). To show the sufficiency, suppose that a divergence $D$ obeys Equation (3.49). Applying $(\partial_i)_p$ to the both sides of the equation and then letting $r = p$, we obtain

$$D((\partial_i)_p \| q) = \eta_i(p) - \eta_i(q). \qquad (3.50)$$

On the other hand, we see that the canonical divergence also satisfies the same equation as above and that a solution of the equation is unique under the initial condition $D(p \| p) = 0$ ($\forall p$). Consequently, $D$ coincides with the canonical divergence. ∎

We may generalize the Pythagorean theorem for the Euclidean distance to the $(g, \nabla)$-divergence $D$ on an arbitrary dually flat space $(S, g, \nabla, \nabla^*)$.

**Theorem 3.8** *Let $p$, $q$, and $r$ be three points in $S$. Let $\gamma_1$ be the $\nabla$-geodesic connecting $p$ and $q$, and let $\gamma_2$ be the $\nabla^*$-geodesic connecting $q$ and $r$. If at the intersection $q$ the curves $\gamma_1$ and $\gamma_2$ are orthogonal (with respect to the inner product $g$), then we have the* **Pythagorean relation**

$$D(p \| r) = D(p \| q) + D(q \| r) \qquad (3.51)$$

*(Figure 3.1).*

## 3.4. CANONICAL DIVERGENCE

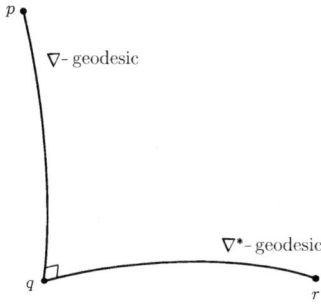

Figure 3.1: The Pythagorean relation for $(g, \nabla)$-divergences.

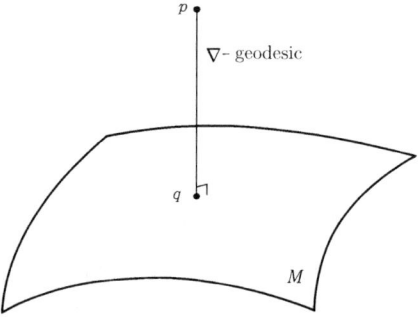

Figure 3.2: The projection theorem of $(g, \nabla)$-divergence

**Proof:** Since a $\nabla$-geodesic is a straight line with respect to $[\theta^i]$, we may parameterize $\gamma_1$ using $t$ as $\theta_t^i = t\theta^i(p) + (1-t)\theta^i(q)$, and obtain $\frac{d}{dt}\theta_t^i \partial_i = \{\theta^i(p) - \theta^i(q)\} \partial_i$ as an expression of the tangent vectors of this curve. Similarly, we may parameterize $\gamma_2$ as $\eta_{ti} = t\eta_i(q) + (1-t)\eta_i(r)$, and obtain as its tangent vectors $\frac{d}{dt}\eta_{ti}\partial^i = \{\eta_i(q) - \eta_i(r)\} \partial^i$. From Equation (3.30) we see that the inner product of these tangent vectors at the intersection point $q$ may be written as $\{\theta^i(p) - \theta^i(q)\} \{\eta_i(q) - \eta_i(r)\}$. Therefore when the curves are orthogonal at $q$, Equation (3.51) follows from Equation (3.49). ∎

The projection theorem given below follows immediately from the theorem above (see Figure 3.2).

**Corollary 3.9** *Let $p$ be a point in $S$ and let $M$ be a submanifold of $S$ which is $\nabla^*$-autoparallel. Then a necessary and sufficient condition for a point $q$ in $M$ to satisfy $D(p \| q) = \min_{r \in M} D(p \| r)$ is for the $\nabla$-geodesic connecting $p$ and $q$ to be orthogonal to $M$ at $q$.*

The point $q$ in the theorem above is called the **$\nabla$-projection of $p$ onto $M$**. More generally, the following holds for any submanifold $M$.

**Theorem 3.10** *Let $p$ be a point in $S$ and let $M$ be a submanifold of $S$. A necessary and sufficient condition for a point $q \in M$ to be a stationary point of the function $D(p \| \cdot) : r \mapsto D(p \| r)$ restricted on $M$ (in other words, the partial derivatives with respect to a coordinate system of $M$ are all 0) is for the $\nabla$-geodesic connecting $p$ and $q$ to be orthogonal to $M$ at $q$.*

**Proof:** Let $\partial_a = \frac{\partial}{\partial u^a}$ be the natural basis of a coordinate system $[u^a]$ of $M$. Then from Equations (3.44) and (3.35) we have

$$\begin{aligned} D(p \| (\partial_a)_q) &= (\partial_a \eta_i)_q D(p \| (\partial^i)_q) \\ &= (\partial_a \eta_i)_q \{\theta^i(q) - \theta^i(p)\} \\ &= \langle (\partial_a)_q, \{\theta^i(q) - \theta^i(p)\}(\partial_i)_q \rangle, \end{aligned} \quad (3.52)$$

from which the theorem follows. ∎

**Corollary 3.11** *Given a point $p$ in $S$ and a positive number $c$, suppose that the "D-sphere" $M = \{q \in S \mid D(p \| q) = c\}$ forms a hypersurface in $S$. Then every $\nabla$-geodesic passing through the center $p$ orthogonally intersects $M$.*

Before concluding this section, let us take another look at the notion of canonical divergence to illustrate how our geometry modifies the usual Riemannian geometry. Let a Riemannian metric $g$ and an affine connection $\nabla$ be given on a manifold $S$ and let $\nabla^*$ be the dual of $\nabla$ with respect to $g$. We do not assume that $(S, g, \nabla, \nabla^*)$ is dually flat, and hence the canonical divergence is not generally defined on $S$. Now, let $\gamma : [a, b] \to S$ ($a < b$) be a smooth curve in $S$ connecting the points $\gamma(a)$ and $\gamma(b)$, on which the dualistic structure $(g_\gamma, \nabla_\gamma, \nabla^*_\gamma)$ is induced from $(g, \nabla, \nabla^*)$ by projection. The coefficients of $g_\gamma$ and $\nabla_\gamma$ corresponding to $g_{ab}$ and $\Gamma^{(\pi)d}_{ab} = \Gamma^{(\pi)}_{ab,c} g^{cd}$ in Equations (1.26) and (1.62) are given by

$$\begin{aligned} g_\gamma(t) &= g_{ij}(\gamma(t)) \dot{\gamma}^i(t) \dot{\gamma}^j(t), \\ \Gamma_\gamma(t) &= \left\{ \dot{\gamma}^i(t) \dot{\gamma}^j(t) \Gamma_{ij,k}(\gamma(t)) + \ddot{\gamma}^j(t) g_{jk}(\gamma(t)) \right\} \dot{\gamma}^k(t) / g_\gamma(t). \end{aligned}$$

Since $\gamma$ is 1-dimensional, $(\gamma, g_\gamma, \nabla_\gamma, \nabla^*_\gamma)$ is always dually flat and the canonical divergence $D_\gamma$ is defined on $\gamma$. We define $D(\gamma) \stackrel{\text{def}}{=} D_\gamma(\gamma(b) \| \gamma(a))$ and call it the **$(g, \nabla)$-divergence of the curve $\gamma$**. Note that $D(\gamma)$ does not depend on the parametrization $t \mapsto \gamma(t)$ of $\gamma$ but its orientation, and that the $(g, \nabla^*)$-divergence of $\gamma$ coincides with the $(g, \nabla)$-divergence of the reversely oriented curve. Some calculation shows that

$$D(\gamma) = \iint_{a \leq s \leq t \leq b} g_\gamma(s) \frac{\mu(t)}{\mu(s)} \, ds \, dt, \quad (3.53)$$

where

$$\mu(t) \stackrel{\text{def}}{=} \exp\left[ \int_a^t \Gamma_\gamma(s) \, ds \right].$$

In particular, if the parameter $t$ is chosen to be $\nabla_\gamma$-affine, or in other words if $\Gamma_\gamma(t) = 0$ for all $t$, then we have (cf. Equation (3.45))

$$D(\gamma) = \iint_{a \leq s \leq t \leq b} g_\gamma(s) \, ds \, dt = \int_a^b (b - s) \, g_\gamma(s) \, ds. \tag{3.54}$$

When $\nabla$ is the Riemannian connection, by applying Equation (3.48) to $D_\gamma$ we see that $D(\gamma) = \frac{1}{2}\|\gamma\|^2$, where $\|\gamma\|$ is the length of $\gamma$ defined by Equation (1.25). In this respect, the $(g, \nabla)$-divergence of a curve gives a modification of the definition of curve length. Now, suppose that $(S, g, \nabla, \nabla^*)$ is a dually flat space, on which the canonical divergence $D$ is defined. In the Riemannian case ($\nabla = \nabla^*$), for any geodesic $\gamma$ we have $\|\gamma\| = d(\gamma(a), \gamma(b))$ and hence $D(\gamma) = D(\gamma(a) \| \gamma(b)) = D(\gamma(b) \| \gamma(a))$ by Equation (3.48). (Note that we are only treating local properties and do not consider cases such as cylinders.) On the other hand, in the general dually flat case we have $D(\gamma) = D(\gamma(b) \| \gamma(a))$ if $\gamma$ is either a $\nabla$-geodesic or a $\nabla^*$-geodesic (see Equation (3.47)). See Henmi and Kobayashi [108] for an interesting physical interpretation of the canonical divergence.

## 3.5 The dualistic structure of exponential families

In this section, we investigate the dually flat structure of an exponential family with respect to the ($\pm 1$)-connections and the Fisher metric, which are shown to be closely linked to some fundamental aspects of statistics. Let us begin with revisiting the question put at the end of §2.6 about the origin of the 1-flatness of an exponential family, which is now easy to answer. For an exponential family $S$, its extension $\tilde{S}$ is an 1-affine manifold and hence is 1-flat, and turns out also to be $(-1)$-flat by the duality. According to Theorem 2.9, $S$ inherits the $(-1)$-flatness from $\tilde{S}$, and turns out to be 1-flat by the duality again.

We showed in §2.3 that with respect to an exponential family

$$p(x; \theta) = \exp\left[C(x) + \theta^i F_i(x) - \psi(\theta)\right], \tag{3.55}$$

the natural parameters $[\theta^i]$ form a 1-affine coordinate system. Now if we define

$$\eta_i = \eta_i(\theta) \stackrel{\text{def}}{=} E_\theta[F_i] = \int F_i(x) p(x; \theta) \, dx, \tag{3.56}$$

then from Equations (2.33) and (2.9) we obtain $\eta_i = \partial_i \psi$. Furthermore, from Equations (2.34) and (2.8) we obtain $\partial_i \partial_j \psi = g_{ij}$. Hence $[\eta_i]$ is a $(-1)$-affine coordinate system dual to $[\theta^i]$, and $\psi$ is the potential of a Legendre transformation. We call this $[\eta_i]$ the **expectation parameters** or the **dual parameters**. For the examples of exponential families given in §2.3, we have the following.

**Example 3.1** *(Example 2.5: Normal Distribution)*

$$\eta_1 = \mu = -\frac{\theta^1}{2\theta^2}, \quad \eta_2 = \mu^2 + \sigma^2 = \frac{(\theta^1)^2 - 2\theta^2}{4(\theta^2)^2}$$

**Example 3.2** *(Example 2.6: Multivariate Normal Distribution)*

$$\eta_i = \mu_i, \quad \eta_{ij} = (\Sigma)_{ij} + \mu_i \mu_j \quad (i \leq j)$$

$$\eta_A = \mu = -\frac{1}{2}(\theta^B)^{-1}\theta^A,$$

$$\eta_B = \Sigma + \mu\mu^t = -\frac{1}{2}(\theta^B)^{-1} + \frac{1}{4}(\theta^B)^{-1}\theta^A(\theta^A)^t(\theta^B)^{-1}$$

**Example 3.3** *(Example 2.7: Poisson Distribution)*

$$\eta = \xi = \exp\theta$$

**Example 3.4** *(Example 2.8: $\mathcal{P}(\mathcal{X})$ for finite $\mathcal{X}$)*

$$\eta_i = p(x_i) = \xi^i = \frac{\exp\theta^i}{1 + \sum_{j=1}^n \exp\theta^j}$$

The dual potential $\varphi$ in Equation (3.36) is then given by

$$\begin{aligned}
\varphi(\theta) &= \theta^i \eta_i(\theta) - \psi(\theta) \\
&= E_\theta[\log p_\theta - C] \\
&= -H(p_\theta) - E_\theta[C],
\end{aligned} \quad (3.57)$$

where $H$ is the **entropy**: $H(p) \stackrel{\text{def}}{=} -\int p(x)\log p(x)\mathrm{d}x$. In addition, from Equation (3.38) we have

$$\varphi(\theta) = \max_{\theta'}\left\{\theta'^i \eta_i(\theta) - \psi(\theta')\right\}, \quad (3.58)$$

where the maximum is attained by $\theta' = \theta$.

From the definition of the Fisher information matrix (i.e., Equation (2.6)) we have

$$g_{ij}(\theta) = E_\theta[(F_i - \eta_i)(F_j - \eta_j)]. \quad (3.59)$$

Now let us regard the function $F_i$ as an estimator for the parameter $\eta_i$ and denote it by $\hat{\eta}_i(x) = F_i(x)$. Then Equation (3.56) means that $\hat{\eta} = [\hat{\eta}_1, \cdots, \hat{\eta}_n]$ is an unbiased estimator for the coordinate system $\eta = [\eta_1, \cdots, \eta_n]$, while Equation (3.59) means that the covariance matrix $V_\eta[\hat{\eta}]$ is equal to $G = [g_{ij}]$. It should be noted that $G$ is the Fisher information matrix for the coordinate system $\theta = [\theta^i]$ and, at the same time, is the inverse of the Fisher information matrix for $\eta = [\eta_i]$

## 3.5. THE DUALISTIC STRUCTURE OF EXPONENTIAL FAMILIES

by Equations (3.31) and (3.32). Hence, $\hat{\eta}$ attains the equality in the Cramér-Rao inequality (Theorem 2.2), or in other words, $\hat{\eta}$ is an efficient estimator. We have thus seen that an m-affine coordinate system of an exponential family always has an efficient estimator.

Conversely, if a coordinate system $\xi = [\xi^i]$ of a model $S = \{p_\xi\}$ has an efficient estimator, then $S$ is an exponential family and $\xi$ is an m-affine coordinate system composed of expectation parameters. We give a geometrical proof to this statement on the assumption that $\mathcal{X}$ is a finite set. Recall that we have already proved in §2.5 that if there is an efficient estimator $\hat{\xi} = [\hat{\xi}^i]$, then $S$ is an exponential family and there are $n(=\dim S)$ linearly independent e-parallel vector fields on $S$, say $\{X^1, \cdots, X^n\}$, such that their e-representations are $(X^i_\xi)^{(e)} = \hat{\xi}^i - \xi^i$. Letting $\partial_i = \frac{\partial}{\partial \xi^i}$, we have (cf. Equation (2.49))

$$\langle \partial_i, X^j \rangle = \partial_i E_\xi\left[ \hat{\xi}^j \right] = \partial_i \xi^j = \delta_i^j. \tag{3.60}$$

In other words, the inner product between $\partial_i$ and an arbitrary e-parallel vector field is always constant on $S$. By the duality of e- and m-connections, this means that $\partial_i$ is m-parallel and consequently $[\xi^i]$ is m-affine. An essentially same argument applies to the case when $\mathcal{X}$ is infinite, and we obtain the following theorem.

**Theorem 3.12** *A necessary and sufficient condition for a coordinate system $\xi$ of a model $S = \{p_\xi\}$ to have an efficient estimator is that $S$ is an exponential family and $\xi$ is m-affine.*

Let us proceed to investigate the canonical divergence. Substituting Equations (3.56) and (3.57) into Equation (3.44) we see that the $(g, \nabla^{(1)})$-divergence on the exponential family $S = \{p_\theta\}$ is given by

$$D^{(1)}(p_\theta \parallel p_{\theta'}) = E_{\theta'}[\log p_{\theta'} - \log p_\theta],$$

which is the 1-divergence defined by Equation (3.26), or in other words, the dual of Kullback divergence, and consequently the $(g, \nabla^{(-1)})$-divergence is the Kullback divergence $D^{(-1)}$. The triangular relation (3.49) in this case is essentially equivalent to the following relation for the Kullback divergence $D = D^{(-1)}$:

$$D(p \parallel q) + D(q \parallel r) - D(p \parallel r)$$
$$= \int \{p(x) - q(x)\}\{\log r(x) - \log q(x)\} \mathrm{d}x, \tag{3.61}$$

which is elementary but often useful in applications.

From Corollary 3.9 and Theorem 3.10, the solutions to the minimization problems

$$\min_{q \in M} D(p \parallel q) \quad \text{and} \quad \min_{q \in M} D(q \parallel p)$$

are respectively given by the $\nabla^{(m)}$-projection and $\nabla^{(e)}$-projection, both of which are important in many applications. The former problem frequently appears in statistics in connection with the maximum likelihood estimation (see Equation (4.38)), while the latter plays a crucial role in the large deviation theory via Sanov's theorem (see §6.2). An example of the latter problem will be given below, for which we begin with a slightly wider setting as follows.

Given $(n+1)$ functions $C, F_1, \ldots, F_n : \mathcal{X} \to \mathbb{R}$, let $S = \{p_\theta \mid \theta \in \Theta\}$ be the $n$-dimensional exponential family represented by Equation (3.55). Then for any $\theta \in \Theta$ and any $q \in \mathcal{P}(\mathcal{X})$ we have

$$H(p_\theta) + E_{p_\theta}[C] + \theta^i E_{p_\theta}[F_i] - H(q) - E_q[C] - \theta^i E_q[F_i]$$
$$= D(q \parallel p_\theta) \geq 0,$$

which leads to

$$\max_{q \in \mathcal{P}(\mathcal{X})} \left\{ H(q) + E_q[C] + \theta^i E_q[F_i] \right\}$$
$$= H(p_\theta) + E_{p_\theta}[C] + \theta^i E_{p_\theta}[F_i] = \psi(\theta). \tag{3.62}$$

Given a vector $\lambda = (\lambda_1, \ldots, \lambda_n) \in \mathbb{R}^n$, let

$$M_\lambda \stackrel{\text{def}}{=} \{q \in \mathcal{P} \mid E_q[F_i] = \lambda_i, \ i = 1, \ldots, n\}. \tag{3.63}$$

Since $M_\lambda$ is defined by a linear constraint on the elements, it is a mixture family. Now let us assume that $S \cap M_\lambda \neq \phi$, or equivalently that there exists an element of $\Theta$, say $\theta_\lambda$, such that $\eta_i(\theta_\lambda) = E_{p_{\theta_\lambda}}[F_i] = \lambda_i$ for $i = 1, \ldots, n$. Then we have:

$$\max_{q \in M_\lambda} \{H(q) + E_q[C]\} = H(p_{\theta_\lambda}) + E_{p_{\theta_\lambda}}[C]$$
$$= \psi(\theta_\lambda) - \theta_\lambda^i \lambda_i$$
$$= \min_{\theta \in \Theta} \{\psi(\theta) - \theta^i \lambda_i\}, \tag{3.64}$$

where the first and second equalities follow from Equation (3.62), while the last follows from Equations (3.57) and (3.58).

When $C = 0$ it follows that $\max_{q \in M_\lambda} H(q) = H(p_{\theta_\lambda})$, which is often referred to as the **principle of maximum entropy**. This has the origin in statistical physics; the thermal equilibrium state which maximizes the thermodynamical entropy $S(p) \stackrel{\text{def}}{=} kH(p)$, where $k(>0)$ is Boltzmann's constant, under the constraint $E_q[\varepsilon] = \bar{\varepsilon}$ on the average of the energy function $\varepsilon$, is given by the Boltzmann-Gibbs distribution

$$p^*(x) = \frac{1}{Z} e^{-\varepsilon(x)/kT},$$

where $T$ is the temperature and $Z$ is the partition function. This corresponds to the previous situation by letting $C = 0$, $n = 1$, $F_i = \varepsilon$, $\lambda = \bar{\varepsilon}$, $\theta_\lambda = -1/kT$ and $\psi(\theta_\lambda) = \log Z$. Assuming $T > 0$, $p^*$ is also characterized as the distribution minimizing Helmholtz's free energy $E_q[\varepsilon] - TS(q)$, which may be regarded as a special case of Equation (3.62).

## 3.5. THE DUALISTIC STRUCTURE OF EXPONENTIAL FAMILIES

When $C(x) = \log p(x)$ for a given distribution $p \in \mathcal{P}$, on the other hand, Equation (3.64) may be rewritten as

$$\min_{q \in M_\lambda} D(q \parallel p) = D(p_{\theta_\lambda} \parallel p)$$
$$= \theta_\lambda^i \lambda_i - \psi(\theta_\lambda) = \max_{\theta \in \Theta}\{\theta^i \lambda_i - \psi(\theta)\}, \quad (3.65)$$

where $p_\theta$ and $\psi$ are now represented as

$$p(x; \theta) = p(x)\exp[\theta^i F_i(x) - \psi(\theta)] \quad \text{and} \quad (3.66)$$

$$\psi(\theta) = \log E_p\left[e^{\theta^i F_i}\right]. \quad (3.67)$$

This $\psi(\theta)$ is commonly called the **logarithmic moment generating function** or the **cumulant generating function** of $p$ with respect to the random variables $F_1, \ldots, F_n$. Note that the mixture family $M_\lambda$ and the exponential family $S = \{p_\theta\}$ intersects orthogonally at $p_{\theta_\lambda}$, and therefore for all $q \in M_\lambda$ and all $\theta$ the following Pythagorean relation holds:

$$D(q \parallel p_\theta) = D(q \parallel p_{\theta_\lambda}) + D(p_{\theta_\lambda} \parallel p_\theta). \quad (3.68)$$

The first equality of Equation (3.65) may be viewed as a consequence of this relation for $\theta = 0$.

Now consider the case when $n = 1$. Given a probability distribution $p \in \mathcal{P}(\mathcal{X})$, a random variable $F: \mathcal{X} \to \mathbb{R}$ and a closed interval $I \subset \mathbb{R}$, Equation (3.65) leads to

$$R(I) \stackrel{\text{def}}{=} \min_{q: E_q[F] \in I} D(q \parallel p) = \min_{\lambda \in I} \max_{\theta}\{\theta\lambda - \psi(\theta)\}, \quad (3.69)$$

where $\psi(\theta) = E_p\left[e^{\theta F}\right]$. The probabilistic meaning of this quantity is given by the large deviation theory, which tells us that

$$\lim_{N \to \infty} \frac{1}{N} \log \Pr\left\{\frac{1}{N}\sum_{t=1}^{N} F(X_t) \in I\right\} = -R(I),$$

where $X_1, X_2, \cdots$ are $\mathcal{X}$-valued random variables that are independent and identically distributed according to $p(x)$. Equation (3.69) may now be considered to be a bridge between two famous large deviation theorems — Sanov's theorem and Cramér's theorem (see e.g. [78]).

We conclude this section by noting that, as examples of Equation (3.54), the Kullback divergence has two mutually dual integral representations. For arbitrary distributions $p_0$ and $p_1$, let us define $p_t^{(\text{m})} \stackrel{\text{def}}{=} (1-t)p_0 + tp_1$ and $p_t^{(\text{e})} \stackrel{\text{def}}{=} p_0^{1-t} p_1^t / Z_t$, where $Z_t$ is the normalizing constant. Then $\{p_t^{(\text{m})}\}$ and $\{p_t^{(\text{e})}\}$ form a mixture family and an exponential family, respectively, which are two different curves connecting $p_0$ and $p_1$. Now, letting $g^{(m)}(t)$ and $g^{(e)}(t)$ respectively denote the Fisher informations of $\{p_t^{(m)}\}$ and $\{p_t^{(e)}\}$, we have

$$D(p_1 \| p_0) = \iint_{0 \le s \le t \le 1} g^{(m)}(s) \, ds \, dt = \int_0^1 (1-s) g^{(m)}(s) \, ds \quad (3.70)$$

$$D(p_0 \| p_1) = \iint_{0 \le s \le t \le 1} g^{(e)}(s) \, ds \, dt = \int_0^1 (1-s) g^{(e)}(s) \, ds. \quad (3.71)$$

## 3.6 The dualistic structure of $\alpha$-affine manifolds and $\alpha$-families

Let us turn our attention to the general $\alpha$-connections and try to extend some of the results in the previous section to their $\alpha$-versions, using the framework of §2.6. Let us fix $\alpha$ to a particular value and let $S = \{p_\theta\} \, (\subset \tilde{\mathcal{P}}(\mathcal{X}))$ be an $\alpha$-affine manifold represented as Equation (2.65). Then as mentioned in §2.6, $S$ is $\alpha$-flat and $[\theta^i]$ forms an $\alpha$-affine coordinate system, while from Equation (3.29) we now see that $S$ is also $(-\alpha)$-flat. In other words, $(S, g, \nabla^{(\alpha)}, \nabla^{(-\alpha)})$ is a dually flat space. Now if we define

$$\eta_i \stackrel{\text{def}}{=} \int F_i(x) \ell^{(-\alpha)}(x; \theta) dx, \quad (3.72)$$

Equations (2.60) and (2.65) lead to

$$\partial_j \eta_i = \int F_i \partial_j \ell^{(-\alpha)} dx = \int \partial_i \ell^{(\alpha)} \partial_j \ell^{(-\alpha)} dx = g_{ij}. \quad (3.73)$$

Since this satisfies Equation (3.32), $[\theta^i]$ and $[\eta_i]$ are mutually dual, and hence $[\eta_i]$ is a $(-\alpha)$-affine coordinate system. In addition, letting for an arbitrary $p \in \tilde{\mathcal{P}}(\mathcal{X})$

$$\Psi^{(\alpha)}(p) \stackrel{\text{def}}{=} \begin{cases} \dfrac{2}{1+\alpha} \int p(x) \, dx & (\alpha \ne -1) \\ \int p(x) \{\log p(x) - 1\} \, dx & (\alpha = -1) \end{cases}, \quad (3.74)$$

we may easily confirm that the potential functions of the Legendre transformation satisfying Equations (3.33) through (3.36) are given by

$$\psi(\theta) = \Psi^{(\alpha)}(p_\theta), \quad (3.75)$$

$$\varphi(\theta) = \Psi^{(-\alpha)}(p_\theta) - \int C(x) \ell^{(-\alpha)}(x; \theta) dx. \quad (3.76)$$

For arbitrary $p, q \in \tilde{\mathcal{P}}(\mathcal{X})$ and $\alpha \in \mathbb{R}$, let

$$D^{(\alpha)}(p \| q) \stackrel{\text{def}}{=} \Psi^{(\alpha)}(p) + \Psi^{(-\alpha)}(q) - \int L^{(\alpha)}(p(x)) L^{(-\alpha)}(q(x)) dx. \quad (3.77)$$

## 3.6. $\alpha$-AFFINE MANIFOLDS AND $\alpha$-FAMILIES

Since this is represented as

$$D^{(\alpha)}(p \| q) = \frac{4}{1-\alpha^2} \int \left\{ \frac{1-\alpha}{2} p + \frac{1+\alpha}{2} q - p^{\frac{1-\alpha}{2}} q^{\frac{1+\alpha}{2}} \right\} dx \quad (3.78)$$

for $\alpha \neq \pm 1$ and

$$D^{(-1)}(p \| q) = D^{(1)}(q \| p) = \int \left\{ q - p + p \log \frac{p}{q} \right\} dx \quad (3.79)$$

for $\alpha = \pm 1$, we see that $D^{(\alpha)}(p \| q)$ coincides with the $\alpha$-divergence defined in §3.2 when $p$ and $q$ are probability distributions. Now, applying Equations (3.75) (3.76) and

$$\theta^i \eta'_i = \int \left\{ \ell^{(\alpha)}(x; \theta) - C(x) \right\} \ell^{(-\alpha)}(x; \theta') dx$$

to the definition of the canonical divergence

$$D(\theta \| \theta') = \psi(\theta) + \varphi(\theta') - \theta^i \eta'_i,$$

we obtain the following theorem.

**Theorem 3.13** *The $(g, \nabla^{(\alpha)})$- and $(g, \nabla^{(-\alpha)})$-divergences on an $\alpha$-affine manifold $S$ $(\subset \tilde{\mathcal{P}}(\mathcal{X}))$ are given by the restriction of $D^{(\alpha)}$ and $D^{(-\alpha)}$ onto $S \times S$, respectively.*

From Theorem 3.13 it follows that, on any $\alpha$-affine manifold, Theorems 3.7, 3.8, 3.9 and 3.10 hold for $D = D^{(\pm \alpha)}$ and $\nabla = \nabla^{(\pm \alpha)}$. On the other hand, an $\alpha$-family is not $\alpha$-flat unless $\alpha = \pm 1$, and hence these theorems do not apply to it. Nevertheless, the property described in Theorem 3.10 is still valid as follows.

**Theorem 3.14** *Let $S$ be an $\alpha$-family, $M$ a submanifold of $S$, and $p$ a point in $S$. A necessary and sufficient condition for a point $q \in M$ to be a stationary point of the function $r \mapsto D^{(\pm \alpha)}(p \| r)$ restricted on $M$ is for the $(\pm \alpha)$-geodesic connecting $p$ and $q$ to be orthogonal to $M$ at $q$ (with $\pm$ the same sign).*

**Proof:** Given a point $q$ in $M$, let

$$V^{(\pm \alpha)} \stackrel{\text{def}}{=} \left\{ p \in \tilde{S} \, \middle| \, \forall a, \, D^{(\pm \alpha)}(p \| (\partial_a)_q) = 0 \right\} \quad \text{and}$$

$$W^{(\pm \alpha)} \stackrel{\text{def}}{=} V^{(\pm \alpha)} \cap S,$$

where $\tilde{S}$ is the denormalization of $S$, which is an $\alpha$-affine manifold, and $\partial_a$ denotes the natural basis of a coordinate system $[u^a]$ of $M$. It suffices to verify the following statements.

(i) $W^{(\pm \alpha)}$ is $(\pm \alpha)$-autoparallel in $S$.

(ii) $T_q(W^{(\pm\alpha)})$ is the orthogonal complement of $T_q(M)$ in $T_q(S)$.

First, noting that $D^{(\pm\alpha)}$ is the $(g, \nabla^{(\pm\alpha)})$-divergence of the $\alpha$-affine manifold $\tilde{S}$ and recalling Equation (3.52), we see the following.

(iii) $V^{(\pm\alpha)}$ is $(\pm\alpha)$-autoparallel in $\tilde{S}$.

(iv) $T_q(V^{(\pm\alpha)})$ is the orthogonal complement of $T_q(M)$ in $T_q(\tilde{S})$.

Then (ii) is immediate from (iv). Next, Equations (3.78) and (3.79) lead to

$$D^{(\pm\alpha)}\left(\tau p \,\|\, (\partial_a)_q\right) = \tau^{(1\mp\alpha)/2} D^{(\pm\alpha)}\left(p \,\|\, (\partial_a)_q\right)$$

for any $p \in \tilde{S}$ and any $\tau > 0$, which implies that $V^{(\pm\alpha)}$ is the denormalization of $W^{(\pm\alpha)}$. Hence, (i) follows from (iii) by Theorem 2.10. ∎

**Corollary 3.15** *Given a point $p$ in an $\alpha$-family $S$ and a positive number $c$, suppose that $M^{\pm} \stackrel{\text{def}}{=} \{q \in S \,|\, D^{(\pm\alpha)}(p \,\|\, q) = c\}$ forms a hypersurface in $S$. Then every $\pm\alpha$-geodesic passing through the center $p$ orthogonally intersects $M^{\pm}$.*

For instance Theorem 3.14 is applied to $S = \mathcal{P}(\mathcal{X})$ for any finite $\mathcal{X}$, since $S$ is an $\alpha$-family for all $\alpha$ (Example 2.11). This may be generalized to the case when $\mathcal{X}$ is infinite in a natural way (see Appendix of Amari [6]).

As briefly explained in §8.4, Kurose [134] showed that a divergence is canonically defined on a manifold equipped with a dualistic structure of constant curvature and exhibits a modified form of Pythagorean relation. The $\alpha$-divergence on an $\alpha$-family is an example of Kurose's divergence, for which the modified Pythagorean relation is demonstrated in the next theorem. We can see that the theorem provides Theorem 3.14 with another proof.

**Theorem 3.16** *Let $S$ be an $\alpha$-family, $p, q$ and $r$ be three points in $S$, $\gamma_1$ be the $\alpha$-geodesic connecting $p$ and $q$ in $S$, and $\gamma_2$ be the $(-\alpha)$-geodesic connecting $q$ and $r$ in $S$. If at the intersection $q$ the curves $\gamma_1$ and $\gamma_2$ are orthogonal, then*

$$\begin{aligned} D^{(\alpha)}(p \,\|\, r) &= D^{(\alpha)}(p \,\|\, q) + D^{(\alpha)}(q \,\|\, r) \\ &\quad - \frac{1-\alpha^2}{4} D^{(\alpha)}(p \,\|\, q) D^{(\alpha)}(q \,\|\, r). \end{aligned} \tag{3.80}$$

**Proof:** When $\alpha = \pm 1$, the theorem is nothing but the Pythagorean relation for the canonical divergence $D^{(\pm 1)}$ on an e- or m-family $S$. Hence we have only to consider the case when $\alpha \neq \pm 1$. Throughout this proof we use a special notation in which $L^{(\pm\alpha)}(p(x))$ is denoted by $p^{(\pm\alpha)}(x)$ and Equation (2.75) is written as

$$p^{(\alpha)}(x) = \sum_{\lambda=0}^{n} \theta^{\lambda}(p) F_{\lambda}(x), \tag{3.81}$$

## 3.6. $\alpha$-AFFINE MANIFOLDS AND $\alpha$-FAMILIES

which we may assume for the $\alpha$-family $S$. Since the $\alpha$-geodesic $\gamma_1$ is an $\alpha$-family due to Theorem 2.11, it is represented as

$$\gamma_1^{(\alpha)}(x;t) = a(t)p^{(\alpha)}(x) + b(t)q^{(\alpha)}(x). \tag{3.82}$$

Assuming $\gamma_1(x;0) = q(x)$, we have $a(0) = 0$ and $b(0) = 1$. From Equation (2.62) we have

$$\begin{aligned}
0 &= \int q^{(-\alpha)}(x) \left.\frac{d\gamma_1^{(\alpha)}(x;t)}{dt}\right|_{t=0} dx \\
&= \dot{a}(0) \int p^{(\alpha)}(x) q^{(-\alpha)}(x)\, dx + \dot{b}(0) \frac{4}{1-\alpha^2}.
\end{aligned} \tag{3.83}$$

On the other hand, the denormalization $\tilde{\gamma}_2$ of the $(-\alpha)$-geodesic $\gamma_2$ is $(-\alpha)$-autoparallel in $\tilde{S}$ due to Theorem 2.10, and forms a plane in a $(-\alpha)$-affine coordinate system of the $\alpha$-affine manifold $\tilde{S}$ given by Equation (3.72). This means that $\gamma_2$ may be parameterized as

$$\int F_\lambda(x)\, \gamma_2^{(-\alpha)}(x;s)\, dx = c(s) \int F_\lambda(x)\, q^{(-\alpha)}(x)\, dx \\
+ d(s) \int F_\lambda(x)\, r^{(-\alpha)}(x)\, dx. \tag{3.84}$$

We assume that $\gamma_2(x;0) = q(x)$ to yield $c(0) = 1$ and $d(0) = 0$. Then the inner product of the tangent vectors of these two curves at the intersection $q$ is given by

$$\langle \dot{\gamma}_1(0), \dot{\gamma}_2(0) \rangle_q$$

$$\stackrel{(A)}{=} \int \left.\frac{d\gamma_1^{(\alpha)}(x;t)}{dt}\right|_{t=0} \left.\frac{d\gamma_2^{(-\alpha)}(x;s)}{ds}\right|_{s=0} dx$$

$$\stackrel{(B)}{=} \int \left\{ \dot{a}(0)p^{(\alpha)}(x) + \dot{b}(0)q^{(\alpha)}(x) \right\} \left.\frac{d\gamma_2^{(-\alpha)}(x;s)}{ds}\right|_{s=0} dx$$

$$\stackrel{(C)}{=} \int \left\{ \dot{a}(0)p^{(\alpha)}(x) + \dot{b}(0)q^{(\alpha)}(x) \right\} \left\{ \dot{c}(0)q^{(-\alpha)}(x) + \dot{d}(0)r^{(-\alpha)}(x) \right\} dx$$

$$\stackrel{(D)}{=} \int \left\{ \dot{a}(0)p^{(\alpha)}(x) + \dot{b}(0)q^{(\alpha)}(x) \right\} \dot{d}(0) r^{(-\alpha)}(x)\, dx$$

$$\stackrel{(E)}{=} \dot{a}(0)\dot{d}(0) \Big[ \int p^{(\alpha)}(x) r^{(-\alpha)}(x)\, dx \\
- \frac{1-\alpha^2}{4}\left\{\int p^{(\alpha)}(x) q^{(-\alpha)}(x)\, dx\right\}\left\{\int q^{(\alpha)}(x) r^{(-\alpha)}(x)\, dx\right\} \Big],$$

where (A) follows from (2.60), (B) follows from (3.82), (C) follows from (3.84) combined with the fact that both $p^{(\alpha)}$ and $q^{(\alpha)}$ belong to the linear span of $\{F_\lambda\}$ due to (3.81), and (D) and (E) follow from (3.83). This proves the theorem, because the quantity $[\cdots]$ in the last line coincides with the difference (RHS) - (LHS) of Equation (3.80). ∎

The additivity (3.28) for the Kullback divergence is modified for the $\alpha$-divergence as follows:

$$D^{(\alpha)}(p_{12}\|q_{12}) = D^{(\alpha)}(p_1\|q_1) + D^{(\alpha)}(p_2\|q_2) - \frac{1-\alpha^2}{4}D^{(\alpha)}(p_1\|q_1)D^{(\alpha)}(p_2\|q_2),$$

where $p_{12}(x_1, x_2) = p_1(x_1)p_2(x_2)$ and $q_{12}(x_1, x_2) = q_1(x_1)q_2(x_2)$. This relation can be seen as an example of the above theorem.

Finally, we observe that the Cramér-Rao inequality and related results such as Theorems 2.2, 2.7, 2.8 and 3.12 may be extended to their $\alpha$-versions at least from a viewpoint of formal analogy. For $A : \mathcal{X} \to \mathbb{R}$ and $p \in \mathcal{P}$, let

$$E_p^{(-\alpha)}[A] \stackrel{\text{def}}{=} \int A(x)L^{(-\alpha)}(p(x))\mathrm{d}x,$$

$$\mu_p^{(-\alpha)}[A] \stackrel{\text{def}}{=} \int A(x)p(x)^{(1+\alpha)/2}\mathrm{d}x$$

$$(= \frac{1+\alpha}{2}E_p^{(-\alpha)}[A] \text{ if } \alpha \neq -1), \text{ and}$$

$$V_p^{(\alpha)}[A] \stackrel{\text{def}}{=} \int \left\{A(x) - \mu_p^{(-\alpha)}[A] \cdot p(x)^{(1-\alpha)/2}\right\}^2 p(x)^\alpha \mathrm{d}x$$

$$= V_p[A \cdot p^{-(1-\alpha)/2}].$$

Similarly, we can define the vectors $E_p^{(-\alpha)}[\vec{A}]$, $\mu_p^{(-\alpha)}[\vec{A}] \in \mathbb{R}^n$ and the $n \times n$ matrix $V_p^{(\alpha)}[\vec{A}]$ for an $n$-tuple $\vec{A} = (A^1, \ldots, A^n)$ of functions $A^i : \mathcal{X} \to \mathbb{R}$. Now we have the following theorems, whose proofs are almost parallel to those of the original results corresponding to the case when $\alpha = 1$, and are omitted.

**Theorem 3.17** *Let $S$ be a statistical model in $\mathcal{P}(\mathcal{X})$ and let $E^{(-\alpha)}[A]|_S$ denote the function $p \mapsto E_p^{(-\alpha)}[A]$ defined for $p \in S$. Then we have*

$$V_p^{(\alpha)}[A] \geq \left\|(\mathrm{d}E^{(-\alpha)}[A]|_S)_p\right\|_p^2. \tag{3.85}$$

*Here the equality holds if and only if*

$$A - \mu_p^{(-\alpha)}[A] \cdot p^{(1-\alpha)/2} \in T_p^{(\alpha)}(S) \stackrel{\text{def}}{=} \left\{X^{(\alpha)} \,\Big|\, X \in T_p(S)\right\}, \tag{3.86}$$

*where $X^{(\alpha)}$ is the $\alpha$-representation of $X$ defined by Equation (2.59). In particular, if $S = \mathcal{P}(\mathcal{X})$ then the equality always holds.*

**Theorem 3.18** *Let $S = \{p_\xi \,;\, \xi \in \Xi\}$ be an $n$-dimensional statistical model in $\mathcal{P}(\mathcal{X})$. If an $n$-tuple of functions $\vec{A} = (A^1, \ldots, A^n)$ satisfies $E_\xi^{(-\alpha)}[\vec{A}] = \xi$ for all $\xi \in \Xi$, then we have $V_\xi^{(\alpha)}[\vec{A}] \geq G(\xi)^{-1}$, where $G(\xi)$ denotes the Fisher information matrix of $S$ with respect to the coordinate system $\xi$.*

**Theorem 3.19** *A necessary and sufficient condition for an n-dimensional statistical model $S$ to have a coordinate system $\xi = [\xi^i]$ and an n-tuple of functions $\vec{A} = (A^i)$ satisfying $E_\xi^{(-\alpha)}[\vec{A}] = \xi$ and $V_\xi^{(\alpha)}[\vec{A}] = G(\xi)^{-1}$ for all $\xi \in \Xi$ is that $S$ is an $\alpha$-family.*

## 3.7 Mutually dual foliations

Let $[\theta^i]$ and $[\eta_i]$ be mutually dual coordinate systems of an $n$-dimensional dually flat space $(S, g, \nabla, \nabla^*)$. First, let us divide the range of the index $i = 1, \cdots, n$ into the two sections $i = 1, \cdots, k$ and $i = k+1, \cdots, n$, and call these sections I and II, respectively. Now let $M(c_I)$ be the set of points whose coordinates $[\eta_i]$ in Section I are fixed to constants $c_I = (c_{I,i})$, for $i = 1, \cdots, k$:

$$M(c_I) = \{p \in S \mid \eta_1(p) = c_{I,1}, \cdots, \eta_k(p) = c_{I,k}\}.$$

This forms a $(n-k)$-dimensional $\nabla^*$-autoparallel submanifold. The set $M(c_I)$ varies with $c_I$, and if $c_I \neq c_{I'}$ then

$$M(c_I) \cap M(c_{I'}) = \emptyset,$$

and also

$$\bigcup_{c_I} M(c_I) = S.$$

In other words, $S$ is partitioned into the submanifolds $\{M(c_I)\}$. We call such a partition a **foliation** of $S$.

In a similar manner, consider the set $E(d_{II})$ of points whose coordinates $\theta^{k+1}, \cdots, \theta^n$ in the $\theta$-coordinate system are fixed to values $d_{II}^{k+1}, \cdots, d_{II}^n$:

$$E(d_{II}) = \{p \in S \mid \theta^{k+1}(p) = d_{II}^{k+1}, \cdots, \theta^n(p) = d_{II}^n\}.$$

This is a $\nabla$-autoparallel submanifold, and the set of $\{E(d_{II})\}$ form another foliation of $S$.

A point $p$ determines the $M(c_I)$ and the $E(d_{II})$ containing it. Then $p$ is the intersection of these $M(c_I)$ and $E(d_{II})$. The tangent space $T_p(M)$ of $M(c_I)$ at this point is the $k$-dimensional space spanned by $\{\partial_1, \cdots, \partial_k\}$, and the tangent space $T_p(E)$ of $E(d_{II})$ is the $(n-k)$-dimensional space spanned by $\{\partial^{k+1}, \cdots, \partial^n\}$. What is interesting is the fact that

$$\langle \partial_i, \partial^j \rangle = 0 \qquad (i \neq j)$$

and that hence $T_p(E)$ and $T_p(M)$ are orthogonal. Such foliations orthogonal to each other are said to be **mutually dual**.

Now divide the $\eta$-coordinates of the point $p$ into the first $k$ dimensions and the remaining $n-k$ dimensions:

$$\eta = (\eta_I, \eta_{II}),$$

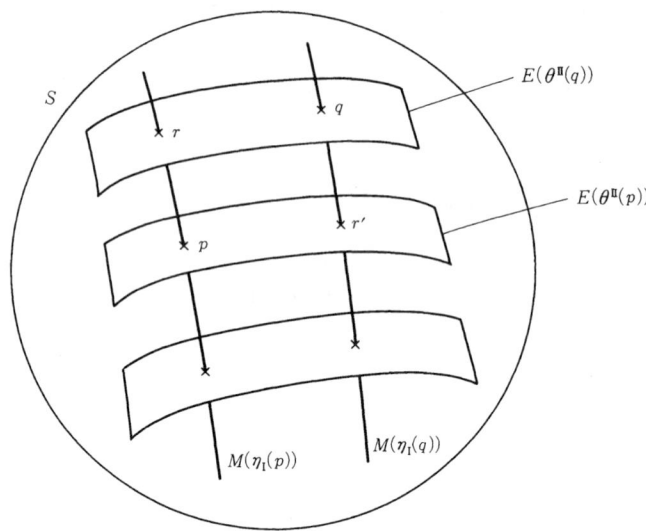

Figure 3.3: Mutually dual foliations of $S$.

and similarly divide the $\theta$-coordinates:

$$\theta = (\theta^I, \theta^{II}).$$

Then the point $p$ is the intersection of $M(\eta_I)$ and $E(\theta^{II})$. If we let

$$\xi = (\eta_I, \theta^{II}),$$

then this defines a coordinate system for $S$. We call this a **mixed coordinate system**.

Let $p$ and $q$ be two points whose mixed coordinates are $(\eta_I(p), \theta^{II}(p))$ and $(\eta_I(q), \theta^{II}(q))$, respectively. Now consider the points whose mixed coordinates are $r : (\eta_I(p), \theta^{II}(q))$ and $r' : (\eta_I(q), \theta^{II}(p))$. Then $r$ is the $\nabla^*$-projection of $q$ onto $M(\eta_I(p))$, and $r'$ is the $\nabla$-projection of $q$ onto $E(\theta^{II}(p))$ (Figure 3.3). From this we see that

$$\begin{aligned} D(p \,\|\, q) &= D(p \,\|\, r) + D(r \,\|\, q) \quad \text{and} \\ D(q \,\|\, p) &= D(q \,\|\, r') + D(r' \,\|\, p), \end{aligned}$$

where we denote by $D$ the $(g, \nabla^*)$-divergence.

In the case when $(S, g, \nabla, \nabla^*) = (\mathcal{P}(\mathcal{X}), g, \nabla^{(e)}, \nabla^{(m)})$, the manifolds $\{M(c_I)\}$ are mixture families, $\{E(d_{II})\}$ are exponential families, and $D$ is the Kullback divergence. An example is given by Equations (3.55) and (3.63). That is, letting $E(C) = \{p_\theta\}$ be the the exponential family defined by Equation (3.55) from a function $C$, we see that the collections of the sets $\{E(C)\}_C$ and $\{M_\lambda\}_\lambda$ form mutually dual foliations (although the parametrization $C \mapsto E(C)$ is redundant in this case), for which the Pythagorean relation (3.68) holds. Another example will be shown in §6.2.

## 3.8 A further look at the triangular relation

In Theorem 3.7 we observed that the triangular relation (3.49) characterizes the canonical divergence. This characterization assumes, however, the existence of mutually dual affine coordinate systems from the beginning and hence is meaningful only when the underlying space is known to be dually flat. In this section, we shall reformulate the triangular relation in a wider setting where the flatness is not assumed, so that we can fully understand the specialty of the canonical divergence among other divergences.

We begin with introducing some important notions from differential geometry. Suppose that an affine connection $\nabla$ is given on a manifold $S$. For a point $q$ and a tangent vector $X \in T_q(S)$, let $\gamma : t \mapsto \gamma(t)$ be a $\nabla$-geodesic with an affine parameter $t$ such that $\gamma(0) = q$ and $\dot{\gamma}(0) = X$. We write $\mathcal{E}_q(X) = \gamma(1)$ when $\gamma(1)$ lies in $S$, and call the mapping $\mathcal{E}_q : X \mapsto \mathcal{E}_q(X)$ the **exponential map** for $\nabla$ at $q$. It can be shown that there exist a neighborhood $U$ of $0$ in $T_q(S)$ and a neighborhood $V$ of $q$ in $S$ such that $\mathcal{E}_q|_U$ is a diffeomorphism from $U$ onto $V$. We assume $U = T_q(S)$ and $V = S$ to simplify the description.

When a basis $\{u_1, \cdots, u_n\}$ of $T_q(S)$ is specified, a coordinate system $[\theta^i]$ of $S$ is defined by $\mathcal{E}_q(\theta^i(p)u_i) = p$, $\forall p \in S$. We call $[\theta^i]$ the **normal coordinate system** for $\nabla$ determined by $\{u_i\}$. In this coordinate system, a geodesic $\gamma$ passing the point $q$ is represented as $\gamma^i(t) \stackrel{\text{def}}{=} \theta^i(\gamma(t)) = tc^i$ by an affine parameter $t$ and constants $\{c^i\}$. Conversely, if a coordinate system $[\theta^i]$ satisfies this property, then $[\theta^i]$ is the normal coordinate system determined by $\{\left(\frac{\partial}{\partial \theta^i}\right)_q\}$. In particular, an affine coordinate system $[\theta^i]$ of a flat connection is a normal coordinate system at $q$ if $\theta^i(q) = 0$ for all $i$.

Let $[\theta^i]$ be a normal coordinate system for $\nabla$ at $q$ and $\{\Gamma_{ij}^k\}$ be the connection coefficients of $\nabla$ with respect to $[\theta^i]$. Then we have (Proposition 8.4 in chap. III of [122], vol.1 )

$$\left(\Gamma_{ij}^k\right)_q + \left(\Gamma_{ji}^k\right)_q = 0. \tag{3.87}$$

Indeed, substituting $\gamma^i(t) = tc^i$ into Equation (1.56) and then letting $t = 0$, we have $c^i c^j (\Gamma_{ij}^k)_q = 0$ for all $(c^i)$, which leads to the equation above. In particular, if $\nabla$ is a symmetric connection, we have $(\Gamma_{ij}^k)_q = 0$.

Now we state the main theorems of this section.

**Theorem 3.20** *Let $\nabla$ and $\nabla^*$ be symmetric connections, $g = \langle \, , \, \rangle$ a Riemannian metric and $D$ a divergence on $S$. Then the following conditions (i) and (ii) are equivalent.*

(i) *The triple $(g, \nabla, \nabla^*)$ is a dualistic structure induced from $D$ in the sense described in §3.2.*

(ii) *The following approximation is valid at every point $q$, as other points $p$ and $r$ approach $q$:*

$$D(p \parallel q) + D(q \parallel r) - D(p \parallel r) = \left\langle \mathcal{E}_q^{-1}(p), \mathcal{E}_q^{*-1}(r) \right\rangle_q + o(\Delta^3), \tag{3.88}$$

where $\mathcal{E}_q$ and $\mathcal{E}_q^*$ are respectively the exponential maps for $\nabla$ and $\nabla^*$ at $q$, and
$$\Delta \stackrel{\text{def}}{=} \max\{\,\|\xi(p) - \xi(q)\|,\, \|\xi(r) - \xi(q)\|\,\}$$
for an arbitrary coordinate system $\xi = [\xi^i]$.

**Theorem 3.21** *In the same situation as the previous theorem, the following conditions (i) and (ii) are equivalent.*

*(i)* $(S, g, \nabla, \nabla^*)$ *is dually flat and $D$ is its canonical divergence.*

*(ii) For all $p, q, r \in S$ it holds that*
$$D(p \| q) + D(q \| r) - D(p \| r) = \langle \mathcal{E}_q^{-1}(p), \mathcal{E}_q^{*-1}(r) \rangle_q . \tag{3.89}$$

The rest of the present section is devoted to proving them. Our first goal is to prove the part (i)$\Rightarrow$(ii) of Theorem 3.20. Let $D$ be a divergence on $S$ and $(g, \nabla, \nabla^*)$ the dualistic structure induced from $D$. Let $p, q, r$ be arbitrary three points in $S$ and let $a^i \stackrel{\text{def}}{=} \xi^i(p) - \xi^i(q)$ and $b^i \stackrel{\text{def}}{=} \xi^i(r) - \xi^i(q)$ for a coordinate system $[\xi^i]$ of $S$. Let us expand $D(p \| q) + D(q \| r) - D(p \| r)$ in terms of $[a^i]$ and $[b^i]$ around $p = q = r$ up to the third order of $\Delta = \max\{\|a\|, \|b\|\}$. We start from the expansion

$$\begin{aligned}D(p \| r) - D(q \| r) &= D((\partial_i)_q \| r) a^i + \frac{1}{2} D((\partial_i \partial_j)_q \| r) a^i a^j \\ &\quad + \frac{1}{6} D((\partial_i \partial_j \partial_k)_q \| r) a^i a^j a^k + o(\Delta^3),\end{aligned}$$

into which we substitute the following expansions:

$$\begin{aligned}D((\partial_i)_q \| r) &= -g_{ij} b^j - \frac{1}{2} \Gamma^*_{jk,i} b^j b^k + o(\Delta^2), \\ D((\partial_i \partial_j)_q \| r) &= g_{ij} - \Gamma_{ij,k} b^k + o(\Delta), \\ D((\partial_i \partial_j \partial_k)_q \| r) &= h_{ijk} + o(1),\end{aligned}$$

where all the quantities $g_{ij}, \Gamma_{ij,k}$, etc. are evaluated at the point $q$, and $h_{ijk}$ is $h_{ijk}^{(D)}$ defined by Equation (3.15). Then we have

$$\begin{aligned}D(p \| r) - D(q \| r) &= -g_{ij} a^i b^j + \frac{1}{2} g_{ij} a^i a^j - \frac{1}{2} \Gamma_{ij,k} a^i a^j b^k \\ &\quad - \frac{1}{2} \Gamma^*_{jk,i} a^i b^j b^k + \frac{1}{6} h_{ijk} a^i a^j a^k + o(\Delta^3),\end{aligned}$$

Combining this with

$$D(p \| q) = \frac{1}{2} g_{ij} a^i a^j + \frac{1}{6} h_{ijk} a^i a^j a^k + o(\Delta^3),$$

## 3.8. A FURTHER LOOK AT THE TRIANGULAR RELATION

we obtain

$$D(p \| q) + D(q \| r) - D(p \| r)$$
$$= g_{ij}a^i b^j + \frac{1}{2}\Gamma_{ij,k}a^i a^j b^k + \frac{1}{2}\Gamma^*_{jk,i}a^i b^j b^k + o(\Delta^3). \quad (3.90)$$

Now, let $a(t)$ and $b(t)$ be curves in $M$ satisfying $a(0) = b(0) = q$, and let $a^i(t) = \xi^i(a(t)) - \xi^i(q)$ and $b^i(t) = \xi^i(b(t)) - \xi^i(q)$. Using the expansion above, we have

$$D(a(t) \| q) + D(q \| b(t)) - D(a(t) \| b(t))$$
$$= g_{ij}\dot{a}^i \dot{b}^j t^2 + \frac{1}{2}(g_{ij}\ddot{a}^i \dot{b}^j + g_{ij}\dot{a}^i \ddot{b}^j)t^3 + \frac{1}{2}(\Gamma_{ij,k}\dot{a}^i \dot{a}^j \dot{b}^k + \Gamma^*_{jk,i}\dot{a}^i \dot{b}^j \dot{b}^k)t^3 + o(t^3)$$
$$= \langle \dot{a}, \dot{b}\rangle t^2 + \frac{1}{2}\left(\langle \nabla_{\dot{a}}\dot{a}, \dot{b}\rangle + \langle \dot{a}, \nabla^*_{\dot{b}}\dot{b}\rangle\right)t^3 + o(t^3),$$

where the derivatives of $a$ and $b$ are evaluated at $t = 0$. In particular, if $a(t)$ and $b(t)$ are $\nabla$-geodesic and $\nabla^*$-geodesic, respectively, with $t$ being their affine parameters, then

$$D(a(t) \| q) + D(q \| b(t)) - D(a(t) \| b(t)) = \langle \dot{a}, \dot{b}\rangle t^2 + o(t^3).$$

Noting that $\mathcal{E}_q(\dot{a}t) = a(t)$ and that $\mathcal{E}^*_q(\dot{b}t) = b(t)$, we see that Equation (3.88) holds.

Next, we show the converse part (ii)⇒(i) of Theorem 3.20. Let $\{u_i\}$ and $\{v^j\}$ be two bases of $T_q(S)$ which are mutually dual in the sense that $\langle u_i, v^j\rangle_q = \delta^j_i$, and let $[\theta^i]$ ($[\eta_j]$, respectively) be the normal coordinate system for $\nabla$ (for $\nabla^*$) at $q$ determined by $[u_i]$ (by $[v^j]$). We denote the components of $g, \nabla, \nabla^*$ with respect to $[\theta^i]$ by $g_{ij}, \Gamma_{ij,k}, \Gamma^*_{ij,k}$. Then we have

$$\langle (\partial_i)_q, (\partial^j)_q\rangle_q = \langle u_i, v^j\rangle_q = \delta^j_i, \quad (3.91)$$

where $\partial_i \stackrel{\text{def}}{=} \frac{\partial}{\partial \theta^i}$ and $\partial^j \stackrel{\text{def}}{=} \frac{\partial}{\partial \eta_j}$, which implies that $(\partial_i \eta_j)_q = (g_{ij})_q$. Now Equation (3.88) is represented as

$$D(p \| q) + D(q \| r) - D(p \| r) = \theta^i(p)\eta_i(q) + o(\Delta^3), \quad (3.92)$$

from which we have

$$D((\partial_i)_p \| q) - D((\partial_i)_p \| r) = \eta_i(r) + o(\Delta^2), \quad (3.93)$$
$$-D((\partial_i)_p \| (\partial_j)_r) = (\partial_j \eta_i)_r + o(\Delta), \quad (3.94)$$
$$-D((\partial_i \partial_j)_p \| (\partial_k)_r) = o(1). \quad (3.95)$$

Letting $p = r = q$ in the last two equations, we have $(g^{(D)}_{ij})_q = (\partial_j \eta_i)_q = (g_{ij})_q$ and $(\Gamma^{(D)}_{ij,k})_q = 0 = (\Gamma_{ij,k})_q$, where the last equality follows from the assumption that $[\theta^i]$ is a normal coordinate system at $q$ for the symmetric connection $\nabla$.

Since $q$ is arbitrary, it is concluded that $g^{(D)} = g$ and $\nabla^{(D)} = \nabla$. The equation $\nabla^{(D^*)} = \nabla^*$ can be verified similarly.

Let us proceed to the proof of Theorem 3.21. It is easy to see that Equation (3.89) coincides with the triangular relation (3.49) when $(S, g, \nabla, \nabla^*)$ is dually flat, and hence the part (i)$\Rightarrow$(ii) is straightforward. Assume (ii) to prove (ii)$\Rightarrow$(i). Then the triple $(g, \nabla, \nabla^*)$ is induced from $D$ owing to Theorem 3.20. Furthermore, the remainder term $o(1)$ in Equation (3.95) can be omitted in this case, and letting $p = r$, we have $-D((\partial_i \partial_j)_p \| (\partial_k)_p) = (\Gamma_{ij,k})_p = 0$ for every $p$. This means that $\nabla$ is flat with $[\theta^i]$ being an affine coordinate system, and therefore $(S, g, \nabla, \nabla^*)$ turns out to be dually flat. Now Equation (3.89) is just the triangular relation (3.49), and it follows from Theorem 3.7 that $D$ is the canonical divergence. The proof is thus completed.

Finally, let us take a look at the case when $\nabla$ and $\nabla^*$ may have nonvanishing torsions. In general, for any two affine connections $\nabla$ and $\nabla'$, their difference $Q(X,Y) \stackrel{\text{def}}{=} \nabla_X Y - \nabla'_X Y$ forms a tensor field of type $(1,2)$. When $Q$ is skew-symmetric ($Q(X,Y) = -Q(Y,X)$), we say that these connections are equivalent to each other. It is easy to see that the geodesics, the exponential maps and the normal coordinate systems are kept invariant under this equivalence relation. We also note that there is a unique symmetric connection in each equivalence class. Indeed, given a connection $\nabla$, the symmetric connection $\nabla'$ equivalent to $\nabla$ is obtained by $\nabla'_X Y = \nabla_X Y - \frac{1}{2} T(X, Y)$, where $T$ is the torsion tensor of $\nabla$, or equivalently by $\Gamma'^k_{ij} = \frac{1}{2}(\Gamma^k_{ij} + \Gamma^k_{ji})$. We call this $\nabla'$ the symmetrization of $\nabla$.

In general, for not necessarily symmetric connections $\nabla$ and $\nabla^*$ and a metric $g$, condition (ii) in Theorem 3.20 holds with a divergence $D$ if and only if the symmetrizations of $\nabla$ and $\nabla^*$ are induced from $D$ together with $g$. Therefore a necessary and sufficient condition for $(g, \nabla, \nabla^*)$ to have such a divergence is that the symmetrizations of $\nabla$ and $\nabla^*$ are mutually dual with respect to $g$.

# Chapter 4

# Statistical inference and differential geometry

Suppose that we are given data generated according to some unknown probability distribution. Statistical inference is the process of extracting information concerning the underlying probability distribution from this data. If we have prior knowledge concerning the underlying mechanism generating this data, in other words if we know the shape of the unknown distribution, then the possible candidates may be constrained to a parameterized family of distributions. We call such a family a **statistical model**. The field of statistics has a long history, and many techniques and theories for inference have been developed. Of course, probability theory and analysis serve as the foundation of this development.

On the other hand, a family of probability distributions which constitutes a statistical model has a rich geometric structure as a manifold with a Riemannian metric and dual connections. In order to obtain new insights into the framework of statistical inference, and to develop superior techniques for inference, it is vital that we reconstruct the field of statistics as the geometric reinterpretation of statistical models. It appears likely that this field will see extensive development in the future. In this chapter, we first consider classical inference of estimation and hypothesis testing from the geometric point of view. We then move towards non-parametric and semi-parametric inference, where the shape of the underlying distribution is unknown.

## 4.1 Estimation based on independent observations

Consider a family of probability distributions $S = \{p(x;\xi)\}$ parameterized by $\xi = [\xi^i]$ for $i = 1, \cdots, n$. Under appropriate regularity conditions, $S$ may be viewed as an $n$-dimensional manifold for which $\xi$ is a (local) coordinate system. Now let $x_1, \cdots, x_N$ be $N$ independent observations of the random variable $x$ distributed according to $p(x;\xi)$. Letting $x^N = (x_1, \cdots, x_N)$, the task of statistical inference is to infer the probability distribution $p(x;\xi)$ given the $N$ data points $x^N$. For example, **estimation** is a kind of inference task, where the goal is to find an estimate $\hat{\xi}$ of $\xi$; alternatively, one might consider the task of **testing**, where the goal is to decide if the hypothesis $H_0 : \xi = \xi_0$ is accepted against the alternative hypothesis $H_1 : \xi \neq \xi_0$, or is rejected.

We have already seen how a Riemannian metric based on the Fisher information matrix and one-parameter family of affine connections called the $\alpha$-connections may be introduced on a manifold $S$ representing a statistical model. Since the probability distribution governing $x^N$ can be written using the distribution of a single data point as

$$p_N(x^N;\xi) = \prod_{t=1}^{N} p(x_t;\xi),$$

we also have

$$\log p_N(x^N;\xi) = \sum_{t=1}^{N} \log p(x_t;\xi). \tag{4.1}$$

By viewing $x^N$ as a random variable, we find $S_N = \{p(x^N;\xi)\}$ to be, like $S = S_1$, a manifold with $\xi$ as a coordinate system. However, from Equation (4.1) and the definitions, we find that the geometric structure introduced on $S_N$ is given by

$$\begin{aligned} g_{ij}^N(\xi) &= N g_{ij}(\xi) \quad \text{and} & (4.2) \\ \Gamma_{ij,k}^{(\alpha)N} &= N \Gamma_{ij,k}^{(\alpha)}(\xi). & (4.3) \end{aligned}$$

In other words, the geometry of $S_N$ is simply that of $S$ scaled by a factor of $N$. Another way in which this can be seen is that the natural basis vectors $\partial_i^N$ of $S^N$ are simply those of $S$ under the scaling transformation

$$\partial_i^N = \sqrt{N} \partial_i.$$

Hence distinguishing between the geometries of $S^N$ and $S$ serves no purpose, and it suffices to simply consider the geometry of $S$.

In the theory of estimation, an estimator is defined as a function on the $N$ data points $x^N$:

$$\hat{\xi} = \hat{\xi}(x^N) = \hat{\xi}(x_1, \cdots, x_N). \tag{4.4}$$

Now, if we consider $x^N$ to be a random variable, then we must consider $\hat{\xi}_N$ to be one also. In addition, it is necessary to select a condition that $\hat{\xi}$ must satisfy which ensures that $\hat{\xi}$ is in some way similar to the actual parameter $\xi$ of the underlying distribution $p(x;\xi)$. Unbiasedness is one such condition which states that $\hat{\xi}$ must be distributed around $\xi$. We have already given its definition in §2.2 for the case $N = 1$, and it is straightforwardly extended to the case $N > 1$ as

$$E_\xi\left[\hat{\xi}\right] = \xi \quad \text{for} \quad \forall \xi, \tag{4.5}$$

## 4.1. ESTIMATION BASED ON INDEPENDENT OBSERVATIONS

where $E_\xi$ now denotes the expectation with respect to the distribution $p_N(x^N;\xi)$.

The mean square error is often used to measure the accuracy of an estimator, which may be expressed as the matrix $V_\xi[\hat{\xi}] = [v_\xi^{ij}]$ where

$$v_\xi^{ij} = v_\xi^{ij}[\hat{\xi}] = E_\xi\left[\left(\hat{\xi}^i - \xi^i\right)\left(\hat{\xi}^j - \xi^j\right)\right]. \tag{4.6}$$

When $\hat{\xi}$ is unbiased, $V_\xi[\hat{\xi}]$ equals the variance-covariance matrix. It should be noted that neither the unbiasedness and the mean square error are geometrically invariant criteria. They actually depend on the choice of coordinate system $\xi$. A lower bound on the mean square error of an unbiased estimator $\hat{\xi}$ is given by the Cramér-Rao inequality (Theorem 2.2 in §2.2), which now takes the form

$$[v_\xi^{ij}] \geq \frac{1}{N}[g^{ij}(\xi)], \tag{4.7}$$

where $[g^{ij}(\xi)]$ denotes the inverse of the Fisher information matrix $[g_{ij}(\xi)]$ of the model $S$ for a single observation, and we use $\geq$ to mean that (L.H.S.)−(R.H.S) is positive semidefinite. In addition, Theorem 3.12 in §3.5 claims that a necessary and sufficient condition for the existence of an efficient estimator, i.e., an unbiased estimator achieving the lower bound for all $\xi$, is that $S_N = \{p_N(x^N;\xi)\}$ is an exponential family with $\xi$ being an m-affine coordinate system. It is easy to see that the condition holds if and only if $S = S_1$ is an exponential family with $\xi$ being an m-affine coordinate system. (To verify the "only if" part may be a good exercise.) We shall discuss a more detail about the efficient estimator for an exponential family in the next section.

Next we proceed to the **asymptotic theory** of estimation, in which the issue is the performance of an estimator in the limit of $N \to \infty$. In this case, the unbiasedness is not so meaningful as in the case of a fixed $N$. Instead, we usually require an estimator $\hat{\xi}_N = \hat{\xi}_N(x^N)$, or more precisely, a sequence of estimators $\{\hat{\xi}_N\,;\,N=1,2,\cdots\}$, to be **consistent** in the sense that for any $\xi$, the estimate $\hat{\xi}_N(x^N)$ converges in probability to $\xi$ as $N \to \infty$ when $x^N$ is distributed according to $p_N(x^N;\xi)$. In other words, this states that for all $\xi$ and all $\epsilon > 0$,

$$\lim_{N\to\infty} \Pr_\xi\left\{|\hat{\xi}_N - \xi| > \epsilon\right\} = 0. \tag{4.8}$$

The probability distribution of a consistent estimator $\hat{\xi}_N$ is concentrated at the underlying value $\xi$ as $N \to \infty$. Under some regularity condition, the expectation of $\hat{\xi}_N$ converges to $\xi$ uniformly, so that we have for all $\xi$,

$$\lim_{N\to\infty} E_\xi\left[\hat{\xi}_N\right] = \xi \tag{4.9}$$

and

$$\lim_{N\to\infty} \partial_j E_\xi\left[\hat{\xi}_N^i\right] = \partial_j \xi^i = \delta_j^i. \tag{4.10}$$

The mean square error of such an asymptotically unbiased estimator satisfies the following **asymptotic Cramér-Rao inequality**:

$$\lim_{N\to\infty} N[v_\xi^{ij}[\hat{\xi}_N]] \geq [g^{ij}(\xi)]. \tag{4.11}$$

Indeed, substituting $c_i\hat{\xi}_N^i$ for the random variable $A$ in Equation (2.50) and letting $\mu^i \stackrel{\text{def}}{=} E_\xi\left[\hat{\xi}_N^i\right]$, we obtain

$$c_ic_jv_\xi^{ij} - \{c_i(\mu^i - \xi^i)\}^2 \geq c_ic_j(\partial_k\mu^i)(\partial_l\mu^j)\frac{1}{N}g^{kl},$$

from which Equation (4.11) follows. A consistent estimator which achieves the equality in the equation above for all $\xi$ is called an **asymptotically efficient estimator** or a **first-order efficient estimator**, or is sometimes simply called an **efficient estimator**. Such an estimator is optimal with respect to the mean square error up to differences of order $N^{-1}$. Given this, it seems reasonable to then ask whether there always exists an asymptotically efficient estimator for an arbitrary model. Unlike the case of efficient estimator of finite $N$, the answer is "yes", and an example is given by the maximum likelihood estimator.

Given $x^N$, consider $p_N(x^N;\xi)$ as a function of $\xi$, and call this the **likelihood function**. We call $\hat{\xi}_N$ the **maximum likelihood estimator** $\hat{\xi}_{\text{m.l.e.}}$ if it satisfies

$$\max_\xi p_N(x^N;\xi) = p_N(x^N;\hat{\xi}_N), \tag{4.12}$$

i.e., if it takes a value for $\xi$ which maximizes the likelihood. The following result is well known in statistics.

**Theorem 4.1** *The maximum likelihood estimator $\hat{\xi}_{\text{m.l.e}}$ is asymptotically efficient. In other words, for all $\xi$ we have*

$$\lim_{N\to\infty} Nv_\xi^{ij}[\hat{\xi}_{\text{m.l.e}}] = g^{ij}(\xi). \tag{4.13}$$

In fact, it is possible to prove the stronger claim that asymptotically, $\hat{\xi}$ is a Gaussian with mean $\xi$ and covariance $N^{-1}[g^{ij}]$. In §4.4, we shall see these classical results in the light of geometrical framework for a model embedded in an exponential family.

The mean square error of an asymptotically efficient estimator $\hat{\xi}$ may be written as

$$v_\xi^{ij}[\hat{\xi}] = \frac{1}{N}g^{ij}(\xi) + O\left(\frac{1}{N^2}\right),$$

Given this, the discussion now shifts to considering the order $N^{-2}$ term of the estimator. This is called the **higher-order asymptotic theory** of statistical estimation, and shall be discussed in §4.5.

## 4.2 Exponential families and observed points

In the definition of an exponential family:

$$p(x;\theta) = \exp\left\{C(x) + \theta^i F_i(x) - \psi(\theta)\right\}, \qquad (4.14)$$

the $n$ functions $F_1(x), \cdots, F_n(x)$ are random variables. Hence let us rename them as the $n$ random variables

$$x_i = F_i(x) \qquad (i = 1, \cdots, n),$$

and let $x = [x_1, \cdots, x_n]$. Suppose we also define the probability density functions on the $n$-dimensional random variable $x = [x_i]$ with respect to the dominating measure

$$\mathrm{d}\mu(x) = \exp\{C(x)\}\mathrm{d}x.$$

Then Equation (4.14) may be rewritten without loss of generality as

$$p(x;\theta) = \exp\left\{\theta^i x_i - \psi(\theta)\right\}. \qquad (4.15)$$

We use this representation in the discussion below. As was already seen, the exponential family $S = \{p(x;\theta)\}$ is a dually flat space, with its e-affine coordinate system given by the natural parameters $\theta$, its m-affine coordinate system given by the expectation parameters

$$\eta_i = E_\theta[x_i].$$

We also have

$$E_\theta[(x_i - \eta_i)(x_j - \eta_j)] = g_{ij}(\theta),$$

where $g_{ij}$ is the Fisher information matrix with respect to the natural parameters.

**Example 4.1 (Normal distribution: see Examples 2.1 and 2.5)** *Consider a family of probability distributions parameterized by $[\mu, \sigma]$ of the form*

$$p(x;\mu,\sigma) = \frac{1}{\sqrt{2\pi}\sigma} \exp\left\{-\frac{1}{2\sigma^2}(x-\mu)^2\right\}.$$

*This is the 2-dimensional space formed by normal distributions, and it may be rewritten as*

$$p(x;\mu,\sigma) = \exp\left\{\frac{\mu}{\sigma^2}x - \frac{1}{2\sigma^2}x^2 - \frac{\mu^2}{2\sigma^2} - \log(\sqrt{2\pi}\sigma)\right\}.$$

*Now introduce the coordinate system $\theta = [\theta^1, \theta^2]$ defined by*

$$\theta^1 = \frac{\mu}{\sigma^2}, \qquad \theta^2 = -\frac{1}{2\sigma^2},$$

and the random variable $x = [x_1, x_2]$ defined by

$$x_1 = F_1(x) = x \quad \text{and} \quad x_2 = F_2(x) = x^2.$$

We then see that we have the exponential family given by

$$p(x; \theta) = \exp\{\theta^i x_i - \psi(\theta)\} \quad \text{and}$$

$$\psi(\theta) = \frac{\mu^2}{2\sigma^2} + \log(\sqrt{2\pi}\sigma).$$

The expectation parameters $\eta = [\eta_1, \eta_2]$ are given by

$$\eta_1 = E[x_1] = \mu \quad \text{and}$$
$$\eta_2 = E[x_2] = E[x^2] = \mu^2 + \sigma^2.$$

Consider $N$ observations $x^N = x_1 x_2 \cdots x_N$, each independently distributed according to an element $p_\theta$ in the given exponential family. The probability distribution on this is given by

$$p_N(x^N; \theta) = \prod_{t=1}^{N} p(x_t; \theta) = \exp[N\{\theta^i \bar{x}_i - \psi(\theta)\}], \tag{4.16}$$

where we have let

$$\bar{x} = \frac{1}{N} \sum_{t=1}^{N} x_t. \tag{4.17}$$

(Note that $x_t$ is not the $t^{\text{th}}$ component of the vector $x$, but rather the $t^{\text{th}}$ observed vector; $\bar{x}_i$ is the $i^{\text{th}}$ component of the vector $\bar{x}$.) This shows that $\{p_N(x^N; \theta)\}$ also forms an exponential family with $\theta$ being natural parameters. Here it should be emphasized that the probability distribution $p_N$ on $x^N$, which has $nN$ components, may actually be expressed as a function of the random variable $\bar{x}$, which has $n$ components. This means that $\bar{x}$ is a sufficient statistic with respect to the exponential family (see §2.2). This, then, guarantees that statistical inference based on $x^N$ may be reduced, without compromising the quality of the result, to inference based on only $\bar{x}$, a compressed representation of the data (the Rao-Blackwell Theorem).

Now consider, with respect to the observation $x^N$, the point which has the coordinates

$$\hat{\eta} = \bar{x}. \tag{4.18}$$

In this way, the sufficient statistic $\bar{x}$ of $x^N$ determines a point $\hat{\eta}$ in $S$ (i.e., a point whose coordinates under the $\eta$-coordinate system is $\bar{x}$.) We call this the **observed point**. Then the following clearly hold:

$$E_\theta[\bar{x}] = \eta$$
$$E_\theta[(\bar{x}_i - \eta_i)(\bar{x}_j - \eta_j)] = \frac{1}{N} g_{ij}(\theta),$$

## 4.3. CURVED EXPONENTIAL FAMILIES

which means that $\hat{\eta}$ is an efficient estimator of $\eta$. (Recall that the Fisher information matrix with respect to the $\eta$-coordinate system is $[g^{ij}] = [g_{ij}]^{-1}$; see §3.5. ) From the central limit theorem, we see that, asymptotically, $\hat{\eta}$ is distributed according to the normal distribution with mean $\eta$ and variance $N^{-1}g_{ij}$. We also see that the observed point is the maximum likelihood estimate of the model. Indeed, for any $\theta$ we have

$$\begin{aligned}
\log p_N\left(x^N; \hat{\theta}\right) - \log p_N\left(x^N; \theta\right) &= N\{(\hat{\theta}^i - \theta^i)\bar{x}_i - \psi(\hat{\theta}) + \psi(\theta)\} \\
&= ND\left(p_{\hat{\theta}} \,\|\, p_\theta\right) \geq 0,
\end{aligned}$$

where $\hat{\theta}$ is the $\theta$-coordinates of the observed point and $D$ is the Kullback divergence.

## 4.3 Curved exponential families

By a **curved exponential family** we mean a set of probability distributions which forms a smooth submanifold within an exponential family; put another way, it is a family of distributions which is smoothly embedded in an exponential family. Letting $n$ and $m$ respectively denote the dimensions of the space of the exponential family and the curved exponential family, we call this an $(n, m)$-curved exponential family.

Let $M$ be a curved exponential family with the coordinate system $u = [u^a]$, where $a = 1, \cdots, m$. Then since the probability distribution denoted by $u$ belongs also to $S$, we may write their $\theta$-coordinates as

$$\theta = \theta(u). \tag{4.19}$$

We may consider this as the parameterization of the submanifold $M$ within $S$. Then the probability distributions in $M$ may be written as

$$p(x; u) = \exp\left\{\theta^i(u)x_i - \psi(\theta(u))\right\}. \tag{4.20}$$

We may also write this in terms of $\eta$-coordinates as

$$\eta = \eta(u). \tag{4.21}$$

In the sequel, we chiefly use the $\eta$-coordinate system to represent a point in $S$, so that Equation (4.21) is primarily preferred.

**Example 4.2** *Let $\varepsilon$ be a random variable distributed according to the standard normal distribution $N(0,1)$ with mean 0 and variance 1. Suppose we observe a signal of strength 1 with noise $\varepsilon$, both scaled by a factor of $u$:*

$$x = u(1 + \varepsilon).$$

*Then $x$ is distributed according to a normal distribution with mean $u$ and variance $u^2$. Hence the candidates we wish to consider, within the 2-dimensional*

88    4. STATISTICAL INFERENCE AND DIFFERENTIAL GEOMETRY

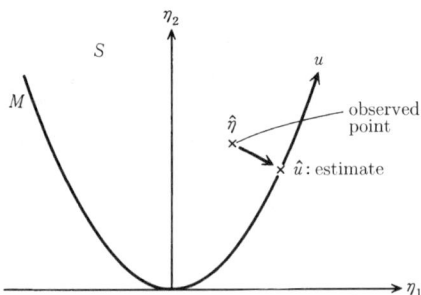

Figure 4.1: An example of a curved exponential family $M$.

space of the normal distributions $S$, are those probability distributions $M = \{p(x; u)\}$ such that
$$\mu = u \quad \text{and} \quad \sigma^2 = u^2.$$

Now $M$ is a $(2,1)$-curved exponential family parameterized by the scalar $u$, and within the space of the exponential family $S$ it is a curve whose defining equations with respect to the $\theta$-coordinate system are

$$\theta^1 = \frac{1}{u} \quad \text{and} \quad \theta^2 = -\frac{1}{2u^2},$$

while with respect to the $\eta$-coordinate system they are

$$\eta_1 = u, \quad \text{and} \quad \eta_2 = 2u^2$$

*(Figure 4.1).*

Suppose the data points $x_1, \cdots, x_N$ have been observed. This determines a point $\hat{\eta} = \bar{x}$ within $S$ whose $\eta$-coordinates are $\bar{x}$. We called this the observed point. Since $\bar{x}$ is a sufficient statistic for $S$, and hence for $M$ also, it suffices to simply consider functions of $\hat{\eta}$ for the estimators $\hat{u} = f(\hat{\eta})$ with which to estimate the parameter $u$ in the underlying distribution $p(x; u)$. In other words, we may represent an estimator by a mapping from $S$ to $M$:

$$f : S \to M \quad \text{where} \quad \hat{\eta} \mapsto \hat{u} = f(\hat{\eta}).$$

The inverse of the estimator $f$, i.e., the set of all points in $S$ which map under estimation to the same point $u$, is in general an $(n-m)$-dimensional submanifold of $S$. We let
$$A(u) = f^{-1}(u) = \{\eta \mid f(\eta) = u\} \tag{4.22}$$
denote this space, and call it the **estimating submanifold** corresponding to the point $u$ in $M$. Then we see that selecting an estimator decomposes the space $S$ into a collection of estimating submanifolds (see Figure 4.2).

## 4.4. CONSISTENCY AND FIRST-ORDER EFFICIENCY

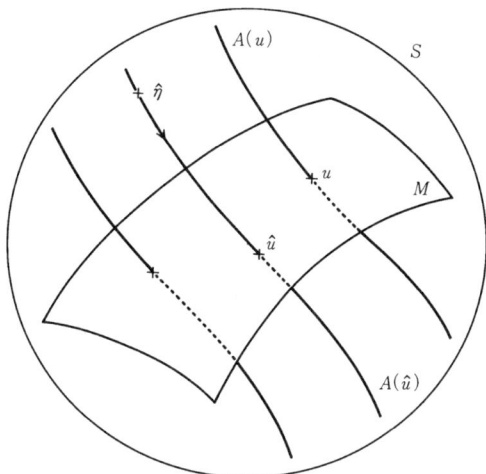

Figure 4.2: Estimating submanifolds $A(u)$.

From this point of view, we see that the characteristics of an estimator are determined entirely by the set of estimating submanifolds $\{A(u)\}$ and the shape of the statistical model $M$. A similar visualization can be made in the case of testing, which we discuss later.

## 4.4 Consistency and first-order efficiency

Asymptotic theory is the analysis of those properties which emerge when the number of samples $N$ is large. Let $p(x;u)$ be the underlying distribution belonging to a curved exponential family described above, and let us consider the accuracy of an estimator $\hat{u}$ with respect to the underlying value $u$.

First let us consider the geometry of a consistent estimator $\hat{u}$. When the underlying distribution is $u$, the expected value of the random variable $x$ is $\eta(u)$. Therefore, by the law of large numbers, $\bar{x}$ converges with probability 1 to the point $\eta(u)$ which represents the underlying distribution under the $\eta$-coordinate system. Hence a necessary and sufficient condition for consistency is for $\eta(u)$, the limit of $\bar{x}$, to be contained in the estimating submanifold $A(u)$ corresponding to $u$. Since the point $\eta(u)$ is indeed contained in $M$, this means that this point $\eta(u)$ is the intersection of $M$ and $A(u)$.

It is possible to consider an estimator for which the estimating submanifold $A(u)$ depends on the number of samples $N$, or equivalently, for which $f(\eta)$ depends on $N$. In this case, we have consistency if $A_N(u) = f_N^{-1}(u)$ contains $\eta(u)$ as $N \to \infty$.

Suppose now that we are given a consistent estimator $\hat{u}$, and let us evaluate its estimation error. Since the observed point $\hat{\eta} = \bar{x}$ converges to the underlying point $\eta(u)$ as $N \to \infty$, $\hat{\eta}$ is distributed close to $\eta(u)$ when $N$ is large. Therefore

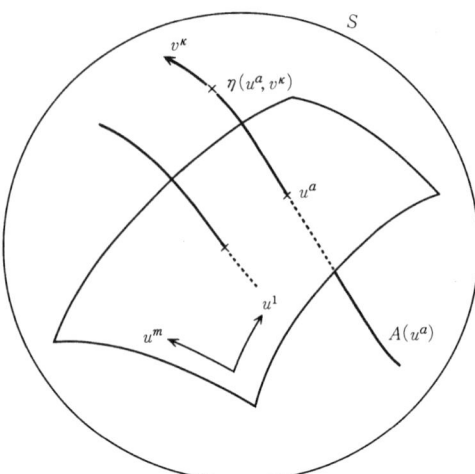

Figure 4.3: The coordinate system $w = (u, v)$ for $S$.

we may consider the tangent space of the point $\eta(u)$ in $S$, and within this space linearize the possible values of $\hat{\eta}$ with respect to the point $\eta(u)$. Now take the difference between $\bar{x}$ and its expected value, scaled by a factor of $\sqrt{N}$:

$$\tilde{x} = \sqrt{N}\{\bar{x} - \eta(u)\} \quad \text{where} \quad \tilde{x} = [\tilde{x}_i]. \tag{4.23}$$

By the central limit theorem, this asymptotically follows the normal distribution with mean 0 and covariance matrix $[g_{ij}(u)]$, where by $g_{ij}(u)$ we mean $g_{ij}(\eta(u))$, the covariance matrix of $x$ under the distribution $p(x; u) = p(x; \eta(u))$. Note that this is also the matrix representing the metric of the tangent space $T_{\eta(u)}(S)$ with respect to the $\theta$-coordinate system.

The space $S$, as can be seen from Figure 4.3, is covered by the family formed by the set of $A(u)$. Since $A(u)$ passes through the point $u$ in $M$ due to the consistency, we may introduce a coordinate system $v$ onto the submanifold $A(u)$ with $u$ as the origin. The dimension of $A(u)$ is $n - m$, and we shall use indices such as $\kappa$ and $\lambda$ ranging over $m+1, m+2, \cdots, n$ for the coordinate system $v$.

Then the point $\eta$ in $S$ may be indexed by $(u, v)$, where $u$ represents the $A(u)$ which contains $\eta$, and $v$ is the index of $\eta$ in $A(u)$. Let us consider $w = (u, v)$ as a new coordinate system for $S$. When using indices, we shall write

$$w = [w^\alpha] = [u^a, v^\kappa],$$

where $\alpha$ ranges from 1 to $n$, $a$ ranges from 1 to $m$, and $\kappa$ ranges from $m+1$ to $n$. Then the $\eta$-coordinates may be written using the $w$-coordinates as

$$\eta = \eta(w) = \eta(u, v). \tag{4.24}$$

This is a coordinate transformation from the $(w = (u, v))$-coordinates to the $\eta$-coordinates. Since the origin $v = 0$ of $A(u)$ is in $M$, the points in $M$ may be written as $\eta(u) = \eta(u, 0)$.

## 4.4. CONSISTENCY AND FIRST-ORDER EFFICIENCY

Let $\hat{w} = (\hat{u}, \hat{v})$ be the $(u, v)$-coordinates of the observed point $\hat{\eta}$; this may be written as

$$\hat{\eta} = \eta(\hat{w}) = \eta(\hat{u}, \hat{v}). \tag{4.25}$$

Since $\hat{u}$ and $\hat{v}$ are close to $u$ and $0$, respectively, let us consider the following normalized quantities:

$$\tilde{u} = \sqrt{N}(\hat{u} - u), \quad \tilde{v} = \sqrt{N}\hat{v}, \quad \tilde{w} = (\tilde{u}, \tilde{v}). \tag{4.26}$$

Now take the Taylor expansion of Equation (4.25) around the point $w = (u, 0)$. Then since $\hat{w} = w + \frac{1}{\sqrt{N}}\tilde{w}$, and expressing in terms of $\eta$-coordinates, we obtain

$$\hat{\eta}_i = \bar{x}_i = \eta_i(u, 0) + \frac{1}{\sqrt{N}} \partial_\alpha \eta_i(u, 0) \tilde{w}^\alpha + \frac{1}{2N}(\partial_\alpha \partial_\beta \eta_i) \tilde{w}^\alpha \tilde{w}^\beta$$
$$+ \frac{1}{6N\sqrt{N}}(\partial_\alpha \partial_\beta \partial_\gamma \eta_i) \tilde{w}^\alpha \tilde{w}^\beta \tilde{w}^\gamma + O\left(\frac{1}{N^2}\right). \tag{4.27}$$

By moving the first term on the right hand side across the equality and scaling by a factor of $\sqrt{N}$, we obtain

$$\tilde{x}_i = B_{\alpha i} \tilde{w}^\alpha + \frac{1}{2\sqrt{N}} C_{\alpha\beta i} \tilde{w}^\alpha \tilde{w}^\beta + \frac{1}{6N} D_{\alpha\beta\gamma i} \tilde{w}^\alpha \tilde{w}^\beta \tilde{w}^\gamma + O\left(\frac{1}{N\sqrt{N}}\right), \tag{4.28}$$

where $B$, $C$, and $D$ are the derivatives of $\eta$ with respect to $w$. For example,

$$B_{\alpha i} = \partial_\alpha \eta_i(u, 0) \quad \text{and} \quad C_{\alpha\beta i} = \partial_\alpha \partial_\beta \eta_i(u, 0).$$

First let us discuss what is typically done in the asymptotic theory of estimation, which is to drop the terms smaller than $O\left(\frac{1}{\sqrt{N}}\right)$ and take the linear approximation. Let us denote the natural basis of $\eta = [\eta_i]$ by

$$e^i = \frac{\partial}{\partial \eta_i} = \partial^i,$$

and that of $w = [w^\alpha]$ by

$$e_\alpha = \frac{\partial}{\partial w^\alpha} = \partial_\alpha.$$

These are bases for the tangent space $T_{\eta(u)}(S)$ of $S$ at the point $\eta(u)$ corresponding to the underlying distribution in $M(\subset S)$. We may decompose $\{e_\alpha\}$ into $\{e_a\} \cup \{e_\kappa\}$, where the tangent space of $M$ is spanned by

$$e_a = \frac{\partial}{\partial u^a} \quad (a = 1, \cdots, m),$$

and the tangent space of $A(u)$ is spanned by

$$e_\kappa = \frac{\partial}{\partial v^\kappa} \quad (\kappa = m+1, \cdots, n).$$

We may also decompose the matrix $B_{\alpha i}$ into the $B_{ai}$ and $B_{\kappa i}$. Since

$$\boldsymbol{e}_a = B_{ai}\boldsymbol{e}^i \quad \text{and} \quad \boldsymbol{e}_\kappa = B_{\kappa i}\boldsymbol{e}^i,$$

we see that they are the components of the tangent vectors of each coordinate axis in $w = (u, v)$ expressed in terms of the basis $\{\boldsymbol{e}^i\}$. Recollect that the inner products of basis vectors determine a metric; with respect to these bases, we have

$$\begin{aligned} g^{ij} &= \langle \boldsymbol{e}^i, \boldsymbol{e}^j \rangle \quad \text{and} \\ g_{\alpha\beta} &= \langle \boldsymbol{e}_\alpha, \boldsymbol{e}_\beta \rangle = B_{\alpha i} B_{\beta j} g^{ij}. \end{aligned} \quad (4.29)$$

The matrix $g_{\alpha\beta}$ representing the metric with respect to the $w$-coordinate system may be decomposed as

$$[g_{\alpha\beta}] = \begin{bmatrix} g_{ab} & g_{a\lambda} \\ g_{\kappa b} & g_{\kappa\lambda} \end{bmatrix}, \quad (4.30)$$

where

$$g_{ab} = \langle \boldsymbol{e}_a, \boldsymbol{e}_b \rangle = B_{ai} B_{bj} g^{ij} \quad (4.31)$$

is the matrix representing the metric on $M$, which is also the Fisher information matrix of the distribution family $M$. When $A(u)$ and $M$ are orthogonal we have

$$g_{a\kappa} = \langle \boldsymbol{e}_a, \boldsymbol{e}_\kappa \rangle = B_{ai} B_{\kappa j} g^{ij} = 0. \quad (4.32)$$

If we ignore the terms smaller than or equal to the order of $\frac{1}{\sqrt{N}}$ in Equation (4.28), we obtain the linear relation

$$\tilde{w}^\alpha = B^{\alpha i} \tilde{x}_i, \quad (4.33)$$

where $[B^{\alpha i}] = [\partial^i w^\alpha]$ is the inverse matrix of $[B_{\alpha i}]$. From Equation (4.29) we have

$$B^{\alpha i} = g^{\alpha\beta} g^{ij} B_{\beta j}, \quad (4.34)$$

where $[g^{\alpha\beta}]$ denotes the inverse matrix of $[g_{\alpha\beta}]$. Since $\tilde{x}_i$ is asymptotically distributed according to the normal distribution with mean 0 and covariance $[g_{ij}]$, its linear transformation $\tilde{w}^\alpha$ is also distributed normally, with mean 0 and covariance $[g^{\alpha\beta}]$. Since $\tilde{u}^a$ is the estimation error $\hat{u}^\alpha - u^\alpha$ scaled by a factor of $\sqrt{N}$, the mean square error scaled by a factor of $N$ is given by

$$\bar{g}^{ab} = E[\tilde{u}^a \tilde{u}^b] = NE[(\hat{u}^a - u^a)(\hat{u}^b - u^b)], \quad (4.35)$$

where $\bar{g}^{ab}$ is the $(a,b)^{\text{th}}$ component of the matrix $[g^{\alpha\beta}]$, the inverse matrix of $[g_{\alpha\beta}]$, and $E$ denotes the expectation with respect to $p(x; u)$. If we decompose the matrix $[g_{\alpha\beta}]$ as Equation (4.30) and take its inverse, we obtain

$$[\bar{g}^{ab}] = \left[g_{ab} - g_{a\kappa} g^{\kappa\lambda} g_{b\lambda}\right]^{-1}. \quad (4.36)$$

Clearly, this is not smaller than $[g^{ab}]$, the inverse of $[g_{ab}]$, and we have

$$\bar{g}^{ab} \geq g^{ab}, \quad (4.37)$$

## 4.4. CONSISTENCY AND FIRST-ORDER EFFICIENCY

where we have used $\geq$ to mean that $[\bar{g}^{ab} - g^{ab}]$ is positive semidefinite. Note that this result just corresponds to the asymptotic Cramér-Rao inequality: Equation (4.11). Furthermore, we see that the equality $\bar{g}^{ab} = g^{ab}$ holds, i.e., the estimator is asymptotically efficient, when and only when $g_{a\kappa} = 0$, or in other words, when and only when $A(u)$ and $M$ are orthogonal.

In summary, we obtain the following theorems.

**Theorem 4.2** *An estimator $\hat{u}$ is consistent if and only if the estimating submanifold $A(u)(= A_N(u))$ contains the point $\eta(u)$ (as $N \to \infty$.)*

**Theorem 4.3** *The asymptotic mean square error:*

$$\bar{g}^{ab} \stackrel{\text{def}}{=} \lim_{N \to \infty} NE\left[\left(\hat{u}^a - u^a\right)\left(\hat{u}^b - u^b\right)\right]$$

*of a consistent estimator $\hat{u}$ is given by*

$$[\bar{g}^{ab}] = \left[g_{ab} - g_{a\kappa} g^{\kappa\lambda} g_{b\lambda}\right]^{-1},$$

*and the estimator is (first-order) asymptotically efficient if and only if $A(u)(= A_N(u))$ and $M$ are orthogonal (as $N \to \infty$).*

Now, let us characterize the maximum likelihood estimator $\hat{u}_{\text{m.l.e}}$ from a geometric point of view, and establish the link between Theorem 4.1 and Theorem 4.3. If we compute the Kullback divergence from the observed point $\hat{\eta}$ to a point $\eta(u)$ in $M$, we obtain

$$\begin{aligned} D\left(\hat{\eta} \| \eta(u)\right) &= \psi(\theta(u)) + \varphi(\hat{\eta}) - \theta^i(u)\hat{\eta}_i \\ &= \varphi(\hat{\eta}) - \frac{1}{N} \log p_N\left(x^N; u\right). \end{aligned} \quad (4.38)$$

Since $\hat{\eta}$ is the result of an observation, $\varphi(\hat{\eta})$ does not depend on $u$. Hence the point $u$ in $M$ which minimizes the divergence with point $\hat{\eta}$ is the point which maximizes the likelihood $p_N\left(x^N; u\right)$, and this is the maximum likelihood estimator $\hat{u}_{\text{m.l.e.}}$. Recollect that the point which minimizes the divergence was the orthogonal projection of the point $\hat{\eta}$ onto $M$ along an m-geodesic (i.e., m-projection, see §3.5). In other words, the estimating submanifold $A(u)$ of $\hat{u}_{\text{m.l.e.}}$ is m-autoparallel and orthogonal to $M$. We thus see that the maximum likelihood estimator is an asymptotically efficient estimator.

More generally, if we consider an arbitrary divergence $D$ on $S$ which induces the Fisher metric (or its constant multiple) in the manner described in §3.2, and if we define the estimator $\hat{u}$ by

$$D\left(\hat{\eta} \| \eta(\hat{u})\right) = \min_{u \in M} D\left(\hat{\eta} \| \eta(u)\right), \quad (4.39)$$

then we see that $\hat{u}$ is asymptotically efficient (Eguchi [84]). Indeed, the estimating submanifold of $\hat{u}$ at the point $u$ is represented (in a neighborhood of $u$) as

$$A(u) = \{\eta \mid D\left(\eta \| (\partial_a)_u\right) = 0, \ \forall a\}. \quad (4.40)$$

from which we have
$$g_{a\kappa} = -D[\partial_\kappa \| \partial_a] = 0.$$

## 4.5 Higher-order asymptotic theory of estimation

Till now, we have been discussing the case when, by making linear approximations, the analysis is contained within a single tangent space. If we do not drop the higher order terms $\frac{1}{\sqrt{N}}$ and $\frac{1}{N}$, then we need to consider not only the tangent space of $A(u)$, but also its curvature. Through the shape of $A(u)$, it is possible to analyze and evaluate various efficient estimators.

Let us outline the higher-order asymptotic theory. Using Equation (4.28), Equation (4.33) may be more accurately written as

$$\begin{aligned}\tilde{w}^\alpha &= B^{\alpha i}\tilde{x}_i - \frac{1}{2\sqrt{N}}C^\alpha_{\beta\gamma}\tilde{w}^\beta\tilde{w}^\gamma \\ &\quad - \frac{1}{6N}D^\alpha_{\beta\gamma\delta}\tilde{w}^\beta\tilde{w}^\gamma\tilde{w}^\delta + \mathrm{O}\!\left(\frac{1}{N\sqrt{N}}\right).\end{aligned} \quad (4.41)$$

We wish to obtain from this the distribution of $\tilde{w}^\alpha$. To do this, let us compute the moments of $\tilde{w}^\alpha$. First, since $E[\tilde{x}_i] = 0$ and $E[\tilde{w}^\beta\tilde{w}^\alpha] = g^{\beta\alpha} + \mathrm{O}(\frac{1}{N})$, we find that the expected value of $\tilde{w}^\alpha$ is given by

$$\begin{aligned}E[\tilde{w}^\alpha] &= -\frac{1}{2\sqrt{N}}C^\alpha + \mathrm{O}\!\left(\frac{1}{N}\right), \quad \text{where} \\ C^\alpha &= C^\alpha_{\beta\gamma}g^{\beta\gamma}.\end{aligned} \quad (4.42)$$

Then in general, even when $\eta(u)$ is contained in $A(u)$, the estimator $\hat{u} = [\hat{u}^a]$ contains a bias of order $\frac{1}{N}$ whose coefficient is $C^a(u)$. This becomes 0 as $N \to \infty$. In order to reduce and compensate for this bias of the estimator, we use $C^a(\hat{u})$ instead of the bias $C^a(u)$, to obtain

$$\hat{u}^{*a} = \hat{u}^a + \frac{1}{2N}C^a(\hat{u}). \quad (4.43)$$

We call this a **bias-corrected estimator**. The bias of $\hat{u}^*$ is

$$E[\hat{u}^*] - u = \mathrm{O}\!\left(\frac{1}{N^2}\right).$$

In order to compute the distribution of $\hat{u}^*$, we first compute the distribution of

$$\hat{w}^* = (\hat{u}^*, \hat{v}^*) = \left[\hat{w}^\alpha + \frac{1}{2N}C^\alpha(\hat{u})\right],$$

and then integrate this over $\hat{v}^*$. To compute the distribution of $\hat{w}^*$, we use the fact that it is asymptotically distributed according to the normal distribution

## 4.5. HIGHER-ORDER ASYMPTOTIC THEORY OF ESTIMATION

$n(\tilde{w}; g^{\alpha\beta})$ with mean 0 and covariance $g^{\alpha\beta}$. Let $A_N$ denote the correction term, and let us write the distribution in the form

$$p(\tilde{w}^*) = n(\tilde{w}^*; g^{\alpha\beta})\{1 + A_N(\tilde{w}^*)\}.$$

Then we may expand $A_N$ as the sum of tensor Hermite polynomials in $\tilde{w}^*$. Each term may be obtained by computing the moments of $\tilde{w}^*$ up to the orders $O\left(\frac{1}{\sqrt{N}}\right)$ and $O\left(\frac{1}{N}\right)$ using Equation (4.28). We call such an expansion the Edgeworth expansion.

We refer the reader interested in the details of these calculation to [6, 9], and instead simply state the results.

**Theorem 4.4** *The mean square error of a bias-corrected first-order efficient estimator is given asymptotically by the expansion:*

$$E\left[(\hat{u}^{*a} - u^a)(\hat{u}^{*b} - u^b)\right] = \frac{1}{N}g^{ab} + \frac{1}{2N^2}K^{ab} + O\left(\frac{1}{N^3}\right). \tag{4.44}$$

$K^{ab}$ *may be decomposed into the sum of positive semidefinite matrices:*

$$K^{ab} = \left(\Gamma_M^{(m)}\right)^{2ab} + 2\left(H_M^{(e)}\right)^{2ab} + \left(H_A^{(m)}\right)^{2ab}, \tag{4.45}$$

*where the terms* $\Gamma_M^{(m)}$, $H_M^{(e)}$, *and* $H_A^{(m)}$ *represent the m-connection coefficients of $M$, the embedding e-curvature of the model $M$, and the embedding m-curvature of $A(u)$, respectively:*

$$\left(\Gamma_M^{(m)}\right)^{2ab} = \Gamma^{(m)a}{}_{cd}\Gamma^{(m)b}{}_{ef}g^{ce}g^{df}, \tag{4.46}$$

$$\left(H_M^{(e)}\right)^{2ab} = H^{(e)\kappa}{}_{ce}H^{(e)\lambda}{}_{df}g_{\kappa\lambda}g^{cd}g^{ea}g^{fb}, \quad \text{and} \tag{4.47}$$

$$\left(H_A^{(m)}\right)^{2ab} = H^{(m)a}{}_{\kappa\lambda}H^{(m)b}{}_{\mu\nu}g^{\kappa\mu}g^{\lambda\nu}, \tag{4.48}$$

*and*

$$\Gamma^{(m)a}{}_{cd} = \left\langle \nabla^{(m)}_{e_c} e_d, e_b \right\rangle g^{ba},$$

$$H^{(e)\kappa}{}_{ce} = \left\langle \nabla^{(e)}_{e_c} e_e, e_\lambda \right\rangle g^{\lambda\kappa}, \quad \text{and}$$

$$H^{(m)a}{}_{\kappa\lambda} = \left\langle \nabla^{(m)}_{e_\kappa} e_\lambda, e_b \right\rangle g^{ba}.$$

This theorem leads to several observations. First, the $\Gamma_M^{(m)}$ term is not a tensor, but rather depends on the choice of coordinate system $u$ for $M$. This reflects the fact that the square error is also not a tensor and depends on the choice of coordinate system. However, once the coordinate system (parameter) $u$ is determined, this value remains invariant on the choice of estimator. Although it is possible to define the coordinate system $u$ so that $\Gamma_M^{(m)}$ is 0 for a particular point (e.g. a normal coordinate system for the m-connection), it is only possible to have $\Gamma_M^{(m)}$ equal to 0 at all points if $M$ is m-flat (and hence e-flat by the duality).

Second, the $H_M^{(e)}$ term represents the e-curvature of $M$. The e-curvature of $M$ is 0 when and only when $M$ is e-autoparallel in $S$, i.e., when and only when $M$ is itself an exponential family (see Theorem 2.5). Hence this term shows that it is not possible to avoid an increase in estimation error given by the e-curvature, which measures the deviation of $M$ from being an exponential family.

Finally, the $H_A^{(m)}$ term shows that the estimation error includes the m-curvature at the point $\eta(u)$ of the estimating submanifold, and this is the only term that depends on the choice of an estimator. Hence a first-order efficient estimator for which the m-curvature of $A(u)$ is 0 is an optimal estimator with respect to the terms up to $\frac{1}{N^2}$, which we refer to as the **second-order efficiency** of the estimator. As we noted before, for the maximum likelihood estimator, the estimating submanifold $A(u)$ is m-autoparallel and orthogonal to $M$. Therefore, we obtain the following.

**Theorem 4.5** *The bias-corrected maximum likelihood estimator $\hat{u}^*_{\text{m.l.e.}}$ is second-order efficient.*

We also see that, for any divergence $D$ which induces $(g, \nabla^{(m)}, \nabla^{(e)})$ on $S$, the bias-corrected version of the minimum $D$ estimator $\hat{u}$ defined by Equation (4.39) is second-order efficient, because Equation (4.40) leads to

$$H^{(m)}_{\kappa\lambda,a} = \left\langle \nabla^{(m)}_{e_\kappa} e_\lambda, e_a \right\rangle = -D\left[\partial_\kappa \partial_\lambda \,\|\, \partial_a\right] = 0.$$

To show that the maximum likelihood estimator is "optimal" was for R. A. Fisher a life-long dream. This dream has been achieved in the asymptotic case by the theorem above. The idea of bias correction was introduced by C. R. Rao, and its higher-order asymptotic optimality was proven by Rao, Ghosh, Pfanzagl, Chibisov, Efron, and Takeuchi-Akahira. Its geometric structure, and in particular its relation to the e-curvature was first noted by B. Efron [83]. The differential geometric framework including both the e-curvature and the m-curvature was introduced by Amari [5, 6]. The link between the asymptotics of the minimum $D$ estimator and the dualistic structure $(g^{(D)}, \nabla^{(D)}, \nabla^{(D^*)})$ was pointed out by Eguchi [84].

It is possible to consider evaluating the estimation error up to the $\frac{1}{N^3}$ term. Recently, Kano [116] calculated the $\frac{1}{N^3}$ terms where the bias of a second-order efficient estimator is corrected up to $\frac{1}{N^2}$. He proved a surprising result that the maximum likelihood estimator does not minimize the $\frac{1}{N^3}$ term, implying that it is not third-order efficient. He gave the third-order efficient estimator explicitly. It should be noted that the expansion in Equation (4.44) of the error is an asymptotic expansion for large $N$, and even when the number of terms is taken to be large, convergence is not necessarily guaranteed.

Finally, we make some remarks on the bias correction. We first emphasize its dependence on the choice of coordinate system. For example, the bias-corrected maximum likelihood estimator $\hat{u}_{\text{m.l.e.}}^*$ depends on the parameterization (coordinate system) $u$ of the model, while the original estimator $\hat{u}_{\text{m.l.e.}}$ does not. The purpose of the bias correction is just to compensate for the dependence of the mean square error on the parameterization. More precisely, the difference of the mean square errors of two bias-corrected first-order efficient estimators turns out to be a tensor, so that we may compare them in a geometrically invariant manner. We should also note that the second-order efficiency defined above is the optimality, up to the order of $\frac{1}{N^2}$, of a bias-corrected estimator compared to all other bias-corrected estimators. In fact, it is sometimes possible to improve a second-order efficient estimator, in the sense of the mean square error up to the order of $\frac{1}{N^2}$, by adding a bias term. However, the superiority of the resulting estimator depends on the manner of parameterization, because the mean square error is not geometrically invariant. The bias correction is thus needed to discuss the asymptotic optimality in terms of the mean square error. Another way to discuss the asymptotic optimality of estimators is to consider a geometrically invariant criterion, rather than the mean square error, to measure the accuracy of estimation. Recently, a remarkable progress in this direction was made by Eguchi and Yanagimoto [88], where some notions from the vector analysis and the harmonic analysis have been introduced in the world of information geometry. See also Komaki [123], where the problem of higher-order optimal prediction including the analysis based on the Bayesian framework is studied from an information geometrical point of view.

## 4.6 Asymptotics of Fisher information

The information contained in the observed data concerning the underlying distribution may be expressed using the Fisher information matrix. The information contained in a single data point $x$ is $g_{ab}$, while that contained in a collection of $N$ data points $x^N$, which is summarized by the sufficient statistic $\bar{x}$, is $Ng_{ab}$. We express them as $g_{ab}(X) = g_{ab}$ and $g_{ab}(X^N) = g_{ab}(\bar{X}) = Ng_{ab}$. Now let

$$y = f(\bar{x})$$

be a statistic which is a function of $\bar{x}$. Unless $f$ is invertible, one cannot recover $\bar{x}$ from $y$, and hence in general the Fisher information contained in $y$ is less than $Ng_{ab}$. The Fisher information contained in $y$ is given by the covariance matrix of $\partial_a \log p(y;u)$,

$$g_{ab}(Y) = E[\partial_a \log p(Y;u)\, \partial_b \log p(Y;u)], \tag{4.49}$$

where $p(y;u)$ is the probability density of $y$, and $u$ is the underlying parameter. According to Theorem 2.1 in §2.2, we have

$$g_{ab}(\bar{X}) = Ng_{ab} = g_{ab}(Y) + \Delta g_{ab}(Y), \tag{4.50}$$

where the information loss due to the contraction from $\bar{x}$ to $y$ is given by the expectation of the conditional covariance:

$$\Delta g_{ab}(Y) = E\big[\operatorname{Cov}[\partial_a \ell(\bar{X}; u), \partial_b \ell(\bar{X}; u) \mid Y]\big]. \tag{4.51}$$

The calculation of an estimator $\hat{u}$ based on $\bar{x}$ is an information loss process. The information loss which a first-order efficient estimator incurs may be computed using the distribution of $\tilde{w}$.

**Theorem 4.6** *(Amari [6, 9]) The information loss of a first-order efficient estimator $\hat{u}$ is*

$$\Delta g_{ab}(\hat{U}) = \left(H_M^{(e)}\right)^2_{ab} + \frac{1}{2}\left(H_A^{(m)}\right)^2_{ab} + \operatorname{O}\left(\frac{1}{N}\right), \tag{4.52}$$

*where*

$$\left(H_M^{(e)}\right)^2_{ab} = H^{(e)\,\kappa}{}_{ca} H^{(e)\,\lambda}{}_{db} g_{\kappa\lambda} g^{cd}, \quad \text{and} \tag{4.53}$$

$$\left(H_A^{(m)}\right)^2_{ab} = H^{(m)}{}_{\kappa\lambda,a} H^{(m)}{}_{\mu\nu,a} g^{\kappa\mu} g^{\lambda\nu}. \tag{4.54}$$

This theorem shows that a first-order efficient estimator preserves a large portion of the order $N$ information $N g_{ab}$ contained in a set of $N$ data points, only incurring an information loss of order 1, and the information loss per data point asymptotically reaches 0. It should be noted that the dominant terms of the information loss are just the covariant versions of the $O(N^{-2})$ terms of the mean-square error given in Theorem 4.4, except that the non-tensorial term $\left(\Gamma_M^{(m)}\right)^2$ does not appear here.

Let us consider where this information is lost, and whether this knowledge could improve inference. First let us examine the whereabouts of the lost information in the case when $\hat{u}$ is the maximum likelihood estimator. Recall that the observed point $\bar{x}$ may be represented as $\bar{x} = \eta(\hat{u}, \hat{v})$ and hence the pair $(\hat{u}, \hat{v})$ contains the whole Fisher information $N g_{ab}$, while $\hat{u}$ has an information loss of $O(1)$. The gap between $(\hat{u}, \hat{v})$ and $\hat{u}$ may be analyzed by a decomposition of $\hat{v}$ given below.

The tangent space $T_u(S)$ of $S$ at the point $\eta(u)$ may be decomposed into a direct sum of the tangent spaces of the model $M$ and the estimating submanifold $A(u)$:

$$T_u(S) = T_u(M) \oplus T_u(A(u)).$$

The bases of $T_u(M)$ and $T_u(A(u))$ are $\{e_a\}$ and $\{e_\kappa\}$, respectively. Let us first extract the curvature directions of $M$ from $T_u(A(u))$. The embedding e-curvature of $M$ is given by the orthogonal component of the change in $e_b$ as it is moved in the direction of $e_a$ with respect to the e-connection:

$$H_{ab}^{(e)\kappa} = \left\langle \nabla_{e_a}^{(e)} e_b, e_\lambda \right\rangle g^{\lambda\kappa}.$$

## 4.6. ASYMPTOTICS OF FISHER INFORMATION

Then
$$e^{(e)}_{ab} = H^{(e)\kappa}_{ab} e_\kappa, \qquad (a,b = 1,\cdots,m) \tag{4.55}$$
is the collection of vectors in $T_u(A(u))$ in the direction of the e-curvature of $M$. Let us denote the linear span of these vectors by $V_u^{(1)}$.

Let us now similarly define even higher order embedding e-curvatures (the change in direction of curvature). To do so, we compute
$$\nabla^{(e)}_{e_a} \nabla^{(e)}_{e_b} e_c,$$
remove the components in $T_u(M)$ and in $V_u^{(1)}$, and let the components orthogonal to them be denoted by
$$e^{(e)}_{abc} = H^{(e)\kappa}_{abc} e_\kappa. \tag{4.56}$$
In other words, $e^{(e)}_{abc}$ is the orthogonal projection of $\nabla^{(e)}_{e_a} \nabla^{(e)}_{e_b} e_c$ onto the orthogonal complement of $T_u(M) \oplus V_u^{(1)}$. We call $\{H^{(e)\kappa}_{abc}\}$ the second-order embedding e-curvature of $M$, and denote the linear span of $\{e^{(e)}_{abc}\}$ by $V_u^{(2)}$. In this way we may successively define the $p^\text{th}$-order e-curvature $\{H^{(e)\kappa}_{a_1\cdots a_{p+1}}\}$ and the space $V_u^{(p)}$ of its direction.

Let $\hat{u}$ be the maximum likelihood estimator. Then, since the estimating submanifolds $\{A(u)\}$ are m-autoparallel in $S$, coordinates $v = [v^\kappa]$ may be chosen so as to form an m-affine coordinate system on each $A(u)$. Now let us consider $\hat{v}$ as a vector in $T_{\hat{u}}(A(\hat{u}))$, representing it by $\hat{v}^\kappa \cdot (e_\kappa)_{\hat{u}}$, and decompose it into the directions of higher-order curvatures. The components of the direction of first-order curvature is given by
$$\hat{r}_{ab} = H^{(e)}_{ab\kappa} \hat{v}^\kappa = H^{(e)\lambda}_{ab} \hat{v}^\kappa g_{\lambda\kappa}, \tag{4.57}$$
the components of the direction of second-order curvature by
$$\hat{r}_{abc} = H^{(e)}_{abc\kappa} \hat{v}^\kappa = H^{(e)\lambda}_{abc} \hat{v}^\kappa g_{\lambda\kappa}, \tag{4.58}$$
and similarly for the higher-order terms, where $H^{(e)}_{ab\kappa}, H^{(e)}_{abc\kappa}, \cdots$ are all evaluated at $\hat{u}$. Restating this in the language of statistics, $\hat{r}_{ab}, \hat{r}_{abc}, \cdots$ can be obtained by taking the higher-order derivatives of the logarithmic likelihood
$$\partial_a \ell(\bar{x}, u), \quad \partial_a \partial_b \ell(\bar{x}, u), \quad \partial_a \partial_b \partial_c \ell(\bar{x}, u), \quad \cdots,$$
orthogonalizing them in order, and evaluating them at the point $\hat{u}$.

Now let
$$\begin{aligned} s^{(1)} &= \hat{u}, \\ s^{(p)} &= \{\hat{u}, \hat{r}_{ab}, \cdots, \hat{r}_{a_1\cdots a_p}\}, \quad p = 2, 3, \cdots. \end{aligned}$$

In addition, let us denote the quadratic form of the $p^\text{th}$-order e-curvature of $M$ by $\left(H^{(e)}_{M,p}\right)^2_{ab}$, which is defined from $H^{(e)\kappa}_{a_1\cdots a_{p+1}}$ in a manner similar to the definition of $(H^{(e)}_{M,1})^2_{ab} = (H^{(e)}_M)^2_{ab}$ in Equation (4.53). We have the following theorem ([9, Theorem 7.10]), which is an extension of Theorem 4.6 in the case of $\hat{u} = \hat{u}_{\text{m.l.e.}}$.

**Theorem 4.7** *When $s^{(p)}$ is maintained as a statistic, the Fisher information lost is*

$$\Delta g_{ab}(s^{(p)}) = N^{-p+1}\left(H^{(e)}_{M,p}\right)^2_{ab} + \mathrm{O}(N^{-p}). \qquad (4.59)$$

The majority of the information is contained in the maximum likelihood estimator $\hat{u}$. Let us now consider whether the lost information, for example the component $\hat{r}_{ab}$ of $\hat{v}$ in the direction of curvature, can still be useful in statistical inference. Of course, use of this information cannot make the estimator $\hat{u}$ itself better. However, $\hat{r}_{ab}$ provides us with information concerning the accuracy of $\hat{u}$.

**Theorem 4.8** *(Amari [7, 9]) The conditional covariance matrix of the estimator $\hat{u}$ when the supplemental information $\hat{r} = \{\hat{r}_{ab}\}$ is known is given by*

$$NE\left[\left(\hat{u}^a - u^a\right)\left(\hat{u}^b - u^b\right) \mid \hat{r}\right] = g^{ab} + \hat{r}^{ab} + \mathrm{O}\left(\frac{1}{N}\right), \qquad (4.60)$$

*where $\hat{r}^{ab} = \hat{r}_{cd} g^{ac} g^{bd}$.*

When $\hat{r}$ was not used, the covariance of $\hat{u}$ (or $\hat{u}^*$) was given by Equation (4.44). In comparison, we see that $\hat{r}_{ab}$ provides a more refined order $\frac{1}{\sqrt{N}}$ measure of the accuracy of $\hat{u}$.

We call statistics such as $\hat{r}_{ab}$ which by themselves contain no (or little) information **ancillary statistics**. Within the statistics community, there is the "folklore" that when ancillary statistics are available, it is useful to classify the situation based on this value, and to then perform conditional inference with respect to them. On close investigation, however, we find that ancillary statistics are useful not because they contain no information, but rather because, although they contain no information by themselves, they contain the most information (conditioned information) supplemental to the information contained in the estimator $\hat{u}$.

## 4.7 Higher-order asymptotic theory of tests

Along with estimation, an important branch of statistical inference is testing. The problem of interval estimation is a manifestation of testing and shall follow from this discussion. In this section we discuss the geometrical theories of testing hypotheses within the framework of a *1-dimensional* curved exponential family $M = \{p(x; u)\}$, where $u$ is supposed to be a *scalar parameter*.

## 4.7. HIGHER-ORDER ASYMPTOTIC THEORY OF TESTS

Let us consider the problem of testing the hypothesis $H_0 : u = u_0$ against the alternative hypothesis $H_1 : u \neq u_0$. Recollect that the observed data $x_1, \cdots, x_N$ may be summarized by the sufficient statistic $\bar{x}$, and that this determines the observed point $\hat{\eta} = \bar{x}$ in the space $S$ of the exponential family. Testing usually consists of choosing a function $\lambda(\bar{x})$ of $\bar{x}$, and then accepting the hypothesis $H_0$ if $\lambda(\bar{x})$ is in some interval, say $I$, and rejecting the hypothesis otherwise. Within the space $S$, those points $\eta$ which satisfy $\lambda(\eta) \in I$ form some region. We call this the **acceptance region** $A$, and if $\bar{x}$ is within this region we accept the hypothesis, and otherwise if it is outside of this region (we call the outside the **critical region**) we reject. The problem lies in properly selecting the region $A$ so as to elicit good testing performance. In general, the interval $I$ and the resulting acceptance region depend on the number of observations $N$, and we denote them by $I_N$ and $A_N$, respectively.

Let us now define the power function of a test $T = \{A_N\,;\,N = 1, 2, \cdots\}$. First let us normalize the parameter $u$, using the Fisher information $g = g_{ab}(u_0)$ ($a = b = 1$) and $N$, as

$$u_t = u_0 + \frac{t}{\sqrt{Ng}}. \tag{4.61}$$

The reason for this normalization is that as $N$ grows larger, it should be possible to detect smaller differences between $u$ and $u_0$. Let the probability of rejecting the hypothesis when the parameter of the underlying distribution is $u_t$ be denoted by

$$\begin{aligned} P_T(t;N) &= \Pr\{\text{reject } H_0 \mid \text{the underlying distribution is } u_t\} \\ &= \Pr\{\bar{x} \notin A_N \mid x_1, \cdots, x_N \sim p(x\,;\,u_t)\}. \end{aligned} \tag{4.62}$$

We call $P_T(t;N)$ when viewed as a function of $t$ the **power function** of $T$.

When $t = 0$, $H_0$ is true, and it would be undesirable to reject the hypothesis. Hence we choose a (small) positive constant $\alpha$, which is called the **level** of the test, and fix

$$P_T(0;N) = \alpha. \tag{4.63}$$

For example, we fix $\alpha$ to values such as 5% (0.05) and 1% (0.01), being regardless of $N$. Since $\lambda(\bar{x})$ converges to $\lambda(\eta(u_0))$ in probability as $N \to \infty$ when $t = 0$, the level condition implies that the interval $I_N$ collapses into this point $\lambda(\eta(u_0))$ in the limit of $N \to \infty$. In addition, as a necessary condition for the power of points for $t \neq 0$ to be greater than that for $t = 0$, we require that

$$P'_T(0;N) = 0, \tag{4.64}$$

where ' denotes the derivative with respect to $t$. We call this the unbiasedness condition. Under these conditions, the power function generally has the limit $\lim_{N \to \infty} P_T(t;N)$.

Among those tests which satisfy the conditions in Equations (4.63) and (4.64), a good test is one whose power at points $t$ ($t \neq 0$) is high. Unfortunately, there does not in general exist a **uniformly most powerful test**, which is a test more powerful at every point $t$ than any other test. Hence given

102    4. STATISTICAL INFERENCE AND DIFFERENTIAL GEOMETRY

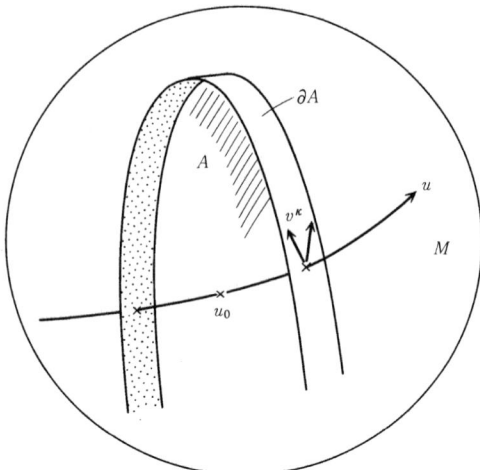

Figure 4.4: Acceptance region $A$.

a test, it is necessary to consider for what values of $t$ it performs well and for what values it performs poorly. To analyze this, let us first expand the power function as

$$P_T(t; N) = P_{T1}(t) + \frac{1}{\sqrt{N}} P_{T2}(t) + \frac{1}{N} P_{T3}(t) + O\left(\frac{1}{N\sqrt{N}}\right). \quad (4.65)$$

In the first-order asymptotic theory, it is assumed that $N$ is large, and that therefore only the term $P_{T1}(t)$ need be considered. In this case there exists a test for which $P_{T1}(t)$ is more (or at least equally) powerful for all $t$ than any other; we call such a test a **first-order uniformly most powerful test**, or an **efficient test**.

We begin our discussion of the geometry of tests by focusing on the boundary of the acceptance region for a test $T$. As shown in Figure 4.4, the acceptance region is bounded by either 1 or 2 hypersurfaces across $M$ which also flank the point $\eta(u_0)$ on the curve $M$.

The hypersurfaces are of the form $\{\eta \mid \lambda(\eta) = c\}$. For each $u$, let us denote the hypersurface with $c = \lambda(\eta(u))$ by $B(u)$. Then in general $B(u)$ transversally intersects with $M$ at $\eta(u)$. Furthermore, taking a (small) open subset of $B(u)$ including $\eta(u)$ and renaming it $B(u)$ if necessary, we may assume that $\{B(u)\}$ forms a family of disjoint submanifolds, just like a family of estimating submanifolds defined for an estimator. Hence we can introduce a coordinate system $v = [v^\kappa]$ on each $B(u)$ so that $(u, v)$ forms a coordinate system of $S$ (or at least of a neighborhood of $\eta(u_0)$ in $S$). Then the observed point $\bar{x}$ is transformed to $(\hat{u}, \hat{v})$ by $\bar{x} = \eta(\hat{u}, \hat{v})$, and the joint probability distribution of $(\hat{u}, \hat{v})$ is determined depending on the geometric shape of the acceptance region.

In the case of the first-order asymptotic theory, it suffices to linearize all the relations around the point $\eta(u_0)$ and analyze within the tangent space $T_{\eta(u_0)}(S)$.

## 4.7. HIGHER-ORDER ASYMPTOTIC THEORY OF TESTS

When the underlying distribution is $u_t$, $\bar{x}$ is asymptotically distributed according to a normal distribution with mean $\eta(u_t, 0)$ and covariance $N^{-1} g_{ij}$. Now let $z_*(\alpha)$ be the two-sided $100\alpha\%$ point of the normal distribution; i.e., let $z_*(\alpha)$ be the point which satisfies

$$\int_{-z_*(\alpha)}^{z_*(\alpha)} n(z) \mathrm{d}z = 1 - \alpha, \tag{4.66}$$

where

$$n(z) \stackrel{\text{def}}{=} \frac{1}{\sqrt{2\pi}} \exp\left\{-\frac{1}{2} z^2\right\}.$$

Also, let

$$\Phi(t) = \int_t^\infty n(z) \mathrm{d}z. \tag{4.67}$$

Note that $\Phi(z_*(\alpha)) = \alpha/2$. The first-order efficiency of a test is characterized by the following theorem (Kumon and Amari [132, 9]).

**Theorem 4.9** *The test $T = \{A_N\}$ is first-order efficient when and only when the bounding hypersurface of the acceptance region $A_N$ is asymptotically orthogonal to $M$ in the limit of $N \to \infty$, or in other words, when and only when the hypersurface $B(u_0) = \{\eta \mid \lambda(\eta) = \lambda(\eta(u_0))\}$ is orthogonal to $M$ at the point $\eta(u_0)$, i.e., $g_{a\kappa}(u_0) = 0$. In this case, its asymptotic power function is given by*

$$P_{T1}(t) = \Phi\left[z_*(\alpha) - t\right] + \Phi\left[z_*(\alpha) + t\right]. \tag{4.68}$$

We now give several examples of tests.

**Example 4.3 (Wald test)** *This is a test based on the maximum likelihood estimator $\hat{u}_{\mathrm{m.l.e.}}$, and uses as its test statistic either*

$$\lambda(\bar{x}) = (\hat{u}_{\mathrm{m.l.e.}}(\bar{x}) - u_0)^2 g(u_0)$$

*or*

$$\lambda(\bar{x}) = (\hat{u}_{\mathrm{m.l.e.}}(\bar{x}) - u_0)^2 g(\hat{u}_{\mathrm{m.l.e.}}(\bar{x})).$$

*Then $\lambda$ is asymptotically distributed according to the $\chi^2$-distribution. The boundary of the acceptance region of this test is given by the constant $\hat{u}_{\mathrm{m.l.e.}}(\bar{x})$; i.e., it is an estimating submanifold of the maximum likelihood estimator. This implies that $B(u)$ is orthogonal to $M$ for all $u$, and therefore the test is efficient.*

**Example 4.4 (Likelihood ratio test)** *The likelihood ratio test is one which uses the statistic*

$$\lambda(\bar{x}) = -2\log \frac{p(\bar{x}; u_0)}{p(\bar{x}; \hat{u}_{\mathrm{m.l.e}}(\bar{x}))}.$$

This is also asymptotically distributed according to a $\chi^2$-distribution. Rewriting this into
$$\lambda(\bar{x}) = 2\{D(\bar{x} \parallel u_0) - D(\bar{x} \parallel \hat{u}_{\text{m.l.e.}}(\bar{x}))\}$$
where $D$ is the Kullback divergence, we see that $B(u_0)$ in this case coincides with the estimating submanifold $A(u_0)$ of $\hat{u}_{\text{m.l.e.}}$. Therefore the test is efficient. Note that $B(u)$ is not generally orthogonal to $M$ for $u \neq u_0$. However, if $M$ is an exponential family, by the Pythagorean theorem we have $\lambda(\bar{x}) = 2D(\hat{u}_{\text{m.l.e.}}(\bar{x}) \parallel u_0)$, and hence this test turns out to be essentially the same as the Wald test.

Besides these, there are well known tests such as the efficient-score (Rao) test, the locally most powerful test, and the conditional test, all of which are first-order efficient and equally powerful if one considers only the term $P_{T1}(t)$. However, when actually testing on a finite number of $N$ samples, these tests behave quite differently. In order to investigate how and where these tests differ in quality, it is necessary to examine the higher-order terms $P_{T2}(t)$ and $P_{T3}(t)$.

The statistics of these first-order efficient tests are all distributed asymptotically according to the $\chi^2$-distribution. However, for finite $N$, it is not reasonable to simply consider the $\chi^2$-distribution when determining the interval $I_N$ with which to define the acceptance region $A_N : \lambda(\bar{x}) \in I_N$ from the level $\alpha$. Instead, it is necessary to consider the difference between the distribution of $\lambda(\bar{x})$ and the $\chi^2$-distribution, and correct the acceptance region given by the $\chi^2$-distribution so that the level $\alpha$ is correct up to order $N^{-3/2}$ and the condition of unbiasedness is satisfied up to order $N^{-1}$. This correction allows us to compare the tests from a common base and to correctly compute the higher-order terms of their power functions.

Examining the $P_{T2}(t)$ component of tests in this manner, it is possible, however, to prove that all first-order efficient tests have the same $P_{T2}(t)$. Hence we say that "first-order efficient tests are second-order efficient." It is necessary, therefore, to examine $P_{T3}(t)$ for each test. As this term contains many components other than such geometrically meaningful quantities as metric or curvature, writing out the results of calculations can be rather tedious. Instead, we introduce a function $\tilde{P}(t; N)$, which does not depend on each test $T$, and compute the difference of the power of $T$ and this function. For each $t$ and $N$, let
$$\tilde{P}(t; N) = \sup_T P_T(t; N),$$
where the supremum is taken over all the efficient tests with the level $\alpha$. In other words, $\tilde{P}(t; N) = P_{\tilde{T}_t}(t; N)$ where $\tilde{T}_t$ is the test which is most powerful at the point $u_t$. Note that there is no test $T$ satisfying $P_T(t; N) = \tilde{P}(t; N)$ for all $t$ in general. Now let
$$\Delta P_T(t) = \lim_{N \to \infty} N\{\tilde{P}(t; N) - P_T(t; N)\}, \tag{4.69}$$

## 4.7. HIGHER-ORDER ASYMPTOTIC THEORY OF TESTS

and call it the **deficiency** of $T$. This quantity measures the number of additional samples which must be obtained in order for the test $T$ to have the same power as $\tilde{T}_t$ at each $t$.

The computation of this quantity succeeded for the first time using differential geometric techniques ([132, 9]). From this the characteristics and differences of a number of tests were clarified.

**Theorem 4.10** *The deficiency of an efficient test $T$ is given by*

$$\Delta P_T(t) = \xi(t,\alpha)\Big[\frac{1}{2}(H_B^{(m)})^2 + z_*^2(\alpha)g^{\kappa\lambda}g^{-2}$$
$$\times\{Q_{ab\kappa} - \mathcal{J}(t,\alpha)H_{ab\kappa}^{(e)}\}\{Q_{cd\lambda} - \mathcal{J}(t,\alpha)H_{cd\lambda}^{(e)}\}\Big], \qquad (4.70)$$

*where*

$$(H_B^{(m)})^2 = g^{-1}H_{\kappa\lambda a}^{(m)}H_{\nu\mu b}^{(m)}g^{\kappa\nu}g^{\lambda\mu}, \qquad (4.71)$$

$$\xi(t,\alpha) = \frac{t}{2}\{n(z_*(\alpha) - t) - n(z_*(\alpha) + t)\}, \qquad (4.72)$$

$$Q_{ab\kappa} = \partial_a g_{b\kappa}, \qquad (4.73)$$

$$\mathcal{J}(t,\alpha) = 1 - \frac{t}{2z_*(\alpha)\tanh t z_*(\alpha)}, \qquad (4.74)$$

*with $g = g_{ab}$, $H_{ab\kappa}^{(e)} = \left\langle \nabla_{e_a}^{(e)} e_b, e_\kappa \right\rangle$ and $H_{\kappa\lambda a}^{(m)} = \left\langle \nabla_{e_\kappa}^{(m)} e_\lambda, e_a \right\rangle$, all of which are evaluated at $u_0$. (Note that since $M$ is 1-dimensional, the value of $a, b, \cdots$ is always 1.)*

From this theorem, we find the following to hold.

(i) A test for which the embedding m-curvature $H_B^{(m)}$ of $B(u_0)$ at $u_0$ is 0 has higher power than those for which this is not the case. The tests mentioned above all meet this condition.

(ii) Since $g_{a\kappa}(u_0) = 0$ holds due to the efficiency of $T$, we asymptotically have

$$g_{a\kappa}(u_t) = \frac{t}{\sqrt{Ng}}Q_{ab\kappa}. \qquad (4.75)$$

Then the theorem shows that, unless $M$ is an exponential family, it is better for the boundary surface of a test to not be exactly orthogonal to $M$ for finite $N$. Instead, it is better for it to be at an asymptotically orthogonal angle depending on the embedding e-curvature $H_{ab\kappa}^{(e)}$ of $M$.

Let us call an efficient test which satisfies $H_B^{(m)} = 0$ and

$$Q_{ab\kappa} = kH_{ab\kappa}^{(e)} \qquad (4.76)$$

# 4. STATISTICAL INFERENCE AND DIFFERENTIAL GEOMETRY

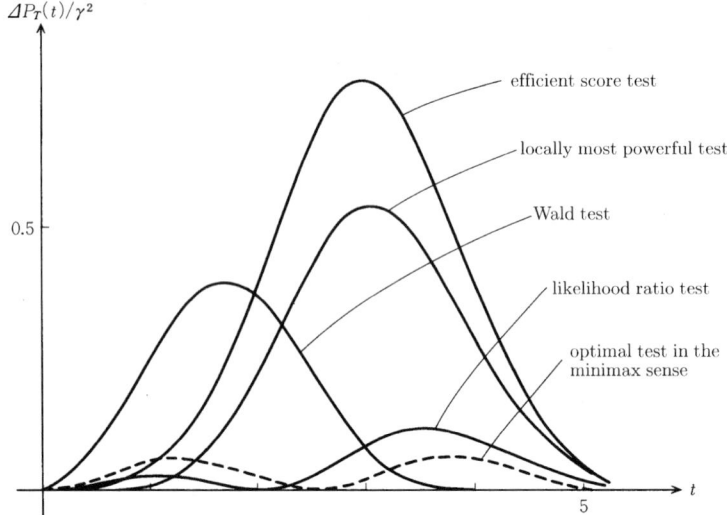

Figure 4.5: Deficiency of two-sided tests, with level $\alpha = 5\%$.

a $k$-test. In addition, let the scalar form of the squared embedding e-curvature be given by

$$\gamma^2 = g^{-2} H^{(e)}_{ab\kappa} H^{(e)}_{cd\lambda} g^{\kappa\lambda}. \tag{4.77}$$

This $\gamma$ is called the Efron curvature.

**Corollary 4.11** *The deficiency of a $k$-test $T(k)$ is given by*

$$\Delta P_{T(k)}(t) = z_*^2(\alpha)\xi(t,\alpha)\{k - \mathcal{J}(t,\alpha)\}^2 \gamma^2. \tag{4.78}$$

From this corollary we see that in order to maximize the power at a point $u_t$, it suffices to choose a $k$-test which satisfies $k = \mathcal{J}(t,\alpha)$. Note that when $\alpha(>0)$ is sufficiently small, $\tanh t z_*(\alpha)$ is nearly equal to the signature of $t$ and therefore the relation $k = \mathcal{J}(t,\alpha)$ is approximately written as $t = \pm 2(1-k)z_*(\alpha)$. All of the first-order efficient tests previously mentioned are $k$-tests for some $k$. We summarize this information below. Also, Figure 4.5 graphs the deficiency of a number of tests as a function of $t$, where the dashed line indicates a $k$-test for which $k$ achieves $\min_k \max_t \Delta P_{T(k)}(t)$.

If the level and unbiasedness are properly corrected, then regardless of the model $M$ used, the properties of the test are determined according to the scalar e-curvature $\gamma$. This means that there is no longer a need to analyze individual models or to conduct simulations in pursuit of a test's power characteristics. In addition, although we have been discussing two-sided tests where $t$ can range over both positive and negative values, one-sided tests, for which the alternative hypothesis is $H_1 : u > u_0$ and $t$ is restricted to $t \geq 0$, may be correspondingly analyzed in a similar manner.

**Theorem 4.12** *Several first-order efficient tests and their form as k-tests are given as follows:*

(i) *The Wald test is a $k = 0$ test, and is most powerful at $t = \pm 2z_*(\alpha)$ when $\alpha$ is sufficiently small. For $\alpha = 0.05$, it is powerful when $t \approx \pm 4$, i.e., at a distance of around $4\sigma$.*

(ii) *The likelihood ratio test is a $k = 0.5$ test, and is most powerful at $t = \pm z_*(\alpha)$ when $\alpha$ is sufficiently small; for instance, $t \approx \pm 2$ for $\alpha = 0.05$. This test uniformly has relatively small deficiency.*

(iii) *The efficient-score test, which uses the test statistic*

$$\lambda(\bar{x}) = g^{ab} \partial_a \ell(\bar{x}; u_0) \partial_b \ell(\bar{x}; u_0),$$

*is a $k = 1$ test. Since $1 = \mathcal{J}(t, \alpha)$ has no solution, this test is nowhere most powerful, whereas the one-sided test using the same statistic turns out to be third-order locally most powerful in the sense that the derivative $P'_{T3}(0)$ is the largest among those of all the efficient tests.*

(iv) *The conditional test, whose definition is omitted here, is a $k = 0.5$ test, and is equivalent to the likelihood test up to the third-order term.*

(v) *The locally most powerful test, for which the second derivative $P''_T(0; N)$ is the largest among all efficient unbiased tests with level $\alpha$, is $k = 1 - 1/\{2z_*^2(\alpha)\}$ test.*

## 4.8 The theory of estimating functions and fiber bundles

### 4.8.1 The fiber bundle of local exponential families

Thus far our discussion has been in the context of curved exponential families, i.e., distribution families $M$ embedded within the space $S$ of an exponential family. This allowed us to rely on the existence of the sufficient statistic $\bar{x}$, define the observed point $\hat{\eta}$ within $S$, and develop a closed theory wholly contained within $S$, including the data. Of course, it is straightforward to generalize this theory to models such as multivariate regression models and time series models with correlated random variables.

It is possible, however, to generalize the asymptotic theory to the case of a general distribution family $M = \{p(x; u)\}$, which cannot be embedded in an exponential family, and, by properly defining the curvature, obtain the similar results. This is done by augmenting each point in the model $M$ with a high-dimensional exponential family which osculates $M$ in some order.

# 4. STATISTICAL INFERENCE AND DIFFERENTIAL GEOMETRY

Now fix a point $u_0$ in $M$. Around this point, $\ell(x;u) = \log p(x;u)$ may be expanded into

$$\begin{aligned}\ell(x;u) &= \ell(x;u_0) + (u^a - u_0^a)\partial_a \ell(x;u_0) \\ &\quad + \frac{1}{2}(u^a - u_0^a)(u^b - u_0^b)\partial_a\partial_b \ell(x;u_0) + \cdots.\end{aligned} \quad (4.79)$$

Exponentiating the expansion we obtain

$$p(x;u) = p(x;u_0)\exp\left\{(u^a - u_0^a)X_a + \frac{1}{2}(u^a - u_0^a)(u^b - u_0^b)X_{ab} + \cdots\right\}, \quad (4.80)$$

where

$$X_a = \partial_a \ell(x;u_0) \quad X_{ab} = \partial_a\partial_b \ell(x;u_0), \quad \cdots.$$

Now let us consider for the point $u_0$ a new exponential family, parameterized by $\theta$, defined in the following way:

$$\begin{aligned} S(u_0) &= \{q(x,\theta;u_0)\} \\ q(x,\theta;u_0) &= p(x;u_0)\exp\{\theta^a X_a + \theta^{ab} X_{ab} - \psi(\theta)\}. \end{aligned} \quad (4.81)$$

The natural parameters of this distribution family are given by the pair $\theta = (\theta^a, \theta^{ab})$. The point $u = u_0$ in the model $M$ coincides with the point $\theta = (0,0)$ in $S(u_0)$. Now construct within $S(u_0)$ the curved exponential family $\tilde{M}(u_0)$, parameterized by $u$, defined by

$$\begin{aligned} \theta = \theta(u): \quad & \theta^a = u^a - u_0^a, \quad \theta^{ab} = \frac{1}{2}(u^a - u_0^a)(u^b - u_0^b) \\ \tilde{M}(u_0) &= \{q(x,\theta(u);u_0)\}. \end{aligned} \quad (4.82)$$

This $\tilde{M}(u_0)$ is a curved exponential family which is a good approximation to $M$ around $u_0$. If we only consider the first component of the expansion, the tangent spaces of $\tilde{M}(u_0)$ and $M$ are the same at $u_0$, and if we consider up to the second component then the "higher-dimensional tangent spaces" including the directions of e-curvature (see §4.6) are the same, and $M$ and $\tilde{M}(u_0)$ osculate each other in the second-order. This may be continued to higher orders.

The augmented structure of $M$ where each point $u$ is supplemented by the exponential family $S(u)$ forms a fiber bundle with the base manifold $M$ and the fiber $S(u)$.

Given data points $x_1, \cdots, x_N$, we may define within each $S(u)$ the observed point $\bar{X}(u)$ given by

$$\bar{X}_a = \partial_a \ell(\bar{x}, u), \quad \text{and} \quad \bar{X}_{ab} = \partial_a\partial_b \ell(\bar{x}, u). \quad (4.83)$$

Performing statistical inference, say estimation, on $S(u)$ determines for each $S(u)$ an estimator $\hat{u}(u)$, and from among these we select the $u^*$ which satisfies

$$\hat{u}(u^*) = u^* \quad (4.84)$$

as our estimate. The properties of this estimator may be analyzed through the curved exponential families $\tilde{M}(u)$ within each $S(u)$. Then since the e-curvature of $M$ at point $u$ is equal to the embedding e-curvature of $\tilde{M}(u)$, the theory developed previously for curved exponential families may be immediately applied. See Amari [11] and Barndorff-Nielsen and Jupp [46] for more details.

### 4.8.2 Hilbert bundles and estimating functions

Given a family of probability distributions $M = \{p(x; u)\}$, suppose there exists a vector-valued function $\boldsymbol{f}(x; u) = [f_a(x; u)]$ where the vectors have the same dimension as $u$ and the following holds for all $u$:

$$E_u[\boldsymbol{f}(X; u)] = 0 \quad \text{and} \quad (4.85)$$
$$\det E_u[\boldsymbol{f}'(X; u)] \neq 0. \quad (4.86)$$

Here, $\det E_u[\boldsymbol{f}'(X; u)]$ is the determinant of the matrix whose components are

$$E_u[\partial_a f_b(X; u)] = \int \partial_a f_b(x; u) p(x; u) \, dx.$$

Given a large number of observations $x_1, \cdots, x_N$, let $\hat{u}$ denote the solution to the equation

$$\sum_{i=1}^{N} \boldsymbol{f}(x_i; \hat{u}) = 0. \quad (4.87)$$

We would expect that the solution $\hat{u}$ to this equation is a good estimator of the underlying distribution $u$ satisfying Equation (4.85), since the left hand side of the equation above divided by $N$ converges to $E_u[\boldsymbol{f}(X; \hat{u})]$ as $N \to \infty$. Note that the solvability of this equation is ensured by Equation (4.86) when $N$ is sufficiently large. In general, we call a function which satisfies Equations (4.85) and (4.86) an **estimating function**, and call Equation (4.87) the **estimating equation**. For instance, we call the derivative of the logarithm of the likelihood function the **score function**; the score function $\boldsymbol{s}(x; u) = [\partial_a \ell(x; u)]$ is clearly an estimating function, and the estimator thus determined is the maximum likelihood estimator.

Recently, much attention has been focused on non-parametric and semi-parametric models. These models differ from classical statistical models in which the candidate distributions are parameterized by a finite-dimensional parameter $u$. A non-parametric model is one in which the shape of the distribution function is unknown, and a semi-parametric model is one in which the distribution density of the random variable $x$ is parameterized by not only the unknown finite-dimensional parameter $u$, but also a parameter $z$ which has a function's degree of freedom (i.e., an infinite degree of freedom.) In this case, the problem is still to estimate the parameter $u$ given the observation $x_1, \cdots, x_N$, but in addition the unknown parameter $z$ acts as a **nuisance parameter**. Let us give an example. The **location-scale model** is widely used in the problem of finding the mean and variance of the underlying distribution given $N$ scalar observations $x_1, \cdots, x_N$, where the shape of the underlying distribution is unknown (we let this be the unknown parameter $z$.) Letting $u = (\mu, \sigma^2)$ be the 2-dimensional parameter representing the mean and variance, respectively, this model may be written as

$$p(x; u, z) = \frac{1}{\sigma} z\left(\frac{x-\mu}{\sigma}\right). \tag{4.88}$$

We would like to estimate $\mu$ and $\sigma$, while the function $z$ remains unknown. Here we only assume that $z(x)$ a smooth positive function on $\mathbb{R}$ which satisfies

$$\int z(x)\mathrm{d}x = 1 \tag{4.89}$$

$$\int xz(x)\mathrm{d}x = 0 \tag{4.90}$$

$$\int x^2 z(x)\mathrm{d}x = 1. \tag{4.91}$$

When we consider a statistical model with a finite-dimensional nuisance parameter $z$ and if a sufficiently large number of data are available, then we only need to estimate the joint parameter $(u, z)$ by, for instance, the maximum likelihood estimation, followed by discarding the estimate of $z$. Unfortunately, this method cannot be applied to our case where $z$ is infinite-dimensional in general. Even in this case, however, if there exists a vector-valued function $\boldsymbol{f}(x; u) = [f_a(x; u)]$ of $x$ and $u$ which satisfies for all $u$ and all $z$

$$E_{u,z}[\boldsymbol{f}(X; u)] = 0 \quad \text{and} \tag{4.92}$$
$$\det E_{u,z}[\boldsymbol{f}'(X; u)] \neq 0, \tag{4.93}$$

where $E_{u,z}$ is the expectation with respect to $p(x; u, z)$, then we may conveniently solve

$$\sum_{i=1}^{N} \boldsymbol{f}(x_i, u) = 0 \tag{4.94}$$

and obtain an estimator $\hat{u}$ without explicitly addressing the unknown $z$.

Then the questions which need to be addressed are:

(i) Does such an estimating function exist?

(ii) If there exist multiple estimating functions, which would provide the optimal estimator $\hat{u}$?

(iii) Does an estimating function which is uniformly optimal for all $z$ exist?

The differential geometric approach is particularly suited to the task of answering such fundamental questions (Amari and Kumon [29], Amari and Kawanabe [27, 28]).

Returning to the general case, let $p(x)$ be a probability distribution, and consider a curve $p(x; t)$ smoothly parameterized by a scalar parameter $t$ and

## 4.8. ESTIMATING FUNCTIONS AND FIBER BUNDLES

satisfying $p(x;0) = p(x)$ within the space $\mathcal{P}$ of all probability distributions. Let $a(x)$ be defined by

$$a(x) = \frac{\mathrm{d}}{\mathrm{d}t} \log p(x;t)|_{t=0}, \tag{4.95}$$

or equivalently

$$p(x;t) = p(x)\{1 + ta(x) + o(t)\}, \tag{4.96}$$

where $o(t)$ obeys $\lim_{t \to 0} o(t)/t = 0$. Since $p(x;t)$ is a probability distribution, $a(x)$ satisfies

$$E_p[a] = \int a(x)p(x)\mathrm{d}x = 0 \tag{4.97}$$

(cf. Equation (2.9)). Assume in addition that

$$E[a^2] < \infty. \tag{4.98}$$

Let $H_p$ denote the real (i.e., not complex) Hilbert space spanned by such functions $a(x)$ for all the curves passing through $p$, whose inner product is given for all $a(x), b(x) \in H_p$ by

$$\langle a, b \rangle = E_p[ab]. \tag{4.99}$$

Note that $H_p$ is a generalization of $T_p^{(e)}(\mathcal{P})$, the space of the e-representations of tangent vectors at $p$ (cf. Equation (2.41)), and the inner product corresponds to the Fisher metric (cf. Equation (2.42)).

Let us augment each point $(u, z)$ in the model $M = \{p(x; u, z)\}$ with the Hilbert space $H_{u,z} = H_p$, where $p(x) = p(x; u, z)$, defined by all possible directions of change $a(x)$ at $p(x; u, z)$. Such a space has the structure of a fiber. The distribution $p(x; u, z)$ changes with the value of the parameter $u$, and the direction corresponding to this change may be expressed using the score function as

$$s_a(x; u, z) = \partial_a \ell(x; u, z), \tag{4.100}$$

where $\partial_a = \frac{\partial}{\partial u^a}$ and

$$\ell(x; u, z) = \log p(x; u, z).$$

The distribution also changes with the value of $z$. Letting $z(t)$ be a curve parameterized by $t$ denoting the change in $z$, the direction of change in the corresponding probability distribution is represented by

$$r(x; u, z, \dot{z}) = \frac{\mathrm{d}}{\mathrm{d}t} \ell(x; u, z(t))|_{t=0}, \tag{4.101}$$

where $\dot{z} = \frac{\mathrm{d}z(t)}{\mathrm{d}t}$. Note that such $r$ has an infinite degree of freedom in general. The linear space $T_{u,z}$ spanned by such functions $\{s_a, r\}$ is the tangent space of $M$, and is contained in $H_{u,z}$. We call the space spanned by $\{r\}$ the **nuisance tangent space**, and denote it by $T_{u,z}^N$.

Now let us define, for each $a \in H_{u,z}$, the e-parallel translation $\Pi^{(e)}{}^{z'}_{z} a$ and the m-parallel translation $\Pi^{(m)}{}^{z'}_{z} a$ from point $(u, z)$ to $(u, z')$ by

$$\left(\Pi^{(e)}{}^{z'}_{z} a\right)(x) = a(x) - E_{u,z'}[a], \quad \text{and} \tag{4.102}$$

$$\left(\Pi^{(m)}{}^{z'}_{z} a\right)(x) = \frac{p(x; u, z)}{p(x; u, z')} a(x), \tag{4.103}$$

where $E_{u,z'}[a]$ is assumed to be finite. These clearly satisfy $E_{u,z'}\left[\Pi^{(e)}{}^{z'}_{z} a\right] = E_{u,z'}\left[\Pi^{(m)}{}^{z'}_{z} a\right] = 0$. However, they do not necessarily belong to $H_{u,z'}$ because $E_{u,z'}\left[(\Pi^{(e)}{}^{z'}_{z} a)^2\right]$ and $E_{u,z'}\left[(\Pi^{(m)}{}^{z'}_{z} a)^2\right]$ may diverge. If they always belong to $H_{u,z'}$, then the parallel translations define linear mappings from the fiber $H_{u,z}$ to $H_{u,z'}$. These may also be written in the form of covariant derivatives, and may be considered to be a generalization of the e-connection and the m-connection in the finite case (cf. Equations (2.39) and (2.43)). What is important is that the following duality (cf. Equation (3.4)) holds:

$$\langle a, b \rangle_{u,z} = \left\langle \Pi^{(e)}{}^{z'}_{z} a, \Pi^{(m)}{}^{z'}_{z} b \right\rangle_{u,z'}, \tag{4.104}$$

where $\langle \,,\, \rangle_{u,z}$ is the inner product of the fiber at point $p(x; u, z)$.

Now fix $u$, and consider the m-parallel transfer of the nuisance space $T^N_{u,z'}$ to the point $(u, z)$ for all points $z'$. We call the closed linear subspace of $H_{u,z}$ spanned by these transfers,

$$H^N_{u,z} = \text{clspan}\left\{\bigcup_{z'} \Pi^{(m)}{}^{z}_{z'} T^N_{u,z'}\right\}, \tag{4.105}$$

the **nuisance fiber**. In addition, we call the subspace spanned by the component of $s_a$ orthogonal to $H^N_{u,z}$ (i.e., the orthogonal projection of $s_a$ onto the orthocomplement of $H^N_{u,z}$) the **information fiber**, and denote it by $H^I_{u,z}$. Finally, we call the orthocomplement of $H^N_{u,z} \oplus H^I_{u,z}$ the **ancillary fiber**, and denote it by $H^A_{u,z}$. Then the fiber $H_{u,z}$ may be decomposed into a direct sum as follows:

$$H_{u,z} = H^I_{u,z} \oplus H^N_{u,z} \oplus H^A_{u,z}. \tag{4.106}$$

Having prepared this framework, let us return to the discussion of estimating functions. Let $\boldsymbol{f}(x; u) = [f_a(x; u)]$ be an estimating function. Then for any $z'$, it satisfies

$$E_{u,z'}[f_a(X; u)] = 0, \tag{4.107}$$

and hence is invariant under e-parallel translation. In addition, differentiating

$$E_{u,z(t)}[f_a(X; u)] = \int p(x; u, z(t)) f_a(x; u) \mathrm{d}x = 0$$

## 4.8. ESTIMATING FUNCTIONS AND FIBER BUNDLES

with respect to $t$ and letting $t = 0$, where $z(0) = z'$ is assumed, we obtain for any $r' \in T^N_{u,z'}$

$$\langle r', f_a \rangle_{u,z'} = 0. \tag{4.108}$$

Since $f_a$ is e-invariant, it follows from the duality (4.104) that

$$\left\langle \Pi^{(\mathrm{m})z}_{z'} r', f_a \right\rangle_{u,z} = 0. \tag{4.109}$$

From this, we see that each component $f_a$ of an estimating function belongs to $(H^N_{u,z})^\perp = H^I_{u,z} \oplus H^A_{u,z}$. Conversely, if a function $f_a$ belongs to $H^I_{u,z} \oplus H^A_{u,z}$ for some $z$, then it satisfies Equation (4.107) for all $z'$. Thus the first requirement (4.92) for an estimating function is geometrically characterized. On the other hand, differentiation of

$$E_{u,z}[f_a(x; u)] = \int p(x; u, z) f_a(x; u) \mathrm{d}x = 0$$

with respect to $u$ leads to

$$E_{u,z}[\partial_b f_a] + \langle s_b, f_a \rangle_{u,z} = 0,$$

which shows that the second requirement (4.93) is equivalent to the regularity of the matrix $[\langle s_b, f_a \rangle_{u,z}]$. Under the assumption that $f_a \in H^I_{u,z} \oplus H^A_{u,z}$, this means that the result of projecting the components $\{f_a\}$ of an estimating function onto $H^I_{u,z}$ must span $H^I_{u,z}$. These considerations lead to the following theorem.

**Theorem 4.13** (Amari and Kawanabe [27]) *A necessary and sufficient condition for $\boldsymbol{f}(x; u)$ to be an estimating function is for its components $\{f_a(x; u)\}$ to be contained in $H^I_{u,z} \oplus H^A_{u,z}$ and for their orthogonal projections onto $H^I_{u,z}$ to span $H^I_{u,z}$.*

From this we see that if $H^I_{u,z}$ is not degenerate and hence has the same dimension as $u$, then there exists an estimating function. For many problems which occur in practice, the following also holds:

$$\Pi^{(\mathrm{m})z}_{z'} T^N_{u,z'} = T^N_{u,z}. \tag{4.110}$$

In this case, we have $H^N_{u,z} = T^N_{u,z}$, and hence $H^I_{u,z}$ is spanned by the projection of the score function $s_a$ onto the orthocomplement of $T^N_{u,z}$. We call this projected function the **efficient score function**. If the dimension of the efficient score function is equal to that of $u$, then it is an estimating function, and the resulting estimator is shown to be first-order efficient.

Recently, the geometrical theory of estimating function has been applied to **independent component analysis (ICA)**. See Amari and Cardoso [24], Amari [20, 21], Kawanabe and Murata [120]. The theory will also play an important role in non-parametric, semi-parametric, and robust estimation, as well as a new statistics using generalized linear models and pseudo-likelihood functions. It is expected that research in this direction, and in the corresponding mathematical foundations, will continue to develop in the future.

# Chapter 5

# The geometry of time series and linear systems

A linear system is a system which contains a memory structure to maintain state, and linearly transforms an input time series into an output time series. If we input white noise into a stable linear system, then the output time series is stationary. Hence from this point of view, we may analyze systems and the times series that they generate together.

Typically, in both control theory and the theory of time series, a single system or a single time series is selected for analysis. Note, however, that since the set of all $n$-dimensional linear systems and the set of all $(p,q)$-ARMA time series are both finitely parameterizable, they both form manifolds of finite dimension. In order to analyze the similarity between two systems or two times series, and to consider the problems of approximation, estimation, and dimension lowering, it is not sufficient to study single systems at a time, but it is necessary rather to consider the space consisting of all such systems and analyze its geometric structure. In this chapter, we survey the differential geometry of systems and time series, and show the importance of dual connections in the analysis of their properties. The first three sections are mostly based on Amari [12], while the last section summarizes the results of Ohara et al. [172, 171, 173]. As the geometric analysis of time series is a recent development, it is a topic which can be expected to grow in the future.

## 5.1 The space of systems and time series

Consider a discrete-time linear system with one input and one output. Letting $\{\varepsilon_t\}$ be the time series of the input signal and $\{x_t\}$ be the output signal, where $t = \cdots, -2, -1, 0, 1, 2, \cdots$, the input-output relation of a stationary system is

given by
$$x_t = \sum_{i=0}^{\infty} h_i \varepsilon_{t-i}. \tag{5.1}$$

We call the coefficients $\{h_0, h_1, \cdots\}$ the **impulse response** of the system. Letting $z$ denote the operator which increments the time step so that
$$zx_t = x_{t+1} \quad \text{and} \quad z^{-1}x_t = x_{t-1}, \tag{5.2}$$
and using the **transfer function** of the system
$$H(z) = \sum_{i=0}^{\infty} h_i z^{-i}, \tag{5.3}$$
the input-output relation may be rewritten as
$$x_t = H(z)\varepsilon_t. \tag{5.4}$$

We assume below that the system in Equation (5.3) satisfies $\sum_{i=0}^{\infty} |h_i|^2 < \infty$. We call such a system a **stable** system. Then $H(z)$ may be viewed as a regular complex function for $|z| \geq 1$.

If the input $\varepsilon_t$ is white Gaussian noise independently distributed according to the standard normal distribution $N(0,1)$, then the output time series is a (colored) stationary Gaussian time series. If we take the Fourier expansion (in the wider sense) of the time series $\{x_t\}$, we obtain
$$X(\omega) = \lim_{T \to \infty} \frac{1}{\sqrt{2T}} \sum_{t=-T}^{T} x_t e^{-i\omega t}. \tag{5.5}$$

Then $X(\omega)$ is a random variable whose **power spectrum** $S(\omega) = E\bigl[|X(\omega)|^2\bigr]$ converges to
$$S(\omega) = |H(e^{i\omega})|^2. \tag{5.6}$$
In addition, we have $0 < S(\omega) < \infty$, $S(-\omega) = S(\omega)$ and
$$\int_{-\pi}^{\pi} |\log S(\omega)| \, d\omega < \infty. \tag{5.7}$$

Conversely, given a stationary Gaussian time series $\{x_t\}$, if its power spectrum satisfies Equation (5.7) then there exists a stable system $H(z) = \sum_i h_i z^{-i}$ and white noise $\{\varepsilon_t\}$ such that $\{x_t\}$ may be written in the form of Equations (5.1) and (5.4), and Equation (5.6) is satisfied. Although there are many such systems $H(z)$, the **minimal phase** system, i.e. a $H(z)$ for which $H(z) \neq 0$ in the region $|z| > 1$, is uniquely determined from $S(\omega)$.

In this way we may analyze the time series $\{x_t\}$, the power spectrum $S(\omega)$, and the minimal phase system $H(z)$ considered together. We assume below a slightly strengthened version of Equation (5.7):
$$\int_{-\pi}^{\pi} |\log S(\omega)|^2 \, d\omega < \infty,$$

## 5.1. THE SPACE OF SYSTEMS AND TIME SERIES

and call the set of all $S$ which satisfy this the **system space** or the **space of Gaussian time series** $L$.

It is possible to consider systems which are not minimal phase by letting the input signal $\varepsilon_t$ be white non-Gaussian noise, but we do not discuss this here.

Let us turn our attention to finitely parameterizable systems and time series. Let $\xi = [\xi^i]$ denote these parameters, where $i = 1, \cdots, n$. Then the power spectrum of the time series defined by this system may be written as $S(\omega; \xi)$. Under certain regularity conditions, the space formed by such time series may be viewed as an $n$-dimensional manifold for which $\xi$ is a local coordinate system. Let us give several examples.

**Example 5.1 (AR model)** *We call a system for which the output $x_t$ at time $t$ may be expressed in terms of the previous p values $x_{t-1}, \cdots, x_{t-p}$ and the current input $\varepsilon_t$ as*

$$a_0 x_t = -\sum_{i=1}^{p} a_i x_{t-i} + \varepsilon_t, \quad a_0 \neq 0 \tag{5.8}$$

*an **AR (autoregressive) model** of degree $p$. The transfer function for this model is*

$$H(z) = \frac{1}{\sum_{i=0}^{p} a_i z^{-i}}, \tag{5.9}$$

*and the power spectrum is*

$$S(\omega; \boldsymbol{a}) = \left| \sum_{t=0}^{p} a_t e^{i\omega t} \right|^{-2}, \tag{5.10}$$

*where $\boldsymbol{a} = (a_0, a_1, \cdots, a_p)$ are called the AR parameters.*

**Example 5.2 (MA model)** *We call a system for which $x_t$ may be written as a linear combination of the previous q inputs as*

$$x_t = \sum_{i=1}^{q} b_i \varepsilon_{t-i+1} \tag{5.11}$$

*an **MA (moving average) model** of degree $q$. The transfer function is given by*

$$H(z) = \sum_{i=1}^{q} b_i z^{-i+1}. \tag{5.12}$$

**Example 5.3 (ARMA model)** *We call a system for which we may write*

$$x_t = -\sum_{i=1}^{p} a_i x_{t-i} + \sum_{i=1}^{q} b_i \varepsilon_{t-i+1}, \tag{5.13}$$

118     5. THE GEOMETRY OF TIME SERIES AND LINEAR SYSTEMS

i.e., one in which, letting $a_0 = 1$, the transfer function may be written as

$$H(z) = \frac{\sum_{i=1}^{q} b_i z^{-i+1}}{\sum_{i=0}^{p} a_i z^{-i}} \tag{5.14}$$

an **ARMA model of degree $(p,q)$**.

**Example 5.4 (Bloomfield exponential model [50])** Let

$$e_0(\omega) = 1, \quad and \quad e_t(\omega) = \sqrt{2}\cos\omega t \quad for\ t = 1, 2, \cdots. \tag{5.15}$$

We call a system whose power spectrum may be written as

$$S(\omega;\xi) = \exp\left\{\sum_{t=0}^{p} \xi_t e_t(\omega)\right\} \tag{5.16}$$

a **Bloomfield exponential model of degree $p$**.

In general, a finite-dimensional discrete-time linear system may be written using the input vector time series $\boldsymbol{\varepsilon}_t$, the state vector $\boldsymbol{x}_t$, and the output vector $\boldsymbol{y}_t$ as

$$\begin{cases} \boldsymbol{x}_{t+1} &= A\boldsymbol{x}_t + B\boldsymbol{\varepsilon}_t \\ \boldsymbol{y}_{t+1} &= C\boldsymbol{x}_{t+1} \end{cases} \tag{5.17}$$

for appropriately sized constant matrices $A$, $B$, and $C$. For example, the ARMA model may be rewritten in this form. The set of all such systems may be analyzed as a system space parameterized by $\{A, B, C\}$. We discuss this possibility in the context of continuous-time systems at the end of this chapter.

## 5.2 The Fisher metric and the $\alpha$-connection on the system space

Let us define a Riemannian metric and a family of affine connections on the system space $L$. Given a time series $\{x_t\}$ whose spectrum is $S(\omega)$, it is possible to compute the distribution of $\boldsymbol{x}_T$, where

$$\boldsymbol{x}_T = (x_{-T}, x_{-T+1}, \cdots, x_0, x_1, \cdots, x_T).$$

This distribution is Gaussian, and for each $\omega$ its frequency component

$$X_T(\omega) = \frac{1}{\sqrt{2T}} \sum_{t=-T}^{T} x_t e^{-i\omega t}$$

also has a Gaussian distribution. It can be shown that, when $T$ is sufficiently large, $X_T(\omega)$ and $X_T(\omega')$ are asymptotically independent for any distinct pair $(\omega, \omega')$ and the joint distribution of the family of random variables

## 5.2. THE FISHER METRIC AND THE $\alpha$-CONNECTION

$X_T = \{X_T(\omega)\}_\omega$ is approximately (and rather symbolically) given by

$$p(X; S) \approx \exp\left\{-\frac{1}{2}\int_{-\pi}^{\pi} \frac{|X(\omega)|^2}{S(\omega)} d\omega - \psi(S)\right\}. \tag{5.18}$$

This family of distributions may be considered as an exponential family parameterized by the function $S(\omega)$, which has an infinite degree of freedom, as $T \to \infty$. It is possible even in this setting to introduce the Fisher metric and the $\alpha$-connections in a manner similar to the case of distribution families of finite dimension.

Suppose that an infinite-dimensional coordinate system $[\xi^i]$ is given on $L$. For example, by expanding

$$\log S(\omega) = \sum_{t=0}^{\infty} \xi_t e_t(\omega) \tag{5.19}$$

where $\{e_t(\omega)\}$ are the basis functions defined by Equation (5.15), we may take $\xi = [\xi^0, \xi^1, \cdots]$ as the coordinate system. The probability distribution of the truncated time series $\boldsymbol{x}_T$ of length $2T+1$ is then a function of $\xi$ and is denoted by $p^T(\boldsymbol{x}_T; \xi)$. Now the set of the distributions $\{p^T(\boldsymbol{x}_T; \xi)\}$ forms a finite-dimensional statistical model, with $\xi$ being a redundant parameter, on which the Fisher metric $g^T$ and the $\alpha$-connection $\nabla^{(\alpha)T}$ are defined. Let their components with respect to $[\xi^i]$ be denoted by

$$g_{ij}^T(\xi) = E_\xi\left[(\partial_i \ell^T)(\partial_j \ell^T)\right] \quad \text{and}$$
$$\Gamma_{ij,k}^{(\alpha)T}(\xi) = E_\xi\left[\left(\partial_i\partial_j \ell^T + \frac{1-\alpha}{2}\partial_i \ell^T \partial_j \ell^T\right)(\partial_k \ell^T)\right],$$

where $\ell^T = \log p^T(\boldsymbol{x}_T; \xi)$. Although these quantities themselves diverge as $T \to \infty$, the following limits are finite:

$$g_{ij}(\xi) \stackrel{\text{def}}{=} \lim_{T\to\infty} \frac{1}{T} g_{ij}^T(\xi) \quad \text{and}$$
$$\Gamma_{ij,k}^{(\alpha)}(\xi) \stackrel{\text{def}}{=} \lim_{T\to\infty} \frac{1}{T} \Gamma_{ij,k}^{(\alpha)T}(\xi).$$

These quantities define the Fisher metric $g$ and the $\alpha$-connection $\nabla^{(\alpha)}$ on $L$, and are shown to be represented in terms of the power spectrum $S = S(\omega; \xi)$ as

$$g_{ij}(\xi) = \frac{1}{2\pi}\int_{-\pi}^{\pi} (\partial_i \log S)(\partial_j \log S) d\omega \tag{5.20}$$

and

$$\Gamma_{ij,k}^{(\alpha)}(\xi) = \frac{1}{2\pi}\int_{-\pi}^{\pi} \left(\partial_i\partial_j \log S - \alpha \partial_i \log S \partial_j \log S\right)(\partial_k \log S) d\omega. \tag{5.21}$$

A finite-dimensional model $M$ may be construed as a submanifold of the system space $L$. Hence we may induce the Fisher metric and the $\alpha$-connections on $M$ from those on $L$. Equations (5.20) and (5.21) may also be used to define them when a coordinate system of $M$ is given.

First, let us confirm that the $\alpha$-connection and $(-\alpha)$-connection are in fact dual. This may be done by explicitly calculating

$$\partial_k g_{ij} = \Gamma^{(\alpha)}_{ki,j} + \Gamma^{(-\alpha)}_{kj,i}. \tag{5.22}$$

The following theorem states a remarkable property of the system space $L$.

**Theorem 5.1** *The system space $L$ is $\alpha$-flat for all $\alpha$.*

Intuitively, we may understand this theorem as follows. Since each system $S$ in $L$ is identified with the "product" of infinitely many independent Gaussian distributions with mean 0 as seen in Equation (5.18), the space $L$ is identified with the direct product of infinitely many copies of the 1-dimensional statistical model consisting of Gaussian distributions with mean 0 which are parameterized by the variance alone. The $\alpha$-connection on $L$ is then the direct product of the $\alpha$-connections on these 1-dimensional component models. On the other hand, every 1-dimensional manifold is flat with respect to an arbitrary affine connection as noted at the end of §1.7. Therefore $L$ is $\alpha$-flat.

**Proof of Theorem 5.1:** Define the $\alpha$-spectrum by

$$R^{(\alpha)}(\omega) = \begin{cases} -\dfrac{1}{\alpha} S(\omega)^{-\alpha} & (\alpha \neq 0) \\ \log S(\omega) & (\alpha = 0) \end{cases}. \tag{5.23}$$

Then, given a coordinate system $\xi = [\xi^i]$, by differentiating the $\alpha$-spectrum with respect to $\xi$ we obtain

$$\partial_i R^{(\alpha)}(\omega;\xi) = S^{-\alpha} \partial_i \log S, \tag{5.24}$$

and also

$$\partial_i \partial_j R^{\alpha}(\omega;\xi) = S^{-\alpha} \{\partial_i \partial_j \log S - \alpha \, \partial_i \log S \, \partial_j \log S\}. \tag{5.25}$$

Hence the coefficients of the $\alpha$-connection in Equation (5.21) may be rewritten as

$$\begin{aligned} \Gamma^{(\alpha)}_{ij,k}(\xi) &= \frac{1}{2\pi} \int_{-\pi}^{\pi} S^{2\alpha} \partial_i \partial_j R^{(\alpha)} \partial_k R^{(\alpha)} d\omega \\ &= \frac{1}{2\pi} \int_{-\pi}^{\pi} \partial_i \partial_j R^{(\alpha)} \partial_k R^{(-\alpha)} d\omega. \end{aligned} \tag{5.26}$$

Now, let $\left\{c_t^{(\alpha)}\right\}$ denote the coefficients of the Fourier expansion of the $\alpha$-spectrum:

$$R^{(\alpha)}(\omega) = \sum_{t=0}^{\infty} c_t^{(\alpha)} e_t(\omega), \tag{5.27}$$

## 5.2. THE FISHER METRIC AND THE $\alpha$-CONNECTION

or equivalently
$$c_t^{(\alpha)} = \frac{1}{2\pi} \int_{-\pi}^{\pi} R^{(\alpha)}(\omega) e_t(\omega) d\omega. \tag{5.28}$$

Then $\boldsymbol{c}^\alpha = (c_0^{(\alpha)}, c_1^{(\alpha)}, \cdots)$ forms a coordinate system for $L$. Call this the **$\alpha$-coordinate system**. If we take $\boldsymbol{c}^{(\alpha)}$ as the coordinate system $\xi$, then
$$\partial_i \partial_j R^{(\alpha)} = 0, \tag{5.29}$$
and from this we obtain
$$\Gamma_{ij,k}^{(\alpha)}(\boldsymbol{c}^{(\alpha)}) = 0. \tag{5.30}$$
Therefore $L$ is $\alpha$-flat, and $\boldsymbol{c}^{(\alpha)}$ is its $\alpha$-affine coordinate system. ∎

In particular, the 0-flatness of $L$ means that $L$ is flat with respect to the Levi-Civita connection and is, in other words, a Euclidean space. The 0-coordinate system $\boldsymbol{c}^{(0)}$ provides a Euclidean coordinate system. In general, the $\pm\alpha$-coordinate systems $\boldsymbol{c}^{(\alpha)}$ and $\boldsymbol{c}^{(-\alpha)}$ are mutually dual with respect to $g$, which may be seen from the relation

$$g_{ij}(\xi) = \frac{1}{2\pi} \int_{-\pi}^{\pi} \partial_i R^{(\alpha)} \partial_j R^{(-\alpha)} d\omega. \tag{5.31}$$

Let us abbreviate the m-affine coordinate system $\boldsymbol{c}^{(-1)}$ as $\boldsymbol{c}$, and note that $\boldsymbol{c}$ consists of the expansion coefficients of

$$S(\omega) = \sum_{t=0}^{\infty} c_t e_t(\omega). \tag{5.32}$$

Now, define the **autocorrelations** $\{r_t\}$ of the time series $\{x_t\}$ by

$$r_t = E[x_s x_{s+t}], \tag{5.33}$$

which provide the expansion

$$S(\omega) = \sum_{-\infty}^{\infty} r_t e^{-i\omega t}. \tag{5.34}$$

Then we have

$$c_0 = r_0, \quad \text{and} \quad c_t = \sqrt{2}\, r_t, \quad t = 1, 2, \cdots, \tag{5.35}$$

and $\{r_t\}$ also give an m-affine coordinate system. Since $c_t$ and $r_t$ differ only with a constant factor, we refer to $\{c_t\}$ as the autocorrelations in the sequel.

Similarly, let us abbreviate the e-affine coordinate system $\boldsymbol{c}^{(1)}$ as $\tilde{\boldsymbol{c}}$. This consists of the expansion coefficients of

$$-S^{-1}(\omega) = \sum_{t=0}^{\infty} \tilde{c}_t e_t(\omega), \tag{5.36}$$

and gives the autocorrelations of the time series $\{\tilde{x}_t\}$ generated by the inverse of $H(z)$:
$$\tilde{x}_t = H(z)^{-1}\varepsilon_t.$$

For dually flat spaces there exist corresponding potential functions as shown in §3.3. Using the **entropy** per data

$$\begin{aligned} H &= \lim_{T\to\infty} \frac{1}{2T+1} E\big[-\log p^T(\boldsymbol{x}_T)\big] \\ &= \frac{1}{4\pi}\int_{-\pi}^{\pi} \log S(\omega)\,d\omega + \frac{1}{2}\log(2\pi e), \end{aligned} \qquad (5.37)$$

the potential function and the dual potential function of the system space $L$ with respect to the dually flat structure $(g, \nabla^{(\alpha)}, \nabla^{(-\alpha)})$ is given, except for the case when $\alpha = 0$, by $\psi = \psi^{(\alpha)}$ and $\varphi = \psi^{(-\alpha)}$ where

$$\psi^{(\alpha)} = \frac{2}{\alpha}H - \frac{1}{2\alpha^2}. \qquad (5.38)$$

When $\alpha = 0$, we have a Euclidean space and the potential function is

$$\psi^{(0)} = \frac{1}{4\pi}\int_{-\pi}^{\pi}\{\log S(\omega)\}^2\,d\omega, \qquad (5.39)$$

which is the $L^2$-norm of $\log S$.

Using these potential functions, the canonical divergence on the dually flat space $(L, g, \nabla^{(\alpha)}, \nabla^{(-\alpha)})$ may now be given by

$$\begin{aligned} &D^{(\alpha)}(S_1 \| S_2) \\ &= \psi^{(\alpha)}(S_1) + \psi^{(-\alpha)}(S_2) - \sum_{t=0}^{\infty} c_t^{(\alpha)}(S_1) c_t^{(-\alpha)}(S_2) \\ &= \begin{cases} \dfrac{1}{2\pi\alpha^2}\displaystyle\int\left\{\left(\dfrac{S_2}{S_1}\right)^{\alpha} - 1 - \alpha\log\dfrac{S_2}{S_1}\right\}d\omega & (\alpha \neq 0) \\ \dfrac{1}{4\pi}\displaystyle\int (\log S_2 - \log S_1)^2\,d\omega & (\alpha = 0) \end{cases}, \end{aligned} \qquad (5.40)$$

which we call the **$\alpha$-divergence** on $L$. Let $p_i^T(\boldsymbol{x}_T)$ be the probability distribution of the truncated time series $\boldsymbol{x}_T$ generated by $S_i$ ($i = 1, 2$), and let $D^{(\alpha)}(p_1^T \| p_2^T)$ be the $\alpha$-divergence between these distributions which is defined according to Equations (3.25) and (3.26). Then we have

$$D^{(\pm 1)}(S_1 \| S_2) = \lim_{T\to\infty}\frac{1}{T}D^{(\pm 1)}(p_1^T \| p_2^T), \qquad (5.41)$$

while for $\alpha \neq \pm 1$ such a direct relation does not hold between $D^{(\alpha)}(S_1 \| S_2)$ and $D^{(\alpha)}(p_1^T \| p_2^T)$.

## 5.3 The geometry of finite-dimensional models

Let us now analyze the finite-dimensional spaces corresponding to AR, MA, and ARMA models. We chiefly focus our attention on AR models.

Let $\mathrm{AR}_p$ denote the space of AR models of degree $p$. The AR parameters $\boldsymbol{a} = (a_0, a_1, \cdots, a_p)$ form a coordinate system of $\mathrm{AR}_p$. Models of a particular degree clearly contain those of lower degree, and hence the following inclusion relation holds:

$$\mathrm{AR}_0 \subset \mathrm{AR}_1 \subset \mathrm{AR}_2 \subset \cdots.$$

Now since the inverse of the spectrum of a system in $\mathrm{AR}_p$ is

$$S^{-1}(\omega; \boldsymbol{a}) = \left| \sum_{t=0}^{p} a_t e^{i\omega t} \right|^2, \tag{5.42}$$

its Fourier expansion has only finitely many non-zero components, and hence it may be written as a finite sum:

$$S^{-1}(\omega; \boldsymbol{a}) = \sum_{t=0}^{p} \tilde{c}_t e_t(\omega). \tag{5.43}$$

Therefore $\mathrm{AR}_p$ is a $p+1$-dimensional subspace of the system space $L$ which, with respect to the coordinate system $\tilde{\boldsymbol{c}}$, satisfies the linear constraint

$$\tilde{c}_t = 0, \quad \text{for } t = p+1, p+2, \cdots. \tag{5.44}$$

This means that $\mathrm{AR}_p$ is an e-autoparallel submanifold of $L$, and hence it is e-flat. The e-affine parameters for $\mathrm{AR}_p$ are $\tilde{c}_0, \tilde{c}_1, \cdots, \tilde{c}_p$, and these may be written as functions of the AR parameters $a_0, \cdots, a_p$.

The m-affine parameters are the autocorrelations $c_0, c_1, \cdots, c_p$. The higher order autocorrelations $c_{p+1}, c_{p+2}, \cdots$ for a time series in $\mathrm{AR}_p$, which are not 0 in general, are determined as functions of $c_0, \cdots, c_p$. These are not linear functions, however, and hence although $\mathrm{AR}_p$ is itself m-flat from duality, it does not form an m-autoparallel submanifold within $L$.

As an example of a result which this geometrical theory allows us to obtain, let us consider the problem of approximating a given system $S(\omega)$ in $L$ with a system in $\mathrm{AR}_p$. In order to evaluate the quality of approximation, we use the $(-1)$-divergence (Kullback divergence)

$$D(S_1 \| S_2) = D^{(-1)}(S_1 \| S_2) = \frac{1}{2\pi} \int_{-\pi}^{\pi} \left( \frac{S_1}{S_2} - 1 - \log \frac{S_1}{S_2} \right) d\omega. \tag{5.45}$$

Then, according to Corollary 3.9, the optimal approximation $\hat{S}_p(\omega) \in \mathrm{AR}_p$ achieving

$$D\left(S \,\Big\|\, \hat{S}_p\right) = \min_{S' \in \mathrm{AR}_p} D(S \| S')$$

is the projection of $S$ onto $\mathrm{AR}_p$ along the m-geodesic which is orthogonal to $\mathrm{AR}_p$ (Figure 5.1). Since $\mathrm{AR}_p$ is e-autoparallel, this projection is uniquely determined.

# 5. THE GEOMETRY OF TIME SERIES AND LINEAR SYSTEMS

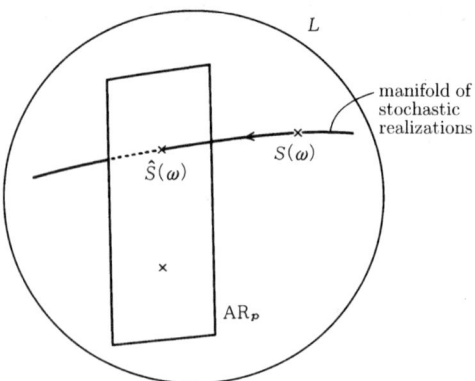

Figure 5.1: The stochastic realization of $S(\omega)$ using AR models.

Let $\boldsymbol{a} = (a_0, a_1, \cdots, a_p)$ be the AR parameters of a system $S'$ in $\mathrm{AR}_p$, and let

$$\bar{a}_i = a_i/a_0 \quad \text{and} \quad \sigma^2 = (1/a_0)^2.$$

Then we have $S'(\omega) = \sigma^2/|1 + \bar{a}_1 e^{-i\omega} + \cdots + \bar{a}_p e^{-i\omega p}|^2$, and some calculation shows that

$$\begin{aligned} D(S \parallel S') &= \frac{1}{\sigma^2} E\big[(x_t + \bar{a}_1 x_{t-1} + \cdots + \bar{a}_p x_{t-p})^2\big] + \log \sigma^2 \\ &\quad -1 - \frac{1}{2\pi} \int_{-\pi}^{\pi} \log S(\omega) \mathrm{d}\omega, \end{aligned}$$

where $E$ denotes the expectation with respect to $S$. From this expression we see that $\hat{S}_p$ is obtained by solving the least-squares problem

$$E\big[(x_t + \bar{a}_1 x_{t-1} + \cdots + \bar{a}_p x_{t-p})^2\big] \longrightarrow \min,$$

and then setting $\sigma^2$ to the resulting minimum value. This procedure is called the AR fitting of the time series $S$.

Let $\boldsymbol{c} = (c_0, c_1, \cdots)$ denote the autocorrelations of $S(\omega)$. Then $\hat{S}_p(\omega) \in \mathrm{AR}_p$ is determined by the point in $\mathrm{AR}_p$ whose m-affine coordinates are $(c_0, \cdots, c_p)$, the first $p+1$ components of $\boldsymbol{c}$. The proof is straightforward: connecting $S$ and $\hat{S}_p$ with an m-geodesic, the tangent vector at point $\hat{S}_p$ is given by

$$\boldsymbol{t} = \sum_{i=0}^{\infty} (c_i - \hat{c}_i) \left(\frac{\partial}{\partial c_i}\right)_{\hat{S}_p},$$

where $\{\hat{c}_i\}$ are the autocorrelations of $\hat{S}_p$. On the other hand, the tangent space of $\mathrm{AR}_p$ at $\hat{S}_p$ is spanned by $\left(\frac{\partial}{\partial \bar{c}_i}\right)_{\hat{S}_p}, i = 0, 1, \cdots, p$. Since $\hat{S}_p$ is the m-projection

## 5.3. THE GEOMETRY OF FINITE-DIMENSIONAL MODELS

of $S$ onto $\text{AR}_p$, and since $\boldsymbol{c} = (c_i)$ and $\tilde{\boldsymbol{c}} = (\tilde{c}_i)$ are mutually dual coordinate systems of $L$, we have

$$0 = \left\langle \boldsymbol{t}, \frac{\partial}{\partial \tilde{c}_i} \right\rangle_{\hat{S}_p} = c_i - \hat{c}_i, \qquad i = 0, 1, \cdots, p.$$

Note that the higher-order terms above $\hat{c}_{p+1}$ differ from those of $\boldsymbol{c}$, and are instead functions of $c_0, \cdots, c_p$.

Given a system $S(\omega)$, with $\boldsymbol{c} = (c_i)$ denoting its autocorrelations, we call a system, say $S'$, a **stochastic realization** of $S(\omega)$ of degree $p$, or of the sequence $c_0, c_1, \cdots, c_p$, when the first $p+1$ autocorrelations of $S'$ agree with $c_0, c_1, \cdots, c_p$. The space of all stochastic realizations forms an m-autoparallel manifold which passes through the stochastic realization $\hat{S}_p$ in $\text{AR}_p$, and is orthogonal to $\text{AR}_p$ at this point.

From Equation (5.45), the divergence between $S(\omega)$ and the white series $S_0(\omega) = 1$ $(\forall \omega)$ is written in terms of the entropy of $S$ as

$$D(S \| S_0) = -2H(S) + \log(2\pi e) + c_0 - 1. \tag{5.46}$$

On the other hand, since $S_0$ belongs to $\text{AR}_p$, we have the Pythagorean relation

$$D(S \| S_0) = D\left(S \| \hat{S}_p\right) + D\left(\hat{S}_p \| S_0\right), \tag{5.47}$$

which leads to

$$D\left(\hat{S}_p \| S_0\right) = \min_S D(S \| S_0), \tag{5.48}$$

where the minimum is taken over all the stochastic realizations of $(c_0, c_1, \cdots, c_p)$. Hence we have

$$H(\hat{S}_p) = \max_S H(S). \tag{5.49}$$

Note also that Equations (5.46) and (5.47) yield

$$D\left(S \| \hat{S}_p\right) = 2\left\{H(\hat{S}_p) - H(S)\right\}, \tag{5.50}$$

from which Equation (5.49) is straightforward. Equation (5.49) means that, of all the possible stochastic realizations of a particular sequence of autocorrelations, the realization $\hat{S}_p$ in $\text{AR}_p$ is the one which maximizes the entropy. This fact is often referred to as the **principle of maximum entropy** (Burg [62]), and now from the geometrical viewpoint we see that this is simply none other than the generalized Pythagorean theorem.

Let $\hat{S}_0, \hat{S}_1, \hat{S}_2, \cdots$ denote the stochastic realizations of $S$ within $\text{AR}_0$, $\text{AR}_1$, $\text{AR}_2, \cdots$, respectively. Then as we increase the degree $p$ of $\text{AR}_p$, the approximation error $D\left(S \| \hat{S}_p\right)$ of the realization decreases correspondingly. By the Pythagorean theorem, the error satisfies the additive relation

$$D\left(S \| \hat{S}_p\right) = D\left(S \| \hat{S}_q\right) + D\left(\hat{S}_q \| \hat{S}_p\right) \tag{5.51}$$

for $q > p$, and is decomposed into

$$D\left(S \,\big\|\, \hat{S}_p\right) = \sum_{i=p}^{\infty} D\left(\hat{S}_{i+1} \,\big\|\, \hat{S}_i\right). \tag{5.52}$$

MA models constitute an m-autoparallel subspace within the system space $L$, and a dual argument similar to that of AR models can be made. Moreover, we may extend the argument to its α-version for an arbitrary α as follows. First, a system is called an **α-model of degree $p$** when its α-coordinates $c^\alpha = (c_0^{(\alpha)}, c_1^{(\alpha)}, \cdots)$ satisfy

$$c_t^{(\alpha)} = 0 \quad \text{for} \quad \forall t \geq p+1,$$

or equivalently, its α-spectrum is represented by the finite sum

$$R^{(\alpha)}(\omega) = \sum_{t=0}^{p} c_t^{(\alpha)} e_t(\omega).$$

In particular, an AR model is a 1-model, an MA model is a $(-1)$-model, and a Bloomfield exponential model is a 0-model. The space $M_p^{(\alpha)}$ of all α-models of degree $p$ forms an α-autoparallel submanifold within $L$. Next, given a system $S$, with $c^\alpha = (c_t^{(\alpha)})$ denoting its α-coordinates, a system $S'$ is called an **α-stochastic realization** of $S$ of degree $p$, or of the sequence $(c_0^{(\alpha)}, c_1^{(\alpha)}, \cdots, c_p^{(\alpha)})$, when its first $p+1$ α-coordinates coincide with $(c_0^{(\alpha)}, c_1^{(\alpha)}, \cdots, c_p^{(\alpha)})$. A $(-1)$-stochastic realization is a stochastic realization in the original sense. Now, we have the following theorems.

**Theorem 5.2** *Given a system $S$, the α-model $\hat{S}_p$ achieving*

$$D^{(-\alpha)}\left(S \,\big\|\, \hat{S}_p\right) = \min_{S' \in M_p^{(\alpha)}} D^{(-\alpha)}\left(S \,\big\|\, S'\right) \tag{5.53}$$

*is a $(-\alpha)$-stochastic realization of $S$ of degree $p$. The approximation error satisfies*

$$D^{(-\alpha)}\left(S \,\big\|\, \hat{S}_p\right) = D^{(-\alpha)}\left(S \,\big\|\, \hat{S}_q\right) + D^{(-\alpha)}\left(\hat{S}_q \,\big\|\, \hat{S}_p\right) \tag{5.54}$$

*for $q > p$ and*

$$D^{(-\alpha)}\left(S \,\big\|\, \hat{S}_p\right) = \sum_{i=p}^{\infty} D^{(-\alpha)}\left(\hat{S}_{i+1} \,\big\|\, \hat{S}_i\right). \tag{5.55}$$

**Theorem 5.3** *Among all the $(-\alpha)$-stochastic realizations of a particular sequence $(c_0^{(-\alpha)}, c_1^{(-\alpha)}, \cdots, c_p^{(-\alpha)})$, the realization $\hat{S}_p$ in $M_p^{(\alpha)}$ has the minimum $(-\alpha)$-potential:*

$$\psi^{(-\alpha)}(\hat{S}_p) = \min_{S} \psi^{(-\alpha)}(S). \tag{5.56}$$

*In particular, when $\alpha > 0$ it has the maximum entropy, and when $\alpha < 0$ it has the minimum entropy.*

## 5.4 Stable systems and stable feedback

Now let us consider the continuous-time linear system

$$\dot{x}(t) = Ax(t) + Bu(t), \tag{5.57}$$

where $x$ is an $n$-dimensional vector representing the state of the system, $u$ is an $m$-dimensional vector representing the input, $A$ is an $n \times n$ matrix, and $B$ is an $n \times m$ matrix. We assume that the pair $(A, B)$ is controllable, and that the rank of $B$ is $m$. For the system to be stable, the real parts of the eigenvalues of $A$ must all be negative.

In this section, we shall first show that the space $\mathcal{S}$ of $n \times n$ stable matrices forms a fiber bundle and that it is possible to introduce a structure of dual connections onto this fiber bundle in a natural manner. Then, by expressing the input $u$ as a linear function of the state $x$, we consider systems which control their own state, i.e. self-controlling state feedback systems. Note that if the input $u$ is the sum of an external input $v$ and the feedback $Fx$ so that

$$u = Fx + v, \tag{5.58}$$

then the system in Equation (5.57) may be described instead as the system

$$\dot{x} = (A + BF)x + Bv. \tag{5.59}$$

We shall study the set $\mathcal{F}(A, B)$ of feedback matrices $F$ for which $A + BF$ is a stable matrix. Furthermore, we shall introduce the space $\mathcal{S}(A, B)$ of stable matrices $A + BF$ realized by feedback matrices $F$, and investigate the geometry of $\mathcal{S}(A, B)$ as a submanifold of $\mathcal{S}$.

First, let us begin by considering the structure of the manifold $\mathcal{S}$ consisting of all stable matrices. $\mathcal{S}$ is clearly a simply connected open set within the Lie group formed by the set of regular matrices $GL(n)$. Let us fix an $n \times n$ positive definite matrix $Q$. Then a stable matrix $A$ may be decomposed using a positive definite matrix $P$ and an skew-symmetric matrix $S$ into

$$A = -\frac{1}{2}QP^{-1} + SP^{-1}. \tag{5.60}$$

This decomposition is unique, and $P$ is the solution to the matrix Lyapunov equation

$$AP + PA^t + Q = 0, \tag{5.61}$$

where $A^t$ is the transposed matrix of $A$. Hence the pair of matrices $(P, S)$ form a coordinate system for $\mathcal{S}$. Since $P$ is symmetric it has $\frac{n(n+1)}{2}$ independent components, and since $S$ is skew-symmetric it has $\frac{n(n-1)}{2}$ independent components.

## 128  5. THE GEOMETRY OF TIME SERIES AND LINEAR SYSTEMS

Adding these we obtain $n^2$, which is the dimension of $\mathcal{S}$. If we decompose the stable matrix $A$ into

$$A_R = -\frac{1}{2}QP^{-1} \quad \text{and} \quad A_I = SP^{-1},$$

then the eigenvalues of $A_R$ are all negative reals, while those of $A_I$ are all purely imaginary.

Following Ohara et al. [172, 171], let us view $\mathcal{S}$ as the direct product of the manifold **PD** formed by the set of positive definite matrices $P$ and the manifold **Skew** formed by the set of skew-symmetric matrices $S$ via Equation (5.60). To investigate the geometric structure of $\mathcal{S} = \mathbf{PD} \times \mathbf{Skew}$, we rely on the following invariance. Consider transformations of bases of the space consisting of all state vectors. Letting $T$ be a matrix representing such a non-singular basis transformation, this transforms $x$ according to

$$\tilde{x} = Tx.$$

Then, in addition, $A$, $Q$, $P$ and $S$ are transformed according to

$$\tilde{A} = TAT^{-1}, \quad \tilde{Q} = TQT^t, \quad \tilde{P} = TPT^t \quad \text{and} \quad \tilde{S} = TST^t, \qquad (5.62)$$

respectively. We require that the structure of the manifold $\mathcal{S}$ remains invariant under this transformation.

Let us first investigate the geometric structure of **PD**. It should be noted that **PD** may be equivalently considered as the space formed by multivariate normal distributions with mean 0 and covariance matrices $P$ in **PD**. Letting $x$ be the random variable, the distributions may be written as

$$p(x; P) = \exp\left\{-\frac{1}{2}x^t P^{-1} x - \psi(P)\right\}. \qquad (5.63)$$

Identifying **PD** with the space of such normal distributions, we may introduce the dualistic structure $(g, \nabla, \nabla^*)$ on **PD** by the Fisher metric,[1] the e-connection and the m-connection. If we use this structure, then **PD** is a dually flat space with $-\frac{1}{2}P^{-1}$ as its $\nabla$-affine $\theta$-coordinate system and $P$ as its $\nabla^*$-affine $\eta$-coordinate system (see Example 2.6 in §2.3 and Example 3.2 in §3.5). Furthermore, the invariance of the structure $(g, \nabla, \nabla^*)$ under the transformation $\tilde{P} = TPT^t$ follows from the invariance of the Fisher metric and the e-, m-connections under the one-to-one transformation $\tilde{x} = Tx$ of the random variables.

Let us consider the space **PD** as an open set in the linear space **Sym** consisting of all $n \times n$ symmetric matrices through the $\eta$-coordinates $P$, so that the tangent space $T_P(\mathbf{PD})$ at each point $P$ is identified with **Sym**. Let $E_{p,q}$ denote the $n \times n$ symmetric matrix whose $(p,q)^{\text{th}}$ and $(q,p)^{\text{th}}$ components are 1, and is 0 elsewhere. Then the natural basis of $T_P(\mathbf{PD}) = \mathbf{Sym}$ corresponding

---
[1] Actually, we define $g$ to be twice the Fisher metric for notational simplicity in this section.

## 5.4. STABLE SYSTEMS AND STABLE FEEDBACK

to the coordinate expression $P = \sum \eta^{p,q} E_{p,q}$ is represented by the set of matrices $\{E_{p,q} \,;\, 1 \leq p \leq q \leq n\}$, whereby a tangent vector $X \in T_P(\mathbf{PD})$ may be expressed as a matrix whose components are $X_{p,q}$:

$$X = \sum_{p \leq q} X_{p,q} E_{p,q}. \tag{5.64}$$

Similarly, the natural basis $\{\tilde{E}^{p,q}\}$ for the $\theta$-coordinates $-\frac{1}{2}P^{-1} = \frac{1}{2}\sum_p \theta_{p,p} E_{p,p} + \frac{1}{4}\sum_{p<q} \theta_{p,q} E_{p,q}$ is represented by

$$\tilde{E}^{p,q} = \frac{1}{2} P E_{p,q} P \quad (p < q) \quad \text{and} \quad \tilde{E}^{p,p} = P E_{pp} P.$$

Then we obtain the following theorem.

**Theorem 5.4** ([171, 173]) **PD** *is a dually flat space with* $-\frac{1}{2}P^{-1}$ *as its $\theta$-coordinate system and $P$ its $\eta$-coordinate system. In terms of $\eta$-coordinates, the Riemannian metric is given by*

$$g_{(p,q)(p',q')} = \langle E_{p,q}, E_{p',q'} \rangle = \operatorname{tr}\left(P^{-1} E_{p,q} P^{-1} E_{p',q'}\right). \tag{5.65}$$

*In other words,*

$$\langle X, Y \rangle = \operatorname{tr}\left(P^{-1} X P^{-1} Y\right). \tag{5.66}$$

*In addition, the covariant derivatives of the pair of dual connections $\nabla$ and $\nabla^*$ are given by*

$$\nabla^*_{E_{p,q}} E_{r,s} = 0 \quad \text{and} \tag{5.67}$$

$$\nabla_{E_{p,q}} E_{r,s} = -E_{p,q} P^{-1} E_{r,s} - E_{r,s} P^{-1} E_{p,q}. \tag{5.68}$$

*We also have*

$$\left\langle \tilde{E}^{p,q}, E_{r,s} \right\rangle = \delta^p_r \delta^q_s, \tag{5.69}$$

$$\nabla_{\tilde{E}^{p,q}} \tilde{E}^{r,s} = 0 \quad \text{and} \tag{5.70}$$

$$\nabla^*_{\tilde{E}^{p,q}} \tilde{E}^{r,s} = \tilde{E}^{p,q} P^{-1} \tilde{E}^{r,s} + \tilde{E}^{r,s} P^{-1} \tilde{E}^{p,q}. \tag{5.71}$$

*For any tangent vector $X \in T_{P_0}(\mathbf{PD})$, the parallel translations from $P_0$ to $P_1$ with respect to the connections $\nabla^*$ and $\nabla$ are given by*

$$\Pi^* X = X \tag{5.72}$$

$$\Pi X = P_1 P_0^{-1} X P_0^{-1} P_1. \tag{5.73}$$

For the space **Skew** defined as the set of skew-symmetric matrices $S$, the invariance under the transformation (5.62) does not suffice to provide us with a means of obtaining a correspondence between the geometric structure of the space and that of a family of probability distributions. Ohara and Amari [171] viewed $\mathcal{S}$ as the fiber bundle (vector bundle) with the base space **PD** and the fiber **Skew**, and introduced the following inner product on the fiber over each $P \in \mathbf{PD}$:

$$\langle S_1, S_2 \rangle = -\operatorname{tr}\left(P^{-1} S_1 P^{-1} S_2\right). \tag{5.74}$$

They also defined two connections on this fiber bundle with the parallel translations

$$\Pi^* S = S \tag{5.75}$$
$$\Pi S = P_1 P_0^{-1} S P_0^{-1} P_1, \tag{5.76}$$

which turn out to be mutually dual in the sense that

$$\langle \Pi^* S_1, \Pi S_2 \rangle_{P_1} = \langle S_1, S_2 \rangle_{P_0}. \tag{5.77}$$

Note that since **Skew** is not the tangent space of **PD** this duality is an extension of the original notion introduced in §3.1.

Finally, we consider the feedback-stabilized manifold $\mathcal{S}(A, B)$. Fix an arbitrary positive definite matrix $Q$, and let $\mathbf{PD}(A, B)$ be the set of positive definite matrices $P$ satisfying

$$(I - BB^+)(AP + PA^t + Q)(I - BB^+) = 0, \tag{5.78}$$

where $B^+$ denotes the pseudo-inverse of $B$. In addition, let $\mathbf{Skew}(B)$ be the set of skew-symmetric matrices $S$ satisfying

$$BB^+ S = S. \tag{5.79}$$

**Theorem 5.5** *([172, 171])* $\mathbf{PD}(A, B)$ *forms an $\frac{m(2n-m+1)}{2}$-dimensional $\nabla^*$-autoparallel submanifold of* **PD** *and hence is dually flat, while* $\mathbf{Skew}(B)$ *forms an $\frac{m(m-1)}{2}$-dimensional linear subspace of* **Skew**. *Those $F$ for which $A + BF$ is stable may be written using $P \in \mathbf{PD}(A, B)$ and $S \in \mathbf{Skew}(B)$ as*

$$F = -B^+(AP + PA^t + Q)\left(I - \frac{1}{2}BB^+\right)P^{-1} - B^+ S P^{-1}. \tag{5.80}$$

This equation gives a diffeomorphism between the space $\mathcal{F}(A, B)$ of stabilizing feedback matrices $F$ and $\mathbf{PD}(A, B) \times \mathbf{Skew}(B)$. In addition, the elements of $\mathcal{S}(A, B)$ are given by

$$A + BF = -\frac{1}{2}QP^{-1} + (S_0(P) - S)P^{-1}, \tag{5.81}$$

where

$$S_0(P) = AP - BB^+\left(AP + PA^t + Q\right)\left(I - \frac{1}{2}BB^+\right) + \frac{1}{2}Q. \tag{5.82}$$

The matrix $S_0(P)$ turns out to be skew-symmetric, and Equation (5.81) shows how $\mathcal{S}(A, B)$ is embedded in $\mathcal{S} \cong \mathbf{PD} \times \mathbf{Skew}$ through Equation (5.60).

## 5.4. STABLE SYSTEMS AND STABLE FEEDBACK

We have not described the meaning of the geometric structure from the point of view of control theory. In this context also, however, it is again possible to introduce two potential functions and a divergence for which the Pythagorean theorem holds. This can be used in designing systems and feedback. For example, it can be helpful in designing feedback systems which best approximate the properties of some underlying system. This is a topic for future research.

# Chapter 6

# Multiterminal information theory and statistical inference

One of the principal problems motivating information theory is the faithful communication of a message given the constraints of channel capacity; the standard approach is to analyze the probabilistic structure of the message and from this construct a code. In contrast, the goal of statistics is to infer the underlying probabilistic structure generating the message. Hence although these two fields share the foundation of probabilistic structures on which they build their theories, because of the differences in their objectives, their analyses tend to follow separate theoretical paths. In multiterminal information theory, however, there appears a problem which binds these two fields together.

We call an information source which has distributed terminals and produces messages which are correlated across these terminals a **multiterminal information source**. Let us suppose that each terminal independently encodes and compresses its message, and that the task of the receiver is not to reconstruct the original message, but rather to infer the probabilistic structure of the underlying multiterminal information source. The Shannon information is relevant to the task of information transfer, while the Fisher information is relevant to the task of statistical inference. Information geometry provides a fundamental framework within which problems that involve both the Shannon and the Fisher information may be solved.

## 6.1 Statistical inference for multiterminal information

Suppose we have the two information sources $X$ and $Y$, and suppose that each source generates a message over a finite alphabet. Let $x^N = x_1 x_2 \cdots x_N$ be the

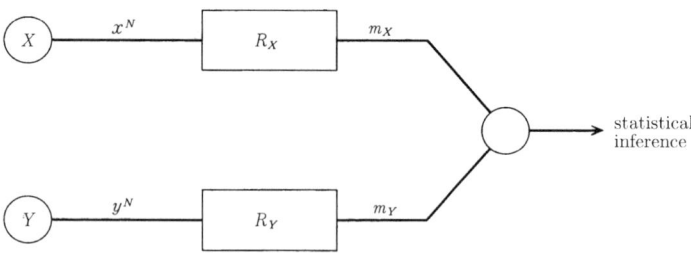

Figure 6.1: Statistical inference of a multiterminal information source.

sequence of characters of length $N$ generated by $X$, and $y^N = y_1 y_2 \cdots y_N$ be the sequence of characters of length $N$ generated by $Y$. Suppose that $x$ and $y$ are correlated, and that for each $i$, the pair $(x_i, y_i)$ is sampled independently from the joint distribution $p(x, y)$. When this distribution is parameterized over $\xi$, we write $p(x, y; \xi)$.

Each information source $X$ and $Y$ independently sends its message to a common receiver, whose task is to infer the probability distribution $p(x, y)$. If $x^N$ and $y^N$ may be directly communicated to the receiver, then the problem reduces to a standard statistical inference problem over the $N$ observations $(x^N, y^N)$. Suppose, however, that the communication paths from $X$ and $Y$ to the receiver have capacities limiting the messages to $R_X$ and $R_Y$ bits per character, respectively. Then it is not possible to send $x^N$ and $y^N$, and it is necessary to use compression. We wish to investigate, with respect to this capacity $(R_X, R_Y)$, what quality of statistical inference is possible and what the appropriate corresponding coding techniques might be (see Figure 6.1.)

Suppose, for example, that $x$ and $y$ are signals which take one of the two values 0 and 1, and that $N = 1000$, $R_X = 0.3$, and $R_Y = 0.4$. Then although $X$ and $Y$ both generates a sequence of 1000 bits, it is only possible for $X$ to send an encoded 300 bits of informations, and $Y$ is limited to 400 bits. The problem is to find the message which allows optimal inference of $p(x, y)$.

This kind of problem is peculiar to multiterminal theory. In order to clarify this, let us consider the case when there is simply a single information source $X$ which generates a sequence of characters according to the distribution $p(x; \xi)$. Then there is no need to send $x^N$ in order to estimate $\xi$. It suffices for the generator to compute the estimator $\hat{\xi} = \hat{\xi}(x_1, \cdots, x_N)$, and send the value of $\hat{\xi}$. In fact, since $\hat{\xi}$ has an estimation error of order $\frac{1}{\sqrt{N}}$, the actual message can consist of only the first $\log N$ digits of $\hat{\xi}$ represented in base 10. Then since the information sent per character

$$\frac{\log N}{N} \to 0$$

as $N$ grows large, the information needed for statistical inference may be sent asymptotically using 0 bits per character. When testing, only a single bit rep-

## 6.1. STATISTICAL INFERENCE

resenting rejection and acceptance need be sent. In the multiterminal case, however, $X$ and $Y$ do not know each other's messages. Hence it is not possible for either to independently compute a good estimator $\hat{\xi}$. In particular, it is not possible to compute information concerning the correlation between the two. Hence it is necessary for the receiver to combine the data from both terminals. The problem, therefore, given that the information sent is limited by the capacities $R_X$ and $R_Y$, is to investigate the possible quality of inference.

Let us begin by formalizing the coding problem. If we have $R_X$ bits per character, then we may send $NR_X$ bits over the entire message. Now let $m_X$ denote the message communicated over the channel, and let $M_X$ denote the set of all possible messages. Let $|M_X|$ denote the total number of such messages. An encoding is a mapping from generated character sequences $x^N$ to messages. This may be represented using a coding function:

$$\hat{m}_X : \{x^N\} \to M_X, \quad \text{where} \quad x^N \mapsto \hat{m}_X(x^N).$$

Then the information rate is

$$R_X = \lim_{N \to \infty} \frac{1}{N} \log |M_X|. \tag{6.1}$$

Restating this simply, if the capacity is $R_X$, then there are

$$|M_X| = 2^{NR_X} \tag{6.2}$$

encodings from which we can choose. If we constrain $|M_X|$ to be only polynomial in size with respect to $N$, then $R_X$ is of order $\frac{\log N}{N}$, and hence $R_X \to 0$ as $N$ grows large. In such a case we call this an asymptotically 0-rate encoding. This discussion applies equally to the information source $Y$.

The role of the receiver is to accept the messages $\hat{m}_X(x^N)$ and $\hat{m}_Y(y^N)$, and perform statistical inference. When the inference problem is estimation, an important goal is to obtain the Fisher information of the random variables $m_X = \hat{m}_X(x^N)$ and $m_Y = \hat{m}_Y(y^N)$. Let $\xi$ denote the parameter of the probability distribution. Then the joint distribution of the code is given by

$$p(m_X, m_Y; \xi) = \sum_{x^N, y^N} p(x^N, y^N; \xi), \tag{6.3}$$

where the sum is over all $x^N$ and $y^N$ which satisfy $\hat{m}_X(x^N) = m_X$ and $\hat{m}_Y(y^N) = m_Y$.

The Fisher information $g_N(\hat{m}_X, \hat{m}_Y)$ may be computed using the distribution over codes. In this case we would like to find

$$g(R_X, R_Y) = \lim_{N \to \infty} \sup \frac{1}{N} g_N(\hat{m}_X, \hat{m}_Y), \tag{6.4}$$

where the sup is taken over all encoding functions $(\hat{m}_X, \hat{m}_Y)$ under the constraint that $|M_X| < 2^{NR_X}$ and $|M_Y| < 2^{NR_Y}$. Then the estimation problem becomes that of finding the value of this $g$ and the corresponding encoding. If this is done, then the maximum likelihood estimate $\hat{\xi}$ obtained from the message would be normally distributed with covariance $\{Ng(R_X, R_Y)\}^{-1}$.

Likewise, in the case of testing, letting the hypothesis $H_0 : \xi = \xi_0$ have as its alternative hypothesis $H_1 : \xi_t = \xi_0 + \frac{t}{\sqrt{Ng}}$, the Fisher information allows us to compute its testing power. Now instead of letting the alternative hypothesis vary with $N$, let us consider the problem of testing the hypothesis $H_0 : p(x, y) = p_0(x, y)$ against $H_1 : p(x, y) = p_1(x, y)$. Then the test accuracy may be increased by correspondingly increasing $N$. Here we impose the restriction that the probability of a Type I error, i.e. the probability of rejecting $H_0$ when the underlying distribution is $p_0$, should be less than a fixed value $\alpha$. It is known that in this case the probability $P_E$ of a Type II error, which is when $H_0$ is accepted although the underlying distribution is $p_1$, decreases exponentially as $2^{-N\beta}$ with respect to $N$ (from the theory of large deviations). We call

$$\beta = \lim_{N\to\infty} \left\{ -\frac{1}{N} \log P_E \right\} \tag{6.5}$$

the **exponent of error**. It is also known by the name of Stein's lemma (see e.g. Dembo and Zeitouni [78]) that if we are sent the actual outputs $x^N$ and $y^N$, then the exponent of error of the optimal test does not depend on $\alpha$, and is given by the Kullback divergence:

$$\beta = D(p_0 \parallel p_1) = D^{(-1)}(p_0 \parallel p_1) \tag{6.6}$$

Then the problem is now to find, given the capacities $R_X$ and $R_Y$, the exponent of error

$$\beta(R_X, R_Y) = \lim_{N\to\infty} \sup \left\{ -\frac{1}{N} \log P_E \right\}, \tag{6.7}$$

where the sup is taken over all encodings consistent with the capacities $R_X$ and $R_Y$.

This problem has not yet been completely solved. Below, we examine the problem from the point of view of information geometry, first in the special case of a 0-rate testing problem, and then in the case of a 0-rate estimation problem. Lastly, we touch upon the general case.

## 6.2 0-rate testing

Let $X$ and $Y$ have as their respective alphabets the $n + 1$ characters $0, 1, \cdots, n$, and the $m + 1$ characters $0, 1, \cdots, m$, and let us write their distribution as

$$p(x, y) = \sum_{i,j} p_{ij} \delta_i(x) \delta_j(y), \tag{6.8}$$

where

$$p_{ij} = \Pr\{x = i, y = j\},$$

## 6.2. 0-RATE TESTING

and $\delta_i(x)$ is a function whose value is 1 when $x = i$ and 0 otherwise, and similarly for $\delta_j(y)$. Since there are $(n+1)(m+1)$ possible pairs $(i,j)$, this is a discrete distribution over $(n+1)(m+1)$ elements, and the set of all such distributions, say $S$, forms an exponential family of dimension $(n+1)(m+1)-1$ (see Example 2.8). By introducing the variables

$$\theta_X^i = \log \frac{p_{i0}}{p_{00}}, \qquad \theta_Y^j = \log \frac{p_{0j}}{p_{00}} \quad \text{and}$$

$$\theta_{XY}^{ij} = \log \frac{p_{ij} p_{00}}{p_{i0} p_{0j}} \quad (i = 1, \cdots, n; j = 1, \cdots, m), \tag{6.9}$$

we may write

$$\begin{aligned} \ell(x,y) &= \log p(x,y) \\ &= \theta_X^i \delta_i(x) + \theta_Y^j \delta_j(y) + \theta_{XY}^{ij} \delta_i(x) \delta_j(y) - \psi(\theta), \end{aligned} \tag{6.10}$$

where the sums for $i$ and $j$ range from 1 to $n$ and 1 to $m$, respectively, and

$$\psi(\theta) = -\log p_{00}. \tag{6.11}$$

Then $\theta = [\theta_X^i, \theta_Y^j, \theta_{XY}^{ij}]$ is an e-affine coordinate system. The corresponding $\eta$-coordinate system is given by

$$\begin{aligned} \eta_i^X &= E[\delta_i(x)] &= p_{i\cdot} &= \sum_{j=0}^m p_{ij}, \\ \eta_j^Y &= E[\delta_j(y)] &= p_{\cdot j} &= \sum_{i=0}^n p_{ij}, \quad \text{and} \\ \eta_{ij}^{XY} &= E[\delta_i(x)\delta_j(y)] &= p_{ij}. \end{aligned} \tag{6.12}$$

Here $\eta^X = [\eta_i^X]$ and $\eta^Y = [\eta_j^X]$ are probability distributions which focus attention on only $x$ and only $y$, respectively. They are called marginal distributions. On the other hand, $\theta_{XY} = [\theta_{XY}^{ij}]$ expresses the correlation between $x$ and $y$. As can be seen from Equation (6.9), if $x$ and $y$ are independent, then there is no correlation,

$$p_{ij} = p_{i\cdot} p_{\cdot j},$$

and hence $\theta_{XY} = 0$.

For simplicity we shall omit the indices $i$ and $j$ and write

$$\theta = (\theta_X, \theta_Y; \theta_{XY}) \quad \text{and} \quad \eta = (\eta^X, \eta^Y; \eta^{XY}).$$

Now if we consider combining parts of the coordinates above to form

$$\xi = (\eta^X, \eta^Y; \theta_{XY}), \tag{6.13}$$

we find that this, too, is a coordinate system for $S$. This is an example of mixed coordinate system described in §3.7.

Now fix the marginal distributions $\eta^X$ and $\eta^Y$, and let us consider the set of probability distributions whose marginals agree with these. Letting

$$M(\eta^X, \eta^Y) = \{\text{distributions whose marginals agree with } \eta^X \text{ and } \eta^Y\}$$

denote this set, $M(\eta^X, \eta^Y)$ forms an m-autoparallel subspace of $S$. Since as we vary $\eta^X$ and $\eta^Y$ this covers the entire space $S$, the set of all $M$ is a foliation of $S$. On the other hand, if we consider the set of all distributions whose values of $\theta_{XY}$ are the same,

$$E(\theta_{XY}) = \{\text{distributions whose } \theta_{XY}^{ij} \text{ coordinates agree with } \theta_{XY}\},$$

this is an e-autoparallel subspace of $S$. The set of all $E$ is also a foliation of $S$. These two foliations are mutually dual in the sense described in §3.7 (see Figure 3.3).

Now let $P_0 = p_0(x, y)$ and $P_1 = p_1(x, y)$ be two probability distributions. Let the mixed coordinate of $P_0$ and $P_1$ be

$$P_0 : (\eta_0^X, \eta_0^Y; \theta_{XY}^0) \quad \text{and} \quad P_1 : (\eta_1^X, \eta_1^Y; \theta_{XY}^1),$$

respectively. Now let us define $\bar{P}$ by the mixed coordinate

$$\bar{P} : (\eta_0^X, \eta_0^Y; \theta_{XY}^1).$$

$\bar{P}$ is a distribution whose marginals agree with those of $P_0$ and whose correlation agrees with $P_1$ (Figure 6.2). Now note that $P_0$ and $\bar{P}$ are in the same m-autoparallel space $M(\eta_0^X, \eta_0^Y)$, and that $\bar{P}$ and $P_1$ are in the same e-autoparallel space $E(\theta_{XY}^1)$, and hence since these two spaces are orthogonal, we have the following from the Pythagorean theorem for the Kullback divergence $D = D^{(-1)}$:

$$D(P_0 \| P_1) = D(P_0 \| \bar{P}) + D(\bar{P} \| P_1). \tag{6.14}$$

Returning to the original topic of discussion let us consider the problem of testing the hypothesis $H_0 : P_0$ against the fixed alternative hypothesis $H_1 : P_1$ when constrained to asymptotically 0-rate encodings. Now the generated outputs $x^N$ and $y^N$ determine an observed point

$$\hat{\eta} = (\hat{\eta}^X, \hat{\eta}^Y, \hat{\eta}^{XY})$$

in $S$. More precisely, we have

$$\hat{\eta}_i^X = \frac{1}{N} \sum_{t=1}^N \delta_i(x_t), \qquad \hat{\eta}_j^Y = \frac{1}{N} \sum_{t=1}^N \delta_j(y_t), \quad \text{and}$$
$$\hat{\eta}_{ij}^{XY} = \frac{1}{N} \sum_{t=1}^N \delta_{ij}(x_t, y_t), \tag{6.15}$$

where these are the $\eta$-coordinates of the empirical distribution based on the observed data.

## 6.2. 0-RATE TESTING

Now the information source $X$ must determine the message $m_X$ as a function $\hat{m}_X(x^N)$ of $x^N$, without knowledge of the data $y^N$. If we consider for $\hat{m}_X(x^N)$ symmetric functions of $x_1, \cdots, x_N$, then the only possibilities are $\hat{\eta}^X = [\hat{\eta}_i^X]$ or a function of $\hat{\eta}^X$. If we send $\hat{\eta}^X$ as the message $m_X$, then since $\hat{\eta}_i^X$ may take the $N+1$ values $\{0, N^{-1}, 2N^{-1}, \cdots, 1\}$, and the index $i$ runs from 1 to $n$, the number $|M_X|$ of possible messages $m_X$ is at most $(N+1)^n$. Hence the information transmission rate is

$$\frac{\log|M_X|}{N} \leq \frac{n\log(N+1)}{N} \sim 0, \tag{6.16}$$

and the encoding satisfies the asymptotically 0-rate constraint. Similarly, let us suppose that the information source $Y$ sends the encoding $m_Y = \hat{\eta}^Y$.

The only portion of the sufficient statistic $\hat{\eta} = (\hat{\eta}^X, \hat{\eta}^Y, \hat{\eta}^{XY})$ available to the receiver in a 0-rate encoding is the pair of marginal distributions $(\hat{\eta}^X, \hat{\eta}^Y)$. Since $\hat{\eta}^{XY}$ cannot be constructed without information being exchanged by $X$ and $Y$, it cannot be obtained in a 0-rate setting. The receiver uses a function $\lambda$ of $(\hat{\eta}^X, \hat{\eta}^Y)$ as its test statistic. The boundary of the acceptance region is given by

$$\lambda(\eta^X, \eta^Y) = c. \tag{6.17}$$

In other words, the boundary consists of the set of all $M(\eta^X, \eta^Y)$ where $(\eta^X, \eta^Y)$ satisfies Equation (6.17).

When $P_0$ is the underlying distribution, $(\hat{\eta}^X, \hat{\eta}^Y)$ converges in probability to the marginal distribution $(\eta_0^X, \eta_0^Y)$ of the underlying distribution as $N \to \infty$. Now consider the acceptance region as a cylinder containing $M(\eta_0^X, \eta_0^Y)$ (Figure 6.2). If the acceptance region $A$ contains an open neighborhood of $M(\eta_0^X, \eta_0^Y)$, then as $N \to \infty$, $(\hat{\eta}^X, \hat{\eta}^Y)$ is contained in $A$ with probability 1. Therefore the probability of an error of Type I converges to 0. In other words, for any $\alpha$, $N$ can be made sufficiently large so that an error of Type I occurs with probability less than $\alpha$.

Before deriving the probability of a Type II error, we give a result from the theory of large deviations. The proof of this result for the problem we are considering is straightforward using a combinatorial argument on multinomial distributions.

**Lemma 6.1 (Sanov's theorem:** *e.g.* **[78])** *Given $N$ independent samples of a random variable whose probability distribution is $P$, the probability that the empirical distribution $\hat{\eta}$ falls in a region $K$ is asymptotically*

$$\Pr\{\hat{\eta} \in K\} \simeq \exp\{-\beta N\}, \tag{6.18}$$

*or more precisely*

$$\lim_{N \to \infty} \frac{1}{N} \log \Pr\{\hat{\eta} \in K\} = -\beta, \tag{6.19}$$

*where*

$$\beta = \inf_{Q \in K} D(Q \parallel P). \tag{6.20}$$

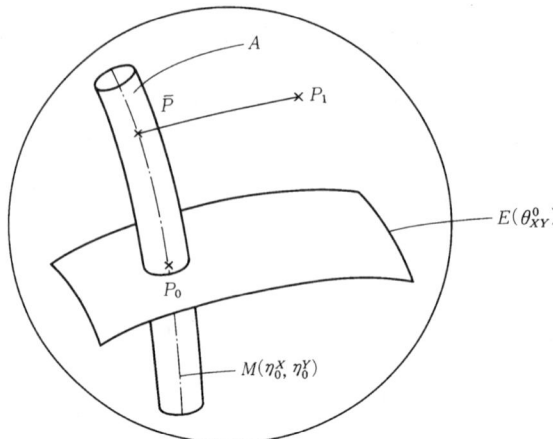

Figure 6.2: The acceptance region $A$ of a 0-rate encoding.

From this lemma we see that the exponent of error can be made arbitrarily close to the supremum of

$$\beta(A) = \inf_{Q \in A} D(P_1 \| Q) \tag{6.21}$$

with respect to $A$. Since as long as $A$ contains $M(\eta_0^X, \eta_0^Y)$ this can be made as small as desired, we have

$$\sup_A \beta(A) = \inf_{Q \in M(\eta_0^X, \eta_0^Y)} D(Q \| P_1) = D(\bar{P} \| P_1). \tag{6.22}$$

We have derived the exponent of error for the case when we send $\hat{\eta}^X$ and $\hat{\eta}^Y$ for $m_X$ and $m_Y$, respectively. The fact that this exponent is achievable was first shown by Han [98], and the present differential geometrical approach was studied by Amari and Han [26]. To show that it is not possible to do better by sending some other message of a 0-rate encoding requires techniques from information theory, which was carried out by Shalaby and Papamarcou [197], and hence we do not prove this here. See also Han and Amari [100].

**Theorem 6.1** *The optimal exponent of error when testing between the hypothesis $H_0 : P_0$ against $H_1 : P_1$ under the assumption of 0-rate encodings is $D(\bar{P} \| P_1)$.*

## 6.3  0-rate estimation

Let $M = \{p(x, y; \xi)\}$ be a statistical model parameterized by the scalar $\xi$. Then the Fisher information of a single data pair $(x, y)$ is

$$g(\xi) = E\left[\{\dot{\ell}(x, y; \xi)\}^2\right], \tag{6.23}$$

## 6.3. 0-RATE ESTIMATION

where we have let $\ell = \log p$ and $\dot{\ell} = \frac{\mathrm{d}}{\mathrm{d}\xi}\ell$. This is the square of the length of the tangent vector of the curve determined by the distribution $M$. If we have $N$ observations $x^N$ and $y^N$, then using the distribution of the sufficient statistic $(\hat{\eta}^X, \hat{\eta}^Y, \hat{\eta}^{XY})$, the Fisher information may be computed to be

$$Ng(\xi) = E_\xi \left[ \left\{ \dot{\ell}\left(\hat{\eta}^X, \hat{\eta}^Y, \hat{\eta}^{XY}; \xi\right) \right\}^2 \right], \tag{6.24}$$

which is $N$ times the information of a single observation.

Suppose, using a 0-rate encoding, that only the information concerning the marginal distributions $m_X = \hat{\eta}^X$ and $m_Y = \hat{\eta}^Y$ is sent, and that the information in $\hat{\eta}^{XY}$ is lost, as in the previous section. Then $\frac{\mathrm{d}}{\mathrm{d}\xi}\log p(m_X, m_Y; \xi)$ (the score function), based on the distributions of $m_X$ and $m_Y$, may be written as a conditional expectation (see Equation (2.19)):

$$\dot{\ell}(m_X, m_Y; \xi) = E_\xi\left[ \dot{\ell}\left(\hat{\eta}^X, \hat{\eta}^Y, \hat{\eta}^{XY}; \xi\right) \,\bigg|\, m_X, m_Y \right]. \tag{6.25}$$

The Fisher information of the statistic $(m_X, m_Y) = (\hat{\eta}^X, \hat{\eta}^Y)$ is given by

$$Ng_M(\xi) = E_\xi\left[ \left\{ \dot{\ell}\left(\hat{\eta}^X, \hat{\eta}^Y; \xi\right) \right\}^2 \right]. \tag{6.26}$$

Let

$$\tilde{x}_i = \sqrt{N}\left(\hat{\eta}^X_i - p_{i\cdot}\right), \qquad \tilde{y}_j = \sqrt{N}\left(\hat{\eta}^Y_j - p_{\cdot j}\right), \quad \text{and}$$
$$\tilde{w}_{ij} = \sqrt{N}\left(\hat{\eta}^{XY}_{ij} - p_{ij}\right). \tag{6.27}$$

By the central limit theorem, these jointly distribute themselves according to a multi-dimensional normal distribution when $N \to \infty$. The score function based on the whole data may be written as

$$\dot{\ell}(\hat{\eta}^X, \hat{\eta}^Y, \hat{\eta}^{XY}; \xi) = \sqrt{N}\left( \dot{\theta}^i_X \tilde{x}_i + \dot{\theta}^j_Y \tilde{y}_j + \dot{\theta}^{ij}_{XY} \tilde{w}_{ij} \right). \tag{6.28}$$

On the other hand, since $\hat{\eta}^X$ and $\hat{\eta}^Y$ belong in the space of random variables spanned by $(\tilde{x}_i, \tilde{y}_j)$, the conditional expectation $\dot{\ell}\left(\hat{\eta}^X, \hat{\eta}^Y; \xi\right)$ is the projection of $\dot{\ell}\left(\hat{\eta}^X, \hat{\eta}^Y, \hat{\eta}^{XY}; \xi\right)$ onto the linear space spanned by $(\tilde{x}_i, \tilde{y}_j)$. Since $\tilde{x}_i$, $\tilde{y}_j$ and $\tilde{w}_{ij}$ are not orthogonal, we may not obtain this projection by simply taking the terms in $\dot{\ell}$ which are linear combinations of $\tilde{x}_i$ and $\tilde{y}_j$.

We compute the projection using a technique from linear algebra. Let us re-index as $\eta_a$ the portion $[\eta^X_i, \eta^Y_j]$ of the $\eta$-coordinates of $S$, so that $a = (i, j)$ is our new index. In addition, let $[g_{ab}]$ denote the corresponding portion of the Fisher information matrix under the $\theta$-coordinate system, and let $[\bar{g}^{ab}]$ denote its inverse.

**Theorem 6.2** *The Fisher information of an asymptotically 0-rate encoding is given by*

$$g_M(\xi) = \dot{\eta}_a \dot{\eta}_b \bar{g}^{ab}. \tag{6.29}$$

Here, using the projected score function, the optimal estimator $\hat{\xi}$ based on the message $(m_X, m_Y)$ is obtained as the solution of

$$E_{\hat{\xi}}\left[\dot{\ell}\left(\hat{\eta}^X, \hat{\eta}^Y, \hat{\eta}^{XY}; \hat{\xi}\right) \mid m_X, m_Y\right] = 0, \quad (6.30)$$

and in addition, $\hat{\xi}$ is asymptotically distributed according to a normal distribution with mean $\xi$ and covariance $g_M(\xi)^{-1} N^{-1}$.

The above result for the achievability of the Fisher information $g_M(\xi)$ is due to Amari [13], while its optimality was recently proven by Han and Amari [100] based on a large deviation approach combined with the argument of [197].

## 6.4 Inference for general multiterminal information

We have until now been considering testing and estimation within the context of an extremely special case, i.e. that of 0-rate encodings. And although we have focused here on the relevant information geometric structures, both the geometric structure and the encoding process are important components of this problem. The problem in the general case when the coding rates $R_X$ and $R_Y$ are unconstrained is still open. It is plausible, however, that information geometry provides a means for solving this problem together with coding techniques from information theory. The following outlines one idea. See Han and Amari [99, 100].

We first recall that the data $(x^N, y^N)$ is a sequence of $N$ independent samples drawn from the joint probability distribution $p(x, y)$ of correlated random variables $X$ and $Y$. Now let $U$ be a random variable whose domain is finite, and assume that the three random variables $U$, $X$, and $Y$ satisfy the Markov relation

$$U - X - Y. \quad (6.31)$$

This is a notation common in information theory which means that $U$ and $Y$ are independent given $X$. From the point of view of information geometry, this means that if we consider the exponential family

$$\begin{aligned}
p(u, x, y) &= \exp\{\theta_U^i \delta_i(u) + \theta_X^i \delta_j(x) + \theta_Y^k \delta_k(y) + \\
&\quad \theta_{UX}^{ij} \delta_{ij}(u, x) + \theta_{UY}^{ik} \delta_{ik}(u, y) + \theta_{XY}^{jk} \delta_{jk}(x, y) + \\
&\quad \theta_{UXY}^{ijk} \delta_{ijk}(u, x, y)\},
\end{aligned} \quad (6.32)$$

then the following hold:

$$\theta_{UY}^{ik} = 0 \quad \text{and} \quad \theta_{UXY}^{ijk} = 0, \quad (6.33)$$

where we have used variables such as

$$\theta_{UXY}^{ijk} = \log \frac{p_{ijk} p_{i00} p_{0j0} p_{00k}}{p_{ij0} p_{0jk} p_{i0k} p_{000}}, \quad (6.34)$$

## 6.4. GENERAL MULTITERMINAL INFORMATION

which are generalizations of those in Equation (6.9). Next, let $V$ be a random variable similarly satisfying the Markov relation

$$V - Y - X. \qquad (6.35)$$

Then, corresponding to the rate constraint specified by $R_X$ and $R_Y$, we may consider a constraint imposed on $U$ and $V$ in terms of some information quantities like entropy or mutual information. Furthermore, this setting has a natural $N$th extension in which we consider random variables $U_N$ and $V_N$ satisfying the Markov relations

$$U_N - X^N - Y^N \quad \text{and} \quad V_N - Y^N - X^N, \qquad (6.36)$$

where $X^N$ and $Y^N$ are the $N$th i.i.d. (independent and identically distributed) extensions of $X$ and $Y$, respectively.

Now, suppose instead of the codes $\hat{m}_X(x^N)$ and $\hat{m}_Y(y^N)$, we send $u_N$ and $v_N$ drawn from the sources $U_N$ and $V_N$ which are correlated with $x^N$ and $y^N$ by the Markov relations above. We may view $u_N$ and $v_N$ as noisy versions of $\hat{m}_X(x^N)$ and $\hat{m}_Y(y^N)$, and hence this is a stochastic encoding. Then, corresponding to the original problem of finding the optimal encoding functions $\hat{m}_X : \{x^N\} \to M_X$ and $\hat{m}_Y : \{y^N\} \to M_Y$ under the rate constraint, an optimization problem for $U_N$ and $V_N$ is naturally formulated, which now we are to pursue.

This is a typical strategy in information theory, where for many classes of problems it is proved that there in fact exists a deterministic encoding system which has asymptotically equivalent characteristics to a stochastic one. The advantage of considering stochastic encodings is that, unlike the original problem for deterministic encodings, the corresponding problem is often reduced to the case when $N = 1$. This is called the **single-letterization** of the problem.

The questions we must address, then, may be decomposed into two parts:

(i) Is the multiterminal statistical inference problem single-letterizable?

If this is the case, then

(ii) How do we solve the problem after single-letterizing?

A geometric consideration of the distribution family $\{p(u,x,y,v)\}$ answers question (ii). In other words, the solution can be obtained by decomposing the fourfold empirical distribution into a communicable portion consisting of marginalized distributions, and an uncommunicable portion consisting of distributions such as $\hat{\eta}^{XY}$.

On the other hand, Ahlswede and Brunashev [3] have made a pessimistic observation on the general single-letterizability of problems of multiterminal statistical inference. However, it might be possible to apply single-letterization to the conditional inference problem where we send $(\hat{\eta}^X, \hat{\eta}^Y)$ at 0-rate as a supplemental message. It is expected that single-letterization or conditional single-letterization provides a key towards new developments.

# Chapter 7

# Information geometry for quantum systems

From a mathematical point of view, quantum mechanics may be construed as an extension of probability theory, and it is possible to generalize many concepts in probability theory to their quantum equivalents. The framework of information geometry for statistical models may also be extended to the quantum mechanical setting. Although a variety of important works related to differential geometrical aspects of quantum mechanics have so far been made by many researchers, we restrict ourselves here to investigating the problem of how the dualistic structure of the Fisher metric and the $\alpha$-connections on statistical models is extended to manifolds of quantum states. We should recognize, however, that study of quantum information geometry has just started, and that we are far from getting its whole perspective at present.

## 7.1 The quantum state space

Let us begin with a very brief introduction of quantum mechanics. (We do not touch upon the quantum time evolution (dynamics) governed by the Schrödinger equation or the state reduction caused by a measurement.) Quantum mechanics describes the physics of a quantum system in terms of linear operators on a complex Hilbert space $\mathcal{H}$. Here we assume that $\mathcal{H}$ is finite-dimensional for mathematical simplicity, and hence we may consider $\mathcal{H}$ as $\mathbb{C}^k$, where $k = \dim \mathcal{H}$, and operators as $k \times k$ matrices with no loss of generality. We denote the inner product of vectors $x$ and $y$ in $\mathcal{H}$ by $\langle x|y \rangle$, and use Dirac's notation in which $|x\rangle\langle x|$ means the operator which maps each element $y$ of $\mathcal{H}$ to $\langle x|y \rangle x$. The Hermitian conjugate (or adjoint) $A^*$ of an operator $A$ is defined by the property $\langle x|Ay \rangle = \langle A^*x|y \rangle$ for all $x$ and $y$. A **quantum state** of the system is then represented by a **density operator** $\rho$ on $\mathcal{H}$, which is a positive semidefinite

Hermitian operator of trace one:

$$\rho = \rho^* \geq 0, \quad \text{and} \quad \text{tr}\,\rho = 1. \tag{7.1}$$

We call $\rho$ a **pure state** when $\rho$ is of rank one and hence is representable as $\rho = |x\rangle\langle x|$ for some unit vector $x$, and a **mixed state** otherwise.

Now suppose that we perform a **measurement**, say $\Pi$, to the quantum system and that the possible outcomes of $\Pi$ form a finite set of events $\{e_1, e_2, \ldots, e_l\}$. When the state of the system is represented by a density operator $\rho$, the probability that a particular event $e_i$ is observed is generally expressed in the form

$$P\{e_i\} = \text{tr}(\rho \pi_i), \tag{7.2}$$

where $\{\pi_i\} = \{\pi_1, \pi_2, \ldots, \pi_l\}$ is a set of operators on $\mathcal{H}$ satisfying

$$\pi_i = \pi_i^* \geq 0, \quad \text{and} \quad \sum_i \pi_i = I, \tag{7.3}$$

with $I$ denoting the identity operator (unit matrix). The set $\{\pi_i\}$ is determined by the measurement $\Pi$ and does not depend on the state $\rho$. Hence we may consider $\{\pi_i\}$ as a mathematical expression of the measurement $\Pi$ and write as $\Pi = \{\pi_i\}$. Note that Equations (7.1), (7.2) and (7.3) qualifies $P$ to be a probability distribution:

$$P\{e_i\} \geq 0 \quad \text{and} \quad \sum_i P\{e_i\} = 1. \tag{7.4}$$

**Note:** More generally, a measurement taking its outcome in a measurable space $(\mathcal{X}, \mathcal{B})$, where $\mathcal{B}$ is a completely additive class ($\sigma$-algebra) consisting of subsets of $\mathcal{X}$, is represented by a mapping $\pi$ which maps each measurable set $B \in \mathcal{B}$ to a nonnegative Hermitian operator $\pi(B) = \pi(B)^* \geq 0$ on $\mathcal{H}$ satisfying $\pi(\mathcal{X}) = I$ and the complete additivity: for any countable family of mutually disjoint measurable sets $\{B_1, B_2, \ldots\} \subset \mathcal{B}$ we have

$$\pi\left(\bigcup_i B_i\right) = \sum_i \pi(B_i).$$

Such a $\pi$ is called a **positive-operator-valued measure, probability operator-valued measure** or **POM** in short. The probability that the measurement outcome lies in a set $B$ is then given by $P(B) = \text{tr}(\rho\,\pi(B))$. See [107] [110] for the detail. When $\mathcal{X} = \{1, 2, \ldots, l\}$ and $\mathcal{B}$ is its power set, Equation (7.3) follows by setting $\pi_i = \pi(\{i\})$.

When $\Pi = \{\pi_i\}$ satisfies the additional condition

$$\pi_i^2 = \pi_i, \quad \text{and} \quad \pi_i \pi_j = 0 \quad \text{if} \quad i \neq j, \tag{7.5}$$

## 7.1. THE QUANTUM STATE SPACE

or in other words, when $\{\pi_i\}$ consists of orthogonal projections onto mutually orthogonal subspaces, it is called a **simple measurement**. This condition may be considered to represent a kind of purity of the measurement.

When the possible outcomes $e_1, e_2, \ldots$ of a simple measurement $\{\pi_i\}$ are labeled with distinct real numbers $a_1, a_2, \ldots$, we call the pair $(\{\pi_i\}, \{a_i\})$ an **observable**. We may represent an observable $(\{\pi_i\}, \{a_i\})$ by the Hermitian operator

$$A = \sum_i a_i \pi_i. \tag{7.6}$$

Note that $\{a_i\}$ are the eigenvalues of $A$ and $\{\pi_i\}$ are the orthogonal projections onto the eigenspaces, and that the correspondence between $A$ and $(\{\pi_i\}, \{a_i\})$ is one-to-one. Then we see that the expectation and the variance of the observable $A$ under the state $\rho$ are given by

$$E_\rho[A] = \sum_i \operatorname{tr}(\rho \pi_i)\, a_i = \operatorname{tr}(\rho A) \quad \text{and} \tag{7.7}$$

$$V_\rho[A] = \sum_i \operatorname{tr}(\rho \pi_i)\, (a_i - E_\rho[A])^2 = \operatorname{tr}\!\left[\rho\, (A - E_\rho[A])^2\right], \tag{7.8}$$

where (and hereafter) a constant multiple $cI$ of the identity $I$ is simply denoted by $c$.

Given a finite-dimensional Hilbert space $\mathcal{H}$, let us denote the set of all Hermitian operators on $\mathcal{H}$ by

$$\mathcal{A} = \{A \mid A = A^*\}, \tag{7.9}$$

and let

$$\mathcal{A}_1 \stackrel{\text{def}}{=} \{A \mid A \in \mathcal{A} \text{ and } \operatorname{tr} A = 1\}. \tag{7.10}$$

Then the totality of density operators

$$\bar{\mathcal{S}} \stackrel{\text{def}}{=} \{\rho \mid \rho = \rho^* \geq 0 \text{ and } \operatorname{tr} \rho = 1\} \tag{7.11}$$

forms a convex subset of $\mathcal{A}_1$. This set is partitioned into $\bar{\mathcal{S}} = \bigcup_{r=1}^{k} \mathcal{S}_r$, where $\mathcal{S}_r \stackrel{\text{def}}{=} \{\rho \in \bar{\mathcal{S}} \mid \operatorname{rank} \rho = r\}$ and $k \stackrel{\text{def}}{=} \dim \mathcal{H}$. In particular, the elements of $\mathcal{S}_1$ are the pure states and are the extreme points (i.e. points which may not be represented as convex combinations of other points) of the convex set $\bar{\mathcal{S}}$, while those of $\mathcal{S}_k$ are strictly positive density operators and constitute the interior of $\bar{\mathcal{S}}$.

Let $\mathcal{U}$ be the set of all unitary operators on $\mathcal{H}$:

$$\mathcal{U} = \{U \mid U^{-1} = U^*\}. \tag{7.12}$$

This forms a Lie group and acts on $\bar{\mathcal{S}}$ by

$$\begin{aligned} \mathcal{U} \times \bar{\mathcal{S}} &\to \bar{\mathcal{S}} \\ (U, \rho) &\mapsto U\rho U^*. \end{aligned} \tag{7.13}$$

Obviously each $\mathcal{S}_r$ is closed under this action. When a state $\rho$, a measurement $\{\pi_i\}$ and an observable $A$ are transformed into

$$\tilde{\rho} = U\rho U^*, \quad \tilde{\pi}_i = U\pi_i U^* \quad \text{and} \quad \tilde{A} = UAU^*$$

by a common unitary operator $U$, the probability distributions and the expectations are kept invariant:

$$\operatorname{tr}(\tilde{\rho}\tilde{\pi}_i) = \operatorname{tr}(\rho\pi_i) \quad \text{and} \quad \operatorname{tr}(\tilde{\rho}\tilde{A}) = \operatorname{tr}(\rho A).$$

The pure state space $\mathcal{S}_1$ may naturally be identified with the complex projective space $\mathbb{C}P^{k-1}$ of complex dimension $k-1$. It is well known that a Riemannian metric on this space which is invariant under the action of the unitary group $\mathcal{U}$ is unique up to a constant. This unique metric is called the Fubini-Study metric and is known to be a Kählerian metric. From an information geometrical point of view, it is important to recognize that the Fubini-Study metric is a quantum version of the Fisher metric and plays an essential role in the statistical estimation theory for pure state models, although we do not further discuss this topic here; see Fujiwara and Nagaoka [92, 94] and Matsumoto [145, 146].

In the rest of the present chapter we focus on the geometry of $\mathcal{S}_k$ which is now denoted by

$$\mathcal{S} \stackrel{\text{def}}{=} \{\rho \mid \rho = \rho^* > 0 \text{ and } \operatorname{tr}\rho = 1\}. \tag{7.14}$$

This space is an open subset of $\mathcal{A}_1$ and hence is naturally regarded as a real (not complex) manifold of dimension $n \stackrel{\text{def}}{=} \dim \mathcal{A}_1 = k^2 - 1$. The tangent space $T_\rho(\mathcal{S})$ of each point $\rho$ may then be identified with

$$\mathcal{A}_0 \stackrel{\text{def}}{=} \{A \mid A \in \mathcal{A} \text{ and } \operatorname{tr} A = 0\}. \tag{7.15}$$

When a tangent vector $X \in T_\rho(\mathcal{S})$ is considered as an element of $\mathcal{A}_0$ by this identification, we denote it by $X^{(\mathrm{m})}$ and call it the **m-representation** of $X$, reflecting that it corresponds to the m-representation in the classical case introduced in §2.5.

For a simple measurement $\Pi = \{\pi_1, \ldots, \pi_l\}$, let

$$\begin{aligned}\mathcal{S}_\Pi &= \mathcal{S} \cap \left\{\sum_{i=1}^l a_i \pi_i \,\bigg|\, (a_1, \ldots, a_l) \in \mathbb{R}^l\right\} \\ &= \left\{\sum_{i=1}^l \frac{p_i}{\operatorname{tr}\pi_i} \pi_i \,\bigg|\, p = (p_1, \ldots, p_l) \in \mathcal{P}_l\right\},\end{aligned} \tag{7.16}$$

where $\mathcal{P}_l$ denotes $\mathcal{P}(\{1, \ldots, l\})$: the totality of positive probability distributions on $\{1, \ldots, l\}$. This forms a $(l-1)$-dimensional submanifold of $\mathcal{S}$. Note that $\mathcal{S}_\Pi$ is commutative in the sense that arbitrary two elements $\rho$ and $\sigma$ of $\mathcal{S}_\Pi$ satisfy $\rho\sigma = \sigma\rho$, and that any commutative model is a submanifold of $\mathcal{S}_\Pi$ for some $\Pi$. When all the projections $\pi_i$ in $\Pi$ are of rank one, we have

$$\mathcal{S} = \{U\rho U^* \mid \rho \in \mathcal{S}_\Pi \text{ and } U \in \mathcal{U}\}. \tag{7.17}$$

## 7.1. THE QUANTUM STATE SPACE

Within the tangent space $T_\rho = T_\rho(\mathcal{S}) \cong \mathcal{A}_0$ at each point $\rho$, let us consider the subspace $T_\rho^u$ consisting of all the tangent vectors corresponding to the action of $\mathcal{U}$. Each element of $T_\rho^u$ is represented as

$$\frac{d}{dt} U_t \rho U_t^* \bigg|_{t=0} = [\Omega, \rho],$$

where $U_t$ is a curve in $\mathcal{U}$ satisfying $U_0 = I$, $\Omega$ is its derivative at $t = 0$ and $[\,,\,]$ denotes the commutator:

$$[A, B] = AB - BA.$$

Hence we have

$$T_\rho^u = \left\{ X \in T_\rho \,\bigg|\, \exists \Omega = -\Omega^* \text{ such that } X^{(m)} = [\Omega, \rho] \right\}. \qquad (7.18)$$

Next, let $T_\rho^c$ denote the subspace of $T_\rho$ corresponding to the change of $\rho$ in the commutative direction. In other words, $T_\rho^c$ is the totality of the derivatives $\frac{d}{dt}\rho_t\big|_{t=0}$ of curves $\rho_t$ satisfying $\rho_0 = \rho$ and $[\rho, \rho_t] = 0$, and is represented as

$$T_\rho^c = \left\{ X \in T_\rho \,\bigg|\, [\rho, X^{(m)}] = 0 \right\}. \qquad (7.19)$$

For a simple measurement $\Pi$ which diagonalizes $\rho$ (i.e. $\rho \in \mathcal{S}_\Pi$), it is clear that $T_\rho(\mathcal{S}_\Pi)$ is a subspace of $T_\rho^c$ and that if $\rho$ has $k$ distinct eigenvalues then $T_\rho^c = T_\rho(\mathcal{S}_\Pi)$. Now the tangent space $T_\rho$ is always decomposed into the direct sum of these two subspaces:

$$T_\rho = T_\rho^c \oplus T_\rho^u. \qquad (7.20)$$

To see this, define the linear mapping $\Lambda : \mathcal{A}_0 \to \mathcal{A}_0$ by $\Lambda(A) = i[\rho, A]$. It then turns out that $\Lambda$ is skew-symmetric with respect to the Hilbert-Schmidt inner product[1] $\langle\!\langle A, B \rangle\!\rangle_{\text{HS}} \stackrel{\text{def}}{=} \text{tr}(AB)$, and that $T_\rho^c$ and $T_\rho^u$ are respectively the kernel and the image of $\Lambda$. Therefore these subspaces are the orthogonal complements of each other, which implies Equation (7.20).

In the subsequent sections we shall introduce some dualistic structures $(g, \nabla, \nabla^*)$ on $\mathcal{S}$ which may be considered as quantum analogues of the Fisher metric and the $\pm \alpha$-connections. We shall see that these dualistic structures satisfy the following common properties.

(i) For an arbitrary simple measurement $\Pi = \{\pi_1, \ldots, \pi_l\}$, the restriction of $(g, \nabla, \nabla^*)$ onto $\mathcal{S}_\Pi$ coincides with the triple of the Fisher metric and the $\pm\alpha$-connections on $\mathcal{P}_l$ for some $\alpha$ under the natural identification $\mathcal{S}_\Pi \cong \mathcal{P}_l$.

(ii) The space $\mathcal{S}_\Pi$ is autoparallel in $\mathcal{S}$ with respect to both $\nabla$ and $\nabla^*$.

---

[1] In the sequel, several kinds of inner products for operators will be introduced, for which we generally use the double bracket $\langle\!\langle\,,\,\rangle\!\rangle$, while the single bracket $\langle\,,\,\rangle$ is used for Riemannian metrics on $\mathcal{S}$ and the inner product on $\mathcal{H}$.

(iii) $g = \langle\,,\,\rangle$, $\nabla$ and $\nabla^*$ are invariant under the action of $\mathcal{U}$. In other words, for an arbitrary unitary operator $U$ and arbitrary vector fields $X$ and $Y$, we have

$$\langle UXU^*, UYU^*\rangle = \langle X, Y\rangle, \tag{7.21}$$
$$\nabla_{UXU^*} UYU^* = U(\nabla_X Y)U^* \quad \text{and} \tag{7.22}$$
$$\nabla^*_{UXU^*} UYU^* = U(\nabla^*_X Y)U^*, \tag{7.23}$$

where $UXU^*$ denotes the vector field[2] such that $(UXU^*)^{(\mathrm{m})}_{U\rho U^*} = UX^{(\mathrm{m})}_\rho U^*$.

(iv) $T^c_\rho$ and $T^u_\rho$ are orthogonal with respect to $g$ at every point $\rho \in \mathcal{S}$.

## 7.2 The geometric structure induced from a quantum divergence

According to the discussion in § 3.2, every torsion-free dualistic structure is induced from some divergence by Equations (3.9)–(3.13). In particular, the dualistic structure consisting of the Fisher metric and the $\pm\alpha$-connections is induced from the $f$-divergence $D_f$ for a smooth convex function $f : \mathbb{R}^+ \to \mathbb{R}$, where $\mathbb{R}^+ \stackrel{\text{def}}{=} \{x \in \mathbb{R} \mid x > 0\}$, satisfying

$$f(1) = 0, \quad f''(1) = 1 \quad \text{and} \quad \alpha = 3 + 2f'''(1), \tag{7.24}$$

which includes the $\alpha$-divergence $D^{(\alpha)}$ as a representative example. Let us introduce a quantum version of the $f$-divergence following Petz [182] and investigate the induced geometry.

Let $\mathcal{B}$ be the totality of (not necessarily Hermitian) operators on $\mathcal{H}$. Given arbitrary strictly positive density operators $\rho, \sigma \in \mathcal{S}$, the **relative modular operator** $\Delta = \Delta_{\sigma,\rho} : \mathcal{B} \to \mathcal{B}$ is defined by

$$\Delta(A) = \sigma A \rho^{-1}, \quad \forall A \in \mathcal{B}, \tag{7.25}$$

or equivalently by

$$\langle\!\langle \Delta(A), B \rangle\!\rangle^-_\rho = \langle\!\langle A, B \rangle\!\rangle^+_\sigma, \quad \forall A, B \in \mathcal{B}, \tag{7.26}$$

where $\langle\!\langle,\rangle\!\rangle^-_\rho$ and $\langle\!\langle,\rangle\!\rangle^+_\sigma$ denote the inner products on $\mathcal{B}$ such that

$$\langle\!\langle A, B \rangle\!\rangle^-_\rho = \mathrm{tr}(\rho A^* B) \quad \text{and} \quad \langle\!\langle A, B \rangle\!\rangle^+_\sigma = \mathrm{tr}(\sigma B A^*). \tag{7.27}$$

Then, since $\Delta$ is Hermitian and positive definite as an operator on the Hilbert space $(\mathcal{B}, \langle\!\langle,\rangle\!\rangle^-_\rho)$, the ordinary operator calculus enables us to define the Hermitian operator $f(\Delta)$ on the same Hilbert space for an arbitrary function $f : \mathbb{R}^+ \to \mathbb{R}$, so that we have the implication:

$$\Delta(A) = \lambda A \ \Rightarrow\ f(\Delta)(A) = f(\lambda)A, \quad \forall \lambda \in \mathbb{R}^+, \forall A \in \mathcal{B}. \tag{7.28}$$

---

[2] Letting $\Phi_U : \rho \mapsto U\rho U^*$, we have $(\mathrm{d}\Phi_U)_\rho(X_\rho) = (UXU^*)_{\Phi_U(\rho)}$.

## 7.2. QUANTUM DIVERGENCE

Now the **quantum $f$-divergence** $D_f$ is defined by

$$D_f(\rho \,\|\, \sigma) = \langle\!\langle I, f(\Delta)(I)\rangle\!\rangle_\rho^- = \text{tr}\bigl[\rho\, f(\Delta)(I)\bigr]. \tag{7.29}$$

Using the spectral representations

$$\rho = \sum_i a_i \mu_i \quad \text{and} \quad \sigma = \sum_j b_j \nu_j, \tag{7.30}$$

where $\{a_i\}, \{b_j\}$ are the eigenvalues[3] and $\{\mu_i\}, \{\nu_j\}$ are simple measurements, we have

$$\Delta(\nu_j \mu_i) = \frac{b_j}{a_i} \nu_j \mu_i \quad \text{and} \quad f(\Delta)(\nu_j \mu_i) = f\!\left(\frac{b_j}{a_i}\right) \nu_j \mu_i, \tag{7.31}$$

which leads to

$$D_f(\rho \,\|\, \sigma) = \sum_{i,j} a_i f\!\left(\frac{b_j}{a_i}\right) \text{tr}(\mu_i \nu_j). \tag{7.32}$$

Let $M = \{M_{ij}\}$ and $N = \{N_{ij}\}$ be the measurements defined by[4]

$$M_{ij} = \mu_i \nu_j \mu_i \quad \text{and} \quad N_{ij} = \nu_j \mu_i \nu_j, \tag{7.33}$$

and define the probability distributions $p = \{p_{ij}\}$ and $q = \{q_{ij}\}$ by

$$p_{ij} = \text{tr}(\rho M_{ij}) = a_i \,\text{tr}(\mu_i \nu_j) \quad \text{and} \quad q_{ij} = \text{tr}(\sigma N_{ij}) = b_j \,\text{tr}(\mu_i \nu_j). \tag{7.34}$$

Then Equation (7.32) indicates that $D_f(\rho \,\|\, \sigma)$ coincides with the classical $f$-divergence of $p, q$. From this, we can observe that $D_f$ for a strictly convex smooth function $f$ satisfying $f(1) = 0$ turns out a divergence on $\mathcal{S}$ in the sense described in § 3.2.

Now assume that $f$ is smooth and satisfies Equation (7.24), and denote the dualistic structure induced from $D_f$ by $(g^{(f)}, \nabla^{(f)}, \nabla^{(f^*)}) = (g^{(D_f)}, \nabla^{(D_f)}, \nabla^{(D_f^*)})$, where we have used the notation of §3.2 in which $D_f^* = D_{f^*}$ holds. This structure is a formal quantum analogue of the triple of the Fisher information and the $\pm\alpha$-connections. Since $D_f$ restricted on $\mathcal{S}_\Pi$ is identical with the classical $f$-divergence on $\mathcal{P}_l$ under the natural identification $\mathcal{S}_\Pi \cong \mathcal{P}_l$, and since $D_f$ is unitarily invariant in the sense that $D_f(U\rho U^* \,\|\, U\sigma U^*) = D_f(\rho \,\|\, \sigma)$, the induced structure satisfies the properties (i) and (iii) in § 7.1. Moreover, some calculation yields that for any vector fields $X, Y$ on $\mathcal{S}_\Pi$ and any $Z$ on $\mathcal{S}$, we have at every point $\rho \in \mathcal{S}_\Pi$

$$\langle X, Z\rangle_\rho = \text{tr}[\rho^{-1} X^{(\text{m})} Z^{(\text{m})}], \tag{7.35}$$

---

[3] We do not care whether the elements of $\{a_i\}$ and $\{b_j\}$ are distinct.

[4] In the situation where the so-called von Neumann's projection hypothesis is applicable to the simple measurement $\mu$ (and to $\nu$, also), which means that the state change $\rho \to \mu_i \rho \mu_i / \text{tr}(\rho \mu_i)$ occurs when $\mu$ is performed to the system in the state $\rho$ to yield the measurement result $i$ with probability $\text{tr}(\rho \mu_i)$, the measurement $M$ ($N$, resp.) may be physically realizable by consecutively performing two measurements $\mu$ and $\nu$ in this order (in the reverse order, resp.).

$$\left\langle \nabla_X^{(f)} Y, Z \right\rangle_\rho = \mathrm{tr}\bigl[\rho^{-1}\{X(Y^{(\mathrm{m})}) - \tfrac{1+\alpha}{2} X^{(\mathrm{m})} Y^{(\mathrm{m})}\} Z^{(\mathrm{m})}\bigr], \tag{7.36}$$

$$\left\langle \nabla_X^{(f^*)} Y, Z \right\rangle_\rho = \mathrm{tr}\bigl[\rho^{-1}\{X(Y^{(\mathrm{m})}) - \tfrac{1-\alpha}{2} X^{(\mathrm{m})} Y^{(\mathrm{m})}\} Z^{(\mathrm{m})}\bigr], \tag{7.37}$$

where $\langle,\rangle$ is the inner product with respect to $g^{(f)}$, and $X(Y^{(\mathrm{m})})$ denotes the derivative of the operator valued function $Y^{(\mathrm{m})} : \rho \mapsto Y_\rho^{(\mathrm{m})}$ in the direction of $X_\rho$, while $X^{(\mathrm{m})} Y^{(\mathrm{m})}$ is the product of the operators $X_\rho^{(\mathrm{m})}$ and $Y_\rho^{(\mathrm{m})}$. We can see that the properties (ii) and (iv) follow from these equations. (Note that $\nabla^{(\pm\alpha)} = \nabla^{(\mathrm{m})} - \tfrac{1\pm\alpha}{2} T$ according to Equation (2.27). ) This means that for the commutative direction the induced structure $(g^{(f)}, \nabla^{(f)}, \nabla^{(f^*)})$ depends on $f$ only through the $\alpha$ determined by Equation (7.24) just as in the classical case. As for the unitary direction, on the other hand, the dependence is more essential. In fact, for $X_1, X_2 \in T_\rho^u$ such that $X_i^{(\mathrm{m})} = [\Omega_i, \rho]$, we have

$$\langle X_1, X_2 \rangle_\rho = 2\mathrm{Re}\,\langle\!\langle \Omega_1, f(\Delta_{\rho,\rho})(\Omega_2) \rangle\!\rangle_\rho^{-}. \tag{7.38}$$

Now, letting $f$ be the $f^{(\alpha)}$ in Equation (3.24), we can define a quantum version of the $\alpha$-divergence, which is called the **quantum $\alpha$-divergence** and is denoted by $D^{(\alpha)}$ as the classical one. Noting that[5]

$$(\Delta_{\sigma,\rho})^r(A) = \sigma^r A \rho^{-r} \quad (\forall r \in \mathbb{R})$$

and

$$(\log \Delta_{\sigma,\rho})(X) = (\log \sigma) X - X(\log \rho),$$

we obtain

$$D^{(\alpha)}(\rho \,\|\, \sigma) = \frac{4}{1-\alpha^2} \left\{ 1 - \mathrm{tr}\left(\rho^{\frac{1-\alpha}{2}} \sigma^{\frac{1+\alpha}{2}}\right) \right\} \quad (\alpha \neq \pm 1) \tag{7.39}$$

$$D^{(-1)}(\rho \,\|\, \sigma) = D^{(1)}(\sigma \,\|\, \rho) = \mathrm{tr}\left[\rho(\log \rho - \log \sigma)\right], \tag{7.40}$$

which just correspond to Equations (3.25) and (3.26). In particular, $D^{(\pm 1)}$ is a quantum equivalent of the Kullback divergence, and is well known in mathematical physics as the **quantum relative entropy**. The general quantum $\alpha$-divergence was introduced by Hasegawa [101] together with the induced dualistic structure $(g^{(\alpha)}(=g^{(-\alpha)}), \nabla^{(\alpha)}, \nabla^{(-\alpha)}) = (g^{(D^{(\alpha)})}, \nabla^{(D^{(\alpha)})}, \nabla^{(D^{(-\alpha)})})$.

Parameterizing the elements of $\mathcal{S}$ as $\rho_\xi$ by a coordinate system $[\xi^i]$ and letting

$$\ell^{(\alpha)}(\xi) \stackrel{\mathrm{def}}{=} \begin{cases} \dfrac{2}{1-\alpha} \rho_\xi^{\frac{1-\alpha}{2}} & (\alpha \neq 1) \\ \log \rho_\xi & (\alpha = 1), \end{cases} \tag{7.41}$$

---

[5] Here the logarithm and the power of an operator are defined in the usual way; when $\rho$ has the eigenvalues $\{p_i\}$ and the orthonormal eigenvectors $\{v_i\}$, then $\log \rho$ and $\rho^r$ have the eigenvalues $\{\log p_i\}$ and $\{p_i^r\}$, respectively, and the same eigenvectors $\{v_i\}$.

## 7.2. QUANTUM DIVERGENCE

the components of $g^{(\alpha)}$ and $\nabla^{(\alpha)}$ are represented as

$$g_{ij}^{(\alpha)} = \text{tr}\left(\partial_i \ell^{(\alpha)} \partial_j \ell^{(-\alpha)}\right) \quad \text{and} \tag{7.42}$$

$$\Gamma_{ij,k}^{(\alpha)} = \text{tr}\left(\partial_i \partial_j \ell^{(\alpha)} \partial_k \ell^{(-\alpha)}\right), \tag{7.43}$$

which correspond to Equations (2.60) and (2.61). In particular, for $X_1, X_2 \in T_\rho^u$ such that $X_i^{(\text{m})} = [\Omega_i, \rho]$, we have $X_i \ell^{(\alpha)} = [\Omega_i, \ell^{(\alpha)}]$ and hence ([101])

$$\langle X_1, X_2 \rangle_\rho = \begin{cases} \dfrac{4}{1-\alpha^2} \text{tr}\left(\left[\Omega_1, \rho^{\frac{1+\alpha}{2}}\right]\left[\Omega_2, \rho^{\frac{1-\alpha}{2}}\right]\right) & (\alpha \neq \pm 1) \\ \text{tr}\left([\Omega_1, \log \rho][\Omega_2, \rho]\right) & (\alpha = \pm 1). \end{cases} \tag{7.44}$$

This follows also from Equation (7.38).

It is not difficult to see that most of the geometrical arguments developed for the Fisher metric and the $\alpha$-connections in §2.6, §3.5 and §3.6 may immediately be applied to this $(g^{(\pm\alpha)}, \nabla^{(\alpha)}, \nabla^{(-\alpha)})$ to yield several parallel results. In particular, $\mathcal{S}$ is dually flat with respect to $(g^{(\pm 1)}, \nabla^{(1)}, \nabla^{(-1)})$ and the canonical divergence of this structure is the relative entropy $D^{(\pm 1)}$. The elements of an arbitrary $\nabla^{(1)}$-autoparallel submanifold $M$ of $\mathcal{S}$, including the case $M = \mathcal{S}$, may be parameterized in the form

$$\rho_\theta = \exp\left[C + \sum_{i=1}^m \theta^i F_i - \psi(\theta)\right], \tag{7.45}$$

where $F_1, \ldots, F_m, C$ are Hermitian operators and $\psi(\theta)$ is an $\mathbb{R}$-valued function, and $[\theta^i]$ turns out a $\nabla^{(1)}$-affine coordinate system, while a $\nabla^{(-1)}$-affine coordinate system dual to $[\theta^i]$ is given by $\eta_i(\theta) \stackrel{\text{def}}{=} \text{tr}(\rho_\theta F_i)$. The dual potential is then represented as

$$\begin{aligned} \varphi(\theta) &= \theta^i \eta_i(\theta) - \psi(\theta) \\ &= \max_{\theta'}\left\{\theta'^i \eta_i(\theta) - \psi(\theta')\right\} \\ &= -H(\rho_\theta) - \text{tr}(\rho_\theta C), \end{aligned} \tag{7.46}$$

where $H$ is the **von Neumann entropy**: $H(\rho) \stackrel{\text{def}}{=} -\text{tr}(\rho \log \rho)$. The structure $(g^{(\pm 1)}, \nabla^{(1)}, \nabla^{(-1)})$ will be revisited in the next section from a different point of view. For an arbitrary $\alpha$, on the other hand, the dualistic structure $(g^{(\pm\alpha)}, \nabla^{(\alpha)}, \nabla^{(-\alpha)})$ on $\mathcal{S}$ may naturally be extended to a dually flat structure on the denormalized manifold

$$\tilde{\mathcal{S}} \stackrel{\text{def}}{=} \{\tau\rho \,|\, \rho \in \mathcal{S} \text{ and } \tau > 0\},$$

which is the set of all positive definite operators on $\mathcal{H}$, and the corresponding canonical divergence gives an extension of $D^{(\alpha)}$ just like Equations (3.78) and (3.79).

Finally, we briefly sketch out a result of Petz [182] concerning how the monotonicity of the classical $f$-divergence in Equation (3.22) is extended to the quantum $f$-divergence. In order to state the result, we need two new notions: **completely positive maps** (**CP maps** for short) and **operator convex functions**. As for the former, we skip the definition here and only note that the CP property is widely believed to characterize the physical realizability of a process of changing quantum states; see Stinespring [203], Kraus [126] and Lindblad [144]. A real valued function $f$ defined on an interval in $\mathbb{R}$ is said to be operator convex when $\lambda f(A) + (1-\lambda)f(B) - f(\lambda A + (1-\lambda)B)$ is positive semidefinite for any $0 \leq \lambda \leq 1$ and any Hermitian operators whose spectra lie in the domain of $f$. The operator concavity is also defined similarly. Every operator convex (or concave) function is a convex (or concave, resp.) function in the usual sense, but the converse is not always true. For instance, it is known that $x \mapsto \log x$ is operator concave, while $x \mapsto e^x$ is not operator convex. In addition, the power function $x \mapsto x^r$ $(x > 0)$ is operator convex iff $1 \leq r \leq 2$ or $-1 \leq r \leq 0$, and is operator concave iff $0 \leq r \leq 1$. See e.g. Bhatia [48] and its references.

Now Petz's theorem claims that if $f$ is operator convex, then the quantum $f$-divergence $D_f$ satisfies the **monotonicity**: $D_f(\rho \,\|\, \sigma) \geq D_f(\Gamma\rho \,\|\, \Gamma\sigma)$ for any (trace-preserving) CP map $\Gamma$. As pointed out by Hasegawa [101], the $f^{(\alpha)}$ used for defining the quantum $\alpha$-divergence $D^{(\alpha)}$ is operator convex iff $-3 \leq \alpha \leq 3$, and therefore the theorem implies the monotonicity of relative entropy $D^{(\pm 1)}$ (Lindblad [144], Araki [33], Uhlmann [209]). Note also that the so-called Wigner-Yanase-Dyson-Lieb concavity (Lieb [143], Uhlmann [209]) is closely related to the monotonicity of $D^{(\alpha)}$ for $-1 \leq \alpha \leq 1$. Special but important cases of the monotonicity of $D_f$ are the **joint convexity** (cf. Equation (3.23)):

$$D_f(\lambda\rho_1 + (1-\lambda)\rho_2 \,\|\, \lambda\sigma_1 + (1-\lambda)\sigma_2)$$
$$\leq \lambda D_f(\rho_1 \,\|\, \sigma_1) + (1-\lambda)D_f(\rho_2 \,\|\, \sigma_2), \quad 0 \leq \lambda \leq 1, \quad (7.47)$$

and the monotonicity with respect to an arbitrary measurement $\Pi = \{\pi_i\}$:

$$D_f(\rho \,\|\, \sigma) \geq D_f(p \,\|\, q), \quad (7.48)$$

where the RHS denotes the classical $f$-divergence between $p_i = \text{tr}(\rho\pi_i)$ and $q_i = \text{tr}(\sigma\pi_i)$.

## 7.3 The geometric structure induced from a generalized covariance

In the classical case we have seen that the dualistic structure $(g, \nabla^{(1)}, \nabla^{(-1)}) = (g, \nabla^{(e)}, \nabla^{(m)})$ is particularly important in most applications among other $(g, \nabla^{(\alpha)}, \nabla^{(-\alpha)})$. In this section we make an attempt to develop a general theory for quantum analogues of $(g, \nabla^{(e)}, \nabla^{(m)})$.

Suppose that we are given a family $\{\langle\!\langle \,\cdot\,,\,\cdot\,\rangle\!\rangle_\rho \mid \rho \in \mathcal{S}\}$ of inner products on $\mathcal{A}$, where $\langle\!\langle A, B \rangle\!\rangle_\rho \in \mathbf{R}$ depends smoothly upon $\rho$ for all $A, B \in \mathcal{A}$, and that it satisfies the following properties:

## 7.3. GENERALIZED COVARIANCE

(a) For all $A, B \in \mathcal{A}$, $\rho \in \mathcal{S}$ and $U \in \mathcal{U}$, it holds that

$$\langle\!\langle UAU^*, UBU^*\rangle\!\rangle_{U\rho U^*} = \langle\!\langle A, B\rangle\!\rangle_\rho. \tag{7.49}$$

(b) If $[\rho, A] = 0$ then

$$\langle\!\langle A, B\rangle\!\rangle_\rho = \mathrm{tr}(\rho AB). \tag{7.50}$$

This may be regarded as a quantum version of the $L^2$-inner product

$$\langle\!\langle A, B\rangle\!\rangle_p = E_p[AB]$$

of real-valued random variables $A$ and $B$ with respect to a probability distribution $p$. Reflecting that $E_p[AB]$ is the covariance of $A$ and $B$ when their expectations vanish, we call $\langle\!\langle\,,\,\rangle\!\rangle = \{\langle\!\langle\,,\,\rangle\!\rangle_\rho \mid \rho \in \mathcal{S}\}$ satisfying the conditions (a) and (b) a **generalized covariance**.

Two important examples of generalized covariances are the **symmetrized inner product**

$$\langle\!\langle A, B\rangle\!\rangle_\rho = \frac{1}{2}\mathrm{tr}(\rho AB + \rho BA) \tag{7.51}$$

and the **Bogoliubov inner product** (also called the **Kubo-Mori inner product** or the **canonical correlation** [127])

$$\langle\!\langle A, B\rangle\!\rangle_\rho = \int_0^1 \mathrm{tr}(\rho^\lambda A \rho^{1-\lambda} B)\mathrm{d}\lambda. \tag{7.52}$$

These examples are unified in the general form ([186])

$$\langle\!\langle A, B\rangle\!\rangle_\rho = \int_0^1 \mathrm{tr}(\rho^\lambda A \rho^{1-\lambda} B)\nu(\mathrm{d}\lambda), \tag{7.53}$$

where $\nu$ is an arbitrary probability measure on $[0, 1]$ satisfying $\nu(\mathrm{d}\lambda) = \nu(1 - \mathrm{d}\lambda)$. Note that the direct analogue $\langle\!\langle A, B\rangle\!\rangle_\rho = \mathrm{tr}(\rho AB)$ ($\forall A, B \in \mathcal{A}$) of the classical $L^2$-product does not yield a generalized covariance, because it may take imaginary values.

Now let us fix a generalized covariance $\langle\!\langle\,,\,\rangle\!\rangle$, and define the **e-representation** of a tangent vector $X \in T_\rho$ as the Hermitian operator $X^{(\mathrm{e})} \in \mathcal{A}$ satisfying

$$\mathrm{tr}\left(X^{(\mathrm{m})} A\right) = \langle\!\langle X^{(\mathrm{e})}, A\rangle\!\rangle_\rho, \quad \forall A \in \mathcal{A}. \tag{7.54}$$

From the condition (b) we see that if $X$ belongs to $T_\rho^c$ then

$$X^{(\mathrm{e})} = \rho^{-1} X^{(\mathrm{m})}, \tag{7.55}$$

which corresponds to Equation (2.40), i.e. the relation between the m- and e-representations in the classical case. When a coordinate system $[\xi^i]$ is given on $\mathcal{S}$ (or on a submanifold of $\mathcal{S}$) so that each state is parameterized as $\rho = \rho_\xi$ and that the m-representation of the natural basis vector is written as

$$(\partial_i)^{(\mathrm{m})} = \partial_i \rho, \tag{7.56}$$

we denote the e-representation by

$$(\partial_i)^{(e)} = L_i. \qquad (7.57)$$

When $\langle\!\langle\,,\,\rangle\!\rangle$ is represented as Equation (7.53), we have

$$\partial_i \rho = \int_0^1 \rho^\lambda L_i \rho^{1-\lambda} \nu(\mathrm{d}\lambda). \qquad (7.58)$$

For instance, when the symmetrized inner product is adopted as $\langle\!\langle\,,\,\rangle\!\rangle$, we have

$$\partial_i \rho = \frac{1}{2}(\rho L_i + L_i \rho), \qquad (7.59)$$

which shows that $L_i$ is the **symmetric logarithmic derivative**, or the **SLD** for short, introduced by C.W. Helstrom [107] to formulate a quantum version of Cramér-Rao inequality (see Theorem 7.5 in the next section). On the other hand, adopting the Bogoliubov inner product as $\langle\!\langle\,,\,\rangle\!\rangle$ leads to

$$L_i = \partial_i \log \rho, \qquad (7.60)$$

or, in other words, we have

$$\partial_i \rho = \int_0^1 \rho^\lambda (\partial_i \log \rho) \rho^{1-\lambda} \mathrm{d}\lambda. \qquad (7.61)$$

This follows from the general formula

$$P - Q = \int_0^1 Q^\lambda (\log P - \log Q) P^{1-\lambda} \mathrm{d}\lambda \qquad (7.62)$$

which holds for arbitrary positive operators $P$ and $Q$. To prove the last equation, it suffices to verify the following identity by differentiating the both sides with $t$:

$$1 - Q^t P^{-t} = \int_0^t Q^\lambda (\log P - \log Q) P^{-\lambda} \mathrm{d}\lambda. \qquad (7.63)$$

Equation (7.60) shows a direct analogy with the e-representation in the classical case which was defined as the derivative of $\log p$.

Using the e-representation, we define the inner product $\langle\,,\,\rangle_\rho$ on the tangent space $T_\rho$ by

$$\langle X, Y \rangle_\rho = \langle\!\langle X^{(e)}, Y^{(e)} \rangle\!\rangle_\rho = \mathrm{tr}\left(X^{(m)} Y^{(e)}\right). \qquad (7.64)$$

Then $g = \langle\,,\,\rangle$ forms a Riemannian metric on $\mathcal{S}$ which may be regarded as a quantum version of the Fisher metric (cf. Equation (2.42)). The components of the metric are given by

$$g_{ij} = \langle\!\langle L_i, L_j \rangle\!\rangle = \mathrm{tr}\left[(\partial_i \rho)(L_j)\right]. \qquad (7.65)$$

## 7.3. GENERALIZED COVARIANCE

We also define quantum versions of the e- and m-connections by

$$\Gamma^{(e)}_{ij,k} = \left\langle \nabla^{(e)}_{\partial_i} \partial_j, \partial_k \right\rangle = \operatorname{tr}\left[(\partial_i L_j)(\partial_k \rho)\right] = \langle\!\langle \partial_i L_j, L_k \rangle\!\rangle, \qquad (7.66)$$

$$\Gamma^{(m)}_{ij,k} = \left\langle \nabla^{(m)}_{\partial_i} \partial_j, \partial_k \right\rangle = \operatorname{tr}\left[(\partial_i \partial_j \rho) L_k\right]. \qquad (7.67)$$

Obviously, $\nabla^{(e)}$ and $\nabla^{(m)}$ are mutually dual with respect to $g$. In addition, it is not difficult to see that the properties (i)–(iv) described in §7.1 hold for this $(g, \nabla^{(e)}, \nabla^{(m)})$ due to the conditions (a) and (b) on $\langle\!\langle \,,\, \rangle\!\rangle$. As observed from Equation (7.60), when the Bogoliubov inner product is adopted as $\langle\!\langle \,,\, \rangle\!\rangle$, the resulting dualistic structure coincides with $(g^{(\pm 1)}, \nabla^{(1)}, \nabla^{(-1)})$ which was defined in the previous section through the quantum relative entropy $D^{(\pm 1)}$.

Most of the arguments of §2.5 for the classical e-, m-connections may be extended to the present setting in a straightforward manner. First, the m-connection $\nabla^{(m)}$ is independent of the choice of generalized covariance, and is the flat affine connection induced from the affine structure of $\mathcal{A}_1$. The parallel translation $\Pi^{(m)}_{\rho,\sigma} : T_\rho \to T_\sigma$ of $\nabla^{(m)}$ is given by

$$\Pi^{(m)}_{\rho,\sigma}(X) = X' \iff X'^{(m)} = X^{(m)}. \qquad (7.68)$$

Next, although the e-representation depends on what generalized covariance is adopted, the set $T^{(e)}_\rho \stackrel{\text{def}}{=} \{X^{(e)} \mid X \in T_\rho\}$ is simply written as

$$T^{(e)}_\rho = \{A \in \mathcal{A} \mid E_\rho[A] = 0\}. \qquad (7.69)$$

This is shown as follows. For an $A \in \mathcal{A}$, let $E[A]$ denote the function $\rho \mapsto E_\rho[A]$ defined on $\mathcal{S}$. Applying a tangent vector $X \in T_\rho$ to this function as a differential operator, we have

$$X(E[A]) = \operatorname{tr}\left(X^{(m)} A\right) = \langle\!\langle X^{(e)}, A \rangle\!\rangle_\rho, \qquad (7.70)$$

where the first equality follows from $X^{(m)} = X\rho$ and the second from Equation (7.54). On the other hand, from the condition (b) on $\langle\!\langle \,,\, \rangle\!\rangle$ we have

$$\langle B, I \rangle_\rho = \operatorname{tr}(\rho B) = E_\rho[B] \qquad (7.71)$$

for all $B \in \mathcal{A}$ including $B = X^{(e)}$. Combining these equations we obtain

$$E_\rho\left[X^{(e)}\right] = X(E[I]) = X(1) = 0, \qquad (7.72)$$

which proves that LHS $\subset$ RHS in Equation (7.69). Since LHS and RHS are both linear spaces of dimension $n = \dim \mathcal{A} - 1$, they must coincide. Now, tracing the proof of Equation (2.43), we see that the parallel translation with respect to $\nabla^{(e)}$ is represented as follows:

$$\Pi^{(e)}_{\rho,\sigma}(X) = X' \iff X'^{(e)} = X^{(e)} - E_\sigma\left[X^{(e)}\right]. \qquad (7.73)$$

We have seen that the parallel translations $\Pi_{\rho,\sigma}^{(m)}$ and $\Pi_{\rho,\sigma}^{(e)}$ are both independent of the choice of curve connecting two points $\rho$ and $\sigma$, and hence $\nabla^{(m)}$ and $\nabla^{(e)}$ have vanishing curvature tensors. In addition, the mixture connection $\nabla^{(m)}$ is always symmetric and hence is flat. How about the e-connection, on the other hand? It should be noted that the symmetry of $\nabla^{(e)}$ is equivalent to the dually flatness of $(\mathcal{S}, g, \nabla^{(e)}, \nabla^{(m)})$ and is also equivalent to the existence of a divergence $D$ on $\mathcal{S}$ which induces $(g, \nabla^{(e)}, \nabla^{(m)})$ in the manner described in §3.2. We have already seen that the e-connection induced from the Bogoliubov inner product is symmetric, and the corresponding canonical divergence is the quantum relative entropy. This is, however, exceptional. In fact, we have:

**Theorem 7.1** *The e-connection induced on $\mathcal{S}$ from a generalized covariance $\langle\!\langle \, , \, \rangle\!\rangle$ is symmetric if and only if $\langle\!\langle \, , \, \rangle\!\rangle$ is the Bogoliubov inner product.*

We shall give a proof after observing some fundamental aspects of the torsion tensor of $\nabla^{(e)}$, which we denote by $\mathrm{Tor}^{(e)}$ to avoid confusing it with the space $T_\rho^{(e)}$ of the e-representations of tangent vectors.

First, from the definition of the torsion tensor we have

$$\left\langle \mathrm{Tor}^{(e)}(\partial_i, \partial_j), \partial_k \right\rangle_\rho = \Gamma_{ij,k}^{(e)} - \Gamma_{ji,k}^{(e)}$$
$$= \langle\!\langle \partial_i L_j - \partial_j L_i, L_k \rangle\!\rangle_\rho. \quad (7.74)$$

In addition, the operator $\partial_i L_j - \partial_j L_i$ belongs to $T_\rho^{(e)}$ at each point $\rho$, or in other words, it holds that

$$E_\rho[\partial_i L_j] = E_\rho[\partial_j L_i], \quad (7.75)$$

and therefore the e-representation of $\mathrm{Tor}^{(e)}(\partial_i, \partial_j)$ is given by

$$\left\{ \mathrm{Tor}^{(e)}(\partial_i, \partial_j) \right\}^{(e)} = \partial_i L_j - \partial_j L_i. \quad (7.76)$$

Equation (7.75) is seen by differentiating $E_\rho[L_j] = 0$ to obtain

$$\mathrm{tr}\,((\partial_i \rho) L_j) + \mathrm{tr}\,(\rho(\partial_i L_j)) = 0,$$

which leads to

$$E_\rho[\partial_i L_j] = -g_{ij}. \quad (7.77)$$

Note that the last equation is an analogue of Equation (2.8).

For each $\rho \in \mathcal{S}$, let us define the mapping $\Phi_\rho : \mathcal{A} \to \mathcal{A}$ by

$$\langle\!\langle A, B \rangle\!\rangle_\rho = \mathrm{tr}(A \Phi_\rho(B)), \quad \forall A, B \in \mathcal{A}. \quad (7.78)$$

Then the relation between the e- and m-representations of a tangent vector $X$ is written as $X^{(m)} = \Phi_\rho(X^{(e)})$, and in particular we have

$$\partial_j \rho = \Phi_\rho(L_j).$$

Operating $\partial_i$ on the both sides of this equation, we obtain

$$\partial_i \partial_j \rho = \Phi_{\partial_i}(L_j) + \Phi_\rho(\partial_i L_j),$$

where $\Phi_{\partial_i}(A)$ for an operator $A$ denotes the derivative of the mapping $\rho \mapsto \Phi_\rho(A)$ by $\partial_i$. Noting that $\partial_i \partial_j \rho = \partial_j \partial_i \rho$, we see that the m-representation of $\mathrm{Tor}^{(\mathrm{e})}(\partial_i, \partial_j)$ is given by

$$\begin{aligned}\left\{\mathrm{Tor}^{(\mathrm{e})}(\partial_i, \partial_j)\right\}^{(\mathrm{m})} &= \Phi_\rho(\partial_i L_j - \partial_j L_i) \\ &= \Phi_{\partial_j}(L_i) - \Phi_{\partial_i}(L_j). \end{aligned} \quad (7.79)$$

When $\langle\!\langle \,,\, \rangle\!\rangle$ is the symmetrized inner product, we have $\Phi_\rho(A) = \frac{1}{2}(\rho A + A\rho)$ and

$$\begin{aligned}\Phi_{\partial_j}(L_i) &= \frac{1}{2}\{(\partial_j \rho)L_i + L_i(\partial_j \rho)\} \\ &= \frac{1}{4}(L_i L_j \rho + L_i \rho L_j + L_j \rho L_i + \rho L_j L_i),\end{aligned}$$

which leads to

$$\left\{\mathrm{Tor}^{(\mathrm{e})}(\partial_i, \partial_j)\right\}^{(\mathrm{m})} = \frac{1}{4}\left[[L_i, L_j], \rho\right]. \quad (7.80)$$

This shows how the nonvanishing torsion appears as a consequence of the noncommutativity of operators.

Now let us prove the 'only if' part of Theorem 7.1, while the 'if' part has already been shown. Assume that $\mathrm{Tor}^{(\mathrm{e})} = 0$. From Equation (7.76), the assumption is equivalent to the existence of a smooth mapping $F : \mathcal{S} \to \mathcal{A}$ such that $\partial_i F = L_i$. Now let $\Pi$ be an arbitrary simple measurement, $\rho$ a point in $\mathcal{S}_\Pi$, and $X$ a tangent vector of $\mathcal{S}_\Pi$ at $\rho$. Then, since $X \log \rho = \rho^{-1}(X\rho) = \rho^{-1} X^{(\mathrm{m})}$ and $[\rho, X \log \rho] = 0$, it follows from the the condition (b) on $\langle\!\langle \,,\, \rangle\!\rangle$ that $\langle\!\langle X \log \rho, A \rangle\!\rangle_\rho = \mathrm{tr}\left(X^{(\mathrm{m})} A\right)$ for all $A \in \mathcal{A}$. Therefore we have $X \log \rho = X^{(\mathrm{e})} = XF(\rho)$, from which we see that the difference $F(\rho) - \log \rho$ is a constant, say $C_\Pi$, on $\mathcal{S}_\Pi$. Moreover, since $\frac{1}{k} I$ ($k = \dim \mathcal{H}$) is a common element of $\mathcal{S}_\Pi$ for all $\Pi$, the constant $C_\Pi$ cannot depend on $\Pi$. Hence we have $F(\rho) = \log \rho + C$, which, combined with Equation (7.60), completes the proof.

## 7.4 Applications to quantum estimation theory

Using the dualistic structure $(g, \nabla^{(\mathrm{e})}, \nabla^{(\mathrm{m})})$ induced from a generalized covariance, we may immediately translate some results in statistics into their quantum versions. Here are the translations of the Cramér-Rao inequality and related results (Theorems 2.7, 2.8, 2.2 and 3.12).

**Theorem 7.2** *Let $A$ be an observable (Hermitian operator), $M$ a submanifold of $\mathcal{S}$, and $\rho$ a point in $M$. Then we have*

$$\langle\!\langle A - E_\rho[A], A - E_\rho[A] \rangle\!\rangle_\rho \geq \|(dE[A]|_M)_\rho\|_\rho^2, \quad (7.81)$$

where the equality holds if and only if

$$A - E_\rho[A] \in T_\rho^{(e)}(M) \stackrel{\text{def}}{=} \left\{ X^{(e)} \,\middle|\, X \in T_\rho(M) \right\}. \tag{7.82}$$

In particular, if $M = \mathcal{S}$ then the equality in Equation (7.81) always holds.

**Theorem 7.3** *Let $M = \{\rho_\xi \,|\, \xi \in \Xi\}$ be an m-dimensional submanifold of $\mathcal{S}$ with a coordinate system $\xi = [\xi^i]$ ranging over $\Xi \subset \mathbb{R}^m$. If an m-tuple $\vec{F} = (F^1, \cdots, F^m) \in \mathcal{A}^m$ of observables is unbiased in the sense that*

$$E_\xi[F^i] \stackrel{\text{def}}{=} \mathrm{tr}(\rho_\xi F^i) = \xi^i, \quad \forall \xi \in \Xi, \, \forall i \in \{1, \ldots, m\}, \tag{7.83}$$

*or more generally, if it is locally unbiased at a point $\xi \in \Xi$ (cf. the note in §2.5) in the sense that*

$$E_\xi[F^i] = \xi^i \quad \text{and} \quad \partial_j E_\xi[F^i] = \delta^i_j, \tag{7.84}$$

*then we have*

$$W_\xi[\vec{F}] \geq G_\xi^{-1}, \tag{7.85}$$

*where $G_\xi$ is the component matrix $[g_{ij}(\xi)]$ of the metric $g$ with respect to $[\xi^i]$, and $W_\xi[\vec{F}]$ is the matrix whose $(i,j)$ element is*

$$w_\xi^{ij} = \langle\!\langle F^i - \xi^i, F^j - \xi^j \rangle\!\rangle_\xi. \tag{7.86}$$

**Theorem 7.4** *Given $(M, [\xi^i])$, there exists an unbiased $\vec{F} = (F^1, \cdots, F^m)$ satisfying $W_\xi[\vec{F}] = G_\xi^{-1}$ for all $\xi$ if and only if $M$ is e-autoparallel in $\mathcal{S}$ and $[\xi^i]$ is an m-affine coordinate system. (Note that an e-autoparallel submanifold is always m-flat, even though it may have nonvanishing e-torsion.)*

From a purely mathematical point of view, all these theorems are natural extensions of the original statistical theorems, whatever generalized covariance may be chosen as $\langle\!\langle \,,\, \rangle\!\rangle$. However this does not ensure that they are as meaningful as the original theorems. We should first note that the quantity $W_\rho[A] \stackrel{\text{def}}{=} \langle\!\langle A - E_\rho[A], A - E_\rho[A] \rangle\!\rangle_\rho$ in Theorem 7.2 is not equal to the variance $V_\rho[A]$ in general. Indeed, $W_\rho[A] = V_\rho[A]$ holds for all $A \in \mathcal{A}$ when and only when the underlying generalized covariance is the symmetrized inner product. It is not clear whether $W_\rho[A]$ has any statistical significance when another generalized covariance is adopted, although a certain kind of equations in statistical physics concerning perturbation from equilibrium states may be regarded as special cases of Theorem 7.2 for the Bogoliubov inner product; see e.g. Kubo et al. [127].

From now on, we assume $\langle\!\langle \,,\, \rangle\!\rangle$ to be the symmetrized inner product and investigate the meaning of Theorems 7.3 and 7.4. When the dimension $m$ of $M$

## 7.4. APPLICATIONS TO QUANTUM ESTIMATION THEORY

is 1, the setting of these theorems properly represents the physically meaningful situation where the unknown value of the scalar parameter $\xi$ is estimated by an observable $F$ and the estimation accuracy is measured by the variance $W_\xi[F] = V_\xi[F]$. This, however, is not the case when $m \geq 2$. In the classical (usual) probability theory, there is no difference between "$m$ $\mathbb{R}$-valued random variables" and "an $\mathbb{R}^m$-valued random variable". On the other hand, in the quantum case, $m$ observables may not naturally be regarded as an $\mathbb{R}^m$-valued measurement unless the observables are commutative, as is well known as the uncertainty principle. Therefore $\vec{F} = (F^1, \cdots, F^m)$ in the theorems does not represent an estimator of the vector parameter $[\xi^1, \cdots, \xi^m]$ in general, even though each $F^i$ may be regarded as an estimator of the scalar parameter $\xi^i$. In order to elucidate the estimation theoretic meaning of the theorems, we need some basic arguments on quantum multiparameter estimation.

As a general representation of estimators for $\xi = [\xi^i]$, we consider a pair $(\Pi, \hat{\xi})$ of a (not necessarily simple) measurement $\Pi = \{\pi(x)\}$ on a finite set $\mathcal{X}$ and a mapping $\hat{\xi} = [\hat{\xi}^i] : \mathcal{X} \to \mathbb{R}^m$, which corresponds to the situation where the unknown value of $\xi$ is estimated by $\hat{\xi}(x)$ based on the result $x$ of the measurement $\Pi$.

**Note:** Since an estimator for an $m$-dimensional parameter is an $\mathbb{R}^m$-valued measurement, it should be represented by a POM (see the note in § 7.1), say $\hat{\pi}$, on $\mathbb{R}^m$, which determines the probability distribution of the estimate under the state $\rho$ as $\Pr\{\hat{\xi} \in B\} = \operatorname{tr}[\rho\hat{\pi}(B)]$. Indeed, the property of a pair $(\Pi, \hat{\xi})$ concerning the estimation may be described in terms of the POM $\hat{\pi}$ defined by

$$\hat{\pi}(B) = \sum_{x:\hat{\xi}(x)\in B} \pi(x).$$

In view of generality, an arbitrary POM on $\mathbb{R}^m$ should be called an estimator. Nevertheless, our elementary and a little redundant representation $(\Pi, \hat{\xi})$ is sufficiently general for the later arguments and has the advantage of being consistent with the description in the previous chapters.

Then, since the $\hat{\xi}(x)$ may be regarded as an estimator for the classical model $M(\Pi) = \{p(x;\xi) \mid \xi \in \Xi\}$ consisting of the probability distributions $p(x;\xi) = \operatorname{tr}[\rho_\xi \pi(x)]$, we are naturally led to the following definitions; the estimator $(\Pi, \hat{\xi})$ is said to be unbiased if

$$E_\xi\left[\Pi, \hat{\xi}^i\right] \stackrel{\text{def}}{=} \sum_x \hat{\xi}^i(x) \operatorname{tr}[\rho_\xi \pi(x)] = \xi^i, \quad \forall \xi \in \Xi, \forall i \in \{1, \ldots, m\}, \quad (7.87)$$

it is said to be locally unbiased at a point $\xi \in \Xi$ if

$$E_\xi\left[\Pi, \hat{\xi}^i\right] = \xi^i \quad \text{and} \quad \partial_j E_\xi\left[\Pi, \hat{\xi}^i\right] = \delta^i_j, \quad \forall i,j \in \{1, \ldots, m\}, \quad (7.88)$$

and the variance-covariance matrix $V_\xi\left[\Pi, \hat{\xi}\right] = [v_\xi^{ij}]$ of $(\Pi, \hat{\xi})$ is defined by

$$v_\xi^{ij} = \sum_x \left\{\hat{\xi}^i(x) - \xi^i\right\}\left\{\hat{\xi}^j(x) - \xi^j\right\} \operatorname{tr}\left[\rho_\xi \pi(x)\right]. \tag{7.89}$$

Now, let
$$F^i \stackrel{\text{def}}{=} \sum_x \hat{\xi}^i(x)\pi(x) \in \mathcal{A}. \tag{7.90}$$

Then we see that the unbiasedness and the locally unbiasedness of $(\Pi, \hat{\xi})$ are equivalent to those of $\vec{F} = (F^1, \cdots, F^m)$ in the sense of Equations (7.83) and (7.84). In addition we have

$$V_\xi\left[\Pi, \hat{\xi}\right] \geq W_\xi\left[\vec{F}\right]. \tag{7.91}$$

To see this, choose $\boldsymbol{c} = (c_i) \in \mathbb{R}^m$ arbitrarily and let

$$\bar{\xi}_{\boldsymbol{c}}(x) \stackrel{\text{def}}{=} c_i\left\{\hat{\xi}^i(x) - \xi^i\right\} \quad \text{and}$$

$$\bar{F}_{\boldsymbol{c}} \stackrel{\text{def}}{=} c_i\left\{F^i - \xi^i\right\} = \sum_x \bar{\xi}_{\boldsymbol{c}}(x)\pi(x).$$

Then we have

$$\begin{aligned}
0 &\leq \sum_x \left\{\bar{\xi}_{\boldsymbol{c}}(x) - \bar{F}_{\boldsymbol{c}}\right\}\pi(x)\left\{\bar{\xi}_{\boldsymbol{c}}(x) - \bar{F}_{\boldsymbol{c}}\right\} \\
&= \sum_x \left\{\bar{\xi}_{\boldsymbol{c}}(x)\right\}^2 \pi(x) - \left\{\bar{F}_{\boldsymbol{c}}\right\}^2
\end{aligned}$$

and hence

$$\sum_x \left\{\bar{\xi}_{\boldsymbol{c}}(x)\right\}^2 \operatorname{tr}\left[\rho_\xi \pi(x)\right] \geq \operatorname{tr}\left[\rho_\xi \left\{\bar{F}_{\boldsymbol{c}}\right\}^2\right],$$

which leads to Equation (7.91). Combining this with Theorem 7.3, we reach the following result.

**Theorem 7.5** *(Helstrom [107]) Let $(\Pi, \hat{\xi})$ be an estimator for $(M, [\xi^i])$ which is unbiased or, more generally, locally unbiased at $\xi$. Then its variance-covariance matrix satisfies*

$$V_\xi\left[\Pi, \hat{\xi}\right] \geq G_\xi^{-1}. \tag{7.92}$$

It can be shown that the bound in the above theorem is "the best" in the following sense ([162]): given a point $\xi \in \Xi$ and a vector $\boldsymbol{c} = (c_1, \cdots, c_m)^t \in \mathbb{R}^m$ arbitrarily, we have

$$\boldsymbol{c}^t G_\xi^{-1} \boldsymbol{c} = \inf_{(\Pi, \hat{\xi})} \boldsymbol{c}^t V_\xi\left[\Pi, \hat{\xi}\right] \boldsymbol{c}, \tag{7.93}$$

## 7.4. APPLICATIONS TO QUANTUM ESTIMATION THEORY

where the infimum is taken over all $(\Pi, \hat{\xi})$ which are locally unbiased at $\xi$. This suggests that it is meaningful to investigate the condition for the existence of an unbiased estimator which achieves the equality in Equation (7.92). Obviously, the equality holds if and only if

$$V_\xi\left[\Pi, \hat{\xi}\right] = W_\xi\left[\vec{F}\right] \quad \text{and} \quad (7.94)$$

$$W_\xi\left[\vec{F}\right] = G_\xi^{-1}. \quad (7.95)$$

Tracing carefully the derivation of Equation (7.91), we see that Equation (7.94) is equivalent to

$$F^i \pi(x) = \hat{\xi}^i(x) \pi(x), \quad \forall i, \forall x. \quad (7.96)$$

This equation implies that the $\{F^i\}$ are commutative (i.e., $[F^i, F^j] = 0$), that $\Pi = \{\pi(x)\}$ is a simple measurement (when $\hat{\xi}$ is assumed to be injective with no loss of generality) and that $\{\hat{\xi}^i(x)\}$ are the eigenvalues of $F^i$. On the other hand, as seen from the proof of Theorem 3.12, Equation (7.95) implies that $\{F^i - \xi^i\}$ form a basis of $T_\xi^{(e)}(M)$ at each $\xi$, or more specifically, that $F^i - \xi^i = g^{ij}(\xi)(L_j)_\xi$ where $\{(L_i)_\xi\}$ are the SLDs at $\xi$ defined by Equation (7.59). Consequently, besides the condition given in Theorem 7.4, it turns out that the additional condition

$$[(L_i)_\xi, (L_j)_{\xi'}] = 0, \quad \forall i, j, \forall \xi, \xi', \quad (7.97)$$

is necessary for the equality in Equation (7.92) to hold for all $\xi$. In general, we say that a manifold $M$ of positive definite density operators is **quasiclassical** if its SLDs satisfy Equation (7.97). Now we have:

**Theorem 7.6** *Given a submanifold $M$ of $\mathcal{S}$ with a coordinate system $\xi = [\xi^i]$, the following three conditions are equivalent.*

(i) *There exists an unbiased estimator $(\Pi, \hat{\xi})$ for $(M, \xi)$ satisfying $V_\xi\left[\Pi, \hat{\xi}\right] = G_\xi^{-1}$ for all $\xi$.*

(ii) *$M$ is quasiclassical and e-autoparallel in $\mathcal{S}$, and $\xi$ is an m-affine coordinate system.*

(iii) *There exist mutually commutative observables $\{F_1, \cdots, F_m\}$, a positive definite operator $P$ (which may be taken to be an arbitrary element $\rho_0$ of $M$), a coordinate system $[\theta^i]$ of $M$ and a function $\psi(\theta)$ such that the elements of $M$ are parameterized as*

$$\rho_\theta = \exp\left[\frac{1}{2}\{\sum_{i=1}^m \theta^i F_i - \psi(\theta)\}\right] \cdot P \cdot \exp\left[\frac{1}{2}\{\sum_{i=1}^m \theta^i F_i - \psi(\theta)\}\right] \quad (7.98)$$

*and that $\xi^i = \text{tr}\,(\rho_\theta F_i)$.*

We have written $F_i$ instead of $F^i$ above for the purpose of emphasizing the similarity of Equation (7.98) to Equation (2.31) (i.e. the definition of exponential families). It is also interesting to compare this with Equation (7.45) which is the general form of e-autoparallel submanifolds when the Bogoliubov inner product is chosen to define the e-connection. The statement (i) ⇒ (ii) has already been shown, while (iii) ⇒ (i) can be verified in a similar manner to the proof that an m-affine coordinate system of an exponential family has an efficient estimator (see §3.5). We shall prove (ii) ⇒ (iii) later.

In order to elucidate the meaning of the quasiclassicality, we need some preliminaries. Given a submanifold $M = \{\rho_\xi\}$ of $\mathcal{S}$ and a measurement $\Pi = \{\pi(x)\}$ on $\mathcal{X}$, let $M(\Pi) = \{p(x;\xi)\}$ be the statistical model on $\mathcal{X}$ consisting of the probability distributions $p(x;\xi) = \mathrm{tr}\,[\rho_\xi \pi(x)]$, and let $G_\xi^\Pi = [g_{ij}^\Pi(\xi)]$ be its Fisher information matrix. Assume further that $\Pi$ is a simple measurement; i.e., $\pi(x)^2 = \pi(x)$ and $\pi(x)\pi(y) = 0$ if $x \neq y$. Then letting

$$L_i^\Pi \stackrel{\mathrm{def}}{=} \sum_x \partial_i \log p(x;\xi)\, \pi(x), \quad (7.99)$$

we have

$$\langle\!\langle L_i, \pi(x)\rangle\!\rangle_\xi = \frac{1}{2} \mathrm{tr}\,[(\rho_\xi L_i + L_i \rho_\xi)\pi(x)]$$
$$= \mathrm{tr}\,[(\partial_i \rho_\xi)\pi(x)] = \partial_i p(x;\xi) = \langle\!\langle L_i^\Pi, \pi(x)\rangle\!\rangle_\xi \quad (7.100)$$

and

$$\langle\!\langle L_i^\Pi, L_j^\Pi \rangle\!\rangle_\xi = g_{ij}^\Pi(\xi). \quad (7.101)$$

Equation (7.100) means that $L_i^\Pi$ is the orthogonal projection of $L_i$ onto the linear subspace

$$\mathcal{A}(\Pi) \stackrel{\mathrm{def}}{=} \left\{ \sum_x a(x)\pi(x) \,\Big|\, a: \mathcal{X} \to \mathbb{R} \right\} \subset \mathcal{A}.$$

This, combined with Equation (7.101), leads to

$$G_\xi \geq G_\xi^\Pi. \quad (7.102)$$

This matrix inequality holds for nonsimple measurements, also. Indeed, using the so-called Naimark extension (see e.g. [110]), an arbitrary measurement $\Pi$ can be realized by a simple measurement, say $\tilde{\Pi}$, in a wider Hilbert space, and the inequality for $\Pi$ is reduced to that for $\tilde{\Pi}$. Another proof is as follows. Let $\Pi$ be a measurement on a set $\mathcal{X}$ for which $G_\xi^\Pi$ is strictly positive. Then, applying Equation (2.54) to the model $M(\Pi)$, we can construct a locally unbiased estimator $\hat{\xi}: \mathcal{X} \to \mathbb{R}^m$ at $\xi$ which satisfies $V_\xi\left[\Pi, \hat{\xi}\right] = (G_\xi^\Pi)^{-1}$. This implies that $(G_\xi^\Pi)^{-1} \geq G_\xi^{-1}$ by Theorem 7.5 and that $G_\xi^\Pi \leq G_\xi$. When $G_\xi^\Pi$ is singular, a limiting procedure is applied to show the inequality.

Now it is not difficult to verify the following propositions.

## 7.4. APPLICATIONS TO QUANTUM ESTIMATION THEORY

(I) For each point $\xi$ and each column vector $\boldsymbol{c} = (c^1, \cdots, c^m)^t \in \mathbb{R}^m$, we have
$$\boldsymbol{c}^t G_\xi \boldsymbol{c} = \max_\Pi \boldsymbol{c}^t G_\xi^\Pi \boldsymbol{c}, \qquad (7.103)$$
where the maximum is attained by a simple measurement $\Pi$ satisfying $c^i(L_i)_\xi \in \mathcal{A}(\Pi)$, or in other words, by the spectral decomposition (i.e., the family of orthogonal projections onto the eigenspaces) of $a^i(L_i)_\xi$ or its refinement.[6]

(II) For each point $\xi$, there exists a (simple) measurement $\Pi$ such that $G_\xi = G_\xi^\Pi$ if and only if $[(L_i)_\xi, (L_j)_\xi] = 0$ for all $i, j$.

(III) There exists a (simple) measurement $\Pi$ such that $G_\xi = G_\xi^\Pi$ for all $\xi$ if and only if $M$ is quasiclassical.

Suppose that $M$ is quasiclassical and that a simple measurement $\Pi$ satisfies $G_\xi = G_\xi^\Pi$ for all $\xi$. In other words, we assume that $(L_i)_\xi \in \mathcal{A}(\Pi)$ for all $i$ and all $\xi$. Then from Equations (7.76) and (7.80) we have $\partial_i L_j = \partial_j L_i$ and hence we can define
$$K_\xi \stackrel{\text{def}}{=} \exp\left[\frac{1}{2}\int_0^\xi (L_i)_\xi \, d\xi^i\right] \in \mathcal{A}(\Pi).$$
This satisfies $[K_\xi, K_{\xi'}] = 0$, and we see that
$$\rho_\xi = K_\xi \rho_0 K_\xi, \qquad (7.104)$$
because both sides of the equation obey the same differential equation with the same initial condition at $\xi = 0$. The converse is also true, and we obtain:

(IV) $M$ is quasiclassical if and only if there exist mutually commutative positive Hermitian operators $\{K_\xi\}$ for which Equation (7.104) holds for all $\xi$.

Given a simple measurement $\Pi = \{\pi(x)\}$ on a set $\mathcal{X}$ and a state $\rho_0$ in $\mathcal{S}$, let
$$\mathcal{S}(\rho_0; \Pi) = \left\{K\rho_0 K \,\middle|\, K \in \mathcal{A}(\Pi), K > 0 \text{ and } \operatorname{tr}(\rho_0 K^2) = 1\right\}. \qquad (7.105)$$
This is the maximal quasiclassical manifold which contains $\rho_0$ and whose SLD Fisher information matrix is preserved by $\Pi$. Note that $\mathcal{S}(\rho_0; \Pi) = \mathcal{S}_\Pi$ (see Equation (7.16)) if and only if $\rho_0 \in \mathcal{A}(\Pi)$. We have:

(V) $\mathcal{S}(\rho_0; \Pi)$ is e-autoparallel in $\mathcal{S}$.

(VI) The mapping $\rho \mapsto p$ from $\mathcal{S}(\rho_0; \Pi)$ to $\mathcal{P}(\mathcal{X})$ defined by $p(x) = \operatorname{tr}(\rho \pi(x))$ turns out to be a diffeomorphism, and the inverse mapping $p \mapsto \rho$ is given by $\rho = K_p \rho_0 K_p$ where
$$K_p \stackrel{\text{def}}{=} \sum_x \sqrt{\frac{p(x)}{p_0(x)}}\, \pi(x), \quad p_0(x) \stackrel{\text{def}}{=} \operatorname{tr}(\rho_0 \pi(x)). \qquad (7.106)$$

---

[6] A measurement $\{\nu(y) \mid y \in \mathcal{Y}\}$ is called a refinement of another measurement $\{\pi(x) \mid x \in \mathcal{X}\}$ when there exists a mapping $f : \mathcal{Y} \to \mathcal{X}$ such that $\pi(x) = \sum_{y \in f^{-1}(x)} \nu(y)$ for all $x \in \mathcal{X}$.

Under this diffeomorphism, the dualistic structure on $\mathcal{S}(\rho_0;\Pi)$ coincides with that on $\mathcal{P}(\mathcal{X})$ consisting of the Fisher metric and the e-, m-connections. In particular, $\mathcal{S}(\rho_0;\Pi)$ is dually flat.

Now we see that the condition (ii) in Theorem 7.6 implies that $M$ is e-autoparallel in $\mathcal{S}(\rho_0,\Pi)$ for some $\Pi$ and forms an exponential family when it is transformed into a statistical model by the diffeomorphism given in proposition (VI). This observation leads to a proof of (ii)$\Rightarrow$(iii) in the theorem.

In the present section we have shown some statistical results as applications of the dualistic geometry based on the symmetrized inner product. It should be stressed, however, that by them we have only visited an entrance to the world of quantum statistical inference. Indeed, a main concern of quantum estimation theory lies in treating models with noncommutative SLDs. A common approach to the parameter estimation problem for such a model $M = \{\rho_\xi\}$ is to look for a good lower bound of $\mathrm{tr}\bigl(QV_\xi\bigl[\Pi,\hat{\xi}\bigr]\bigr)$ for a given positive definite real matrix $Q \in \mathbb{R}^{m\times m}$. The right logarithmic derivatives (RLDs) $\{\tilde{L}_i\}$ defined by $\partial_i \rho_\xi = \rho_\xi \tilde{L}_i$ have been introduced for this purpose (Yuen and Lax [227]; see also Helstrom [107] and Holevo [110]). However, the problem of finding an explicit representation of the best bound

$$C_\xi(Q) \stackrel{\mathrm{def}}{=} \min\Bigl\{\mathrm{tr}\bigl(QV_\xi\bigl[\Pi,\hat{\xi}\bigr]\bigr) \,\Big|\, (\Pi,\hat{\xi}) : \text{locally unbiased at } \xi\Bigr\} \qquad (7.107)$$

is a hard one, and has been solved only in a few special models, among which are the quantum Gaussian model ([227, 110]) and the simplest nontrivial case where both the number of parameters $m$ and $\dim \mathcal{H}$ are 2 (Nagaoka [157, 159]). In the latter case we have

$$C_\xi(Q) = \mathrm{tr}\bigl(QG_\xi^{-1}\bigr) + \frac{\sqrt{\det Q}}{\det G_\xi}\,\mathrm{tr\,abs}\bigl(\rho_\xi\,[(L_1)_\xi,(L_2)_\xi]\bigr), \qquad (7.108)$$

where $\mathrm{tr\,abs}\,(A) = |\lambda_1| + |\lambda_2|$ for an operator having the eigenvalues $\{\lambda_1,\lambda_2\}$. See the references cited in the guide to the bibliography for other important results in quantum estimation theory.

# Chapter 8

# Miscellaneous topics

The field of information geometry developed from the investigation of the natural structures inherent in spaces of probability distributions. The structure of Riemannian spaces with dual connections which emerged from this investigation gives insight not only into fields directly related to probability theory, but also into a wide range of fields by providing a framework within which to analyze the underlying structures of the field. It is expected that new theoretical developments will arise from the application of information geometric ideas to these disparate areas. In particular, problems involving convex functions and Legendre transformations have natural geometric structures as dually flat spaces. In this way, it is meaningful to reconsider such fields as information theory and statistical mechanics from the point of view of dualistic geometry. The same can be said for the field of large deviations in probability theory. In this chapter we present several such topics relevant to the development of information geometry and describe the key mathematical problems which remain to be solved.

## 8.1 The geometry of convex analysis, linear programming and gradient flows

Let $M$ be a convex region in $\mathbb{R}^n$. Let $\theta$ denote coordinates in $\mathbb{R}^n$, and suppose we have a smooth convex function $\psi(\theta)$. This setting allows us to trace the argument of §3.3 in a reverse way. First, a Riemannian metric on $M$ is defined by

$$g_{ij}(\theta) = \partial_i \partial_j \psi(\theta). \tag{8.1}$$

If, in addition, we perform the Legendre transformation

$$\eta_i = \partial_i \psi(\theta), \tag{8.2}$$

then this gives a one-to-one mapping from $\theta$ to $\eta$, and it defines a dual convex function

$$\varphi(\eta) = \max_\theta \left[ \theta^i \eta_i - \psi(\theta) \right] = \theta^i(\eta)\eta_i - \psi\{\theta(\eta)\}. \tag{8.3}$$

The inverse transformation is

$$\theta^i = \partial^i \varphi(\eta), \tag{8.4}$$

and the two function $\psi$ and $\varphi$ satisfy the relation

$$\psi(\theta) + \varphi(\eta) - \theta^i \eta_i = 0. \tag{8.5}$$

Using these two convex functions $\psi$ and $\varphi$ we may introduce on $M$ an analog of the $\alpha$-connection as follows: letting

$$T_{ijk} = \partial_i \partial_j \partial_k \psi(\theta) \tag{8.6}$$

be defined with respect to the $\theta$-coordinate system, we define the $\alpha$-connection by

$$\Gamma^{(\alpha)}_{ij,k}(\theta) = \frac{1-\alpha}{2} T_{ijk} = [ij;k] - \frac{\alpha}{2} T_{ijk}, \tag{8.7}$$

where $[ij;k]$ denotes the coefficients of the Riemannian connection given in Equation (1.69). Then the triple $(g, \nabla^{(\alpha)}, \nabla^{(-\alpha)})$ forms a dualistic structure. In particular, this space is flat for $\alpha = \pm 1$, and has $\theta$ and $\eta$ as its affine coordinate systems. In addition, the canonical divergence between the two points $P$ and $Q$ is given by

$$\begin{aligned} D(P \parallel Q) &= \psi(P) + \varphi(Q) - \theta^i_P \eta_{Qi} \\ &= \psi(P) - \psi(Q) + (\theta^i_Q - \theta^i_P) \partial_i \psi(Q). \end{aligned} \tag{8.8}$$

Now, if the point $P_0$ minimizing $\psi(P)$ is contained in $M$, then its $\eta$-coordinates are

$$\eta_{0i} = \partial_i \psi(P_0) = 0. \tag{8.9}$$

In addition, if $\psi$ is normalized so that $\psi(P_0) = 0$, then

$$D(P \parallel P_0) = \psi(P). \tag{8.10}$$

It is important to realize that underlying all such problems involving convexity and Legendre transformations is a fundamental structure described by geometry of dual connections.

The problem of finding a point $\theta$ which minimizes the linear function

$$V(\theta) = c_i \theta^i \tag{8.11}$$

within a closed convex region $\bar{M}$ appears frequently in numerical programming. Let us consider this problem in our framework. Suppose that a convex region $M$ in $\mathbb{R}^n$ is given in terms of a piecewise smooth convex function $f(\theta)$ as

$$M = \{\theta \mid f(\theta) > 0\}.$$

Letting $\omega$ denote the $(n-1)$-dimensional position vector on the boundary $\partial M$ of the region $M$, define the function $\psi$ on $M$ by

$$\psi(\theta) = -\int_{\partial M} \log \left[ \sum_{i=1}^n \partial_i f(\theta(\omega)) \{\theta^i - \theta^i(\omega)\} \right] d\omega. \tag{8.12}$$

## 8.1. THE GEOMETRY OF CONVEX ANALYSIS

It then turns out that $\psi$ is a convex function on $M$, from which the dually flat structure is introduced on $M$. In particular, when $M$ is a region bounded by $m$ hyper-planes so that the definition of $M$ may be written as

$$M = \left\{ \theta \;\Big|\; \sum_{i=1}^{n} A_i^{\mu} \theta^i - b^{\mu} > 0, \; \mu = 1, \cdots, m \right\}, \tag{8.13}$$

then the convex function above may be rewritten in the form

$$\psi(\theta) = -\sum_{\mu=1}^{m} W_{\mu} \log \left( \sum_{i=1}^{n} A_i^{\mu} \theta^i - b^{\mu} \right), \tag{8.14}$$

where $W_{\mu}$ are some positive constants. The problem of minimizing a linear function $\sum c_i \theta^i$ on $M$ is called a **linear programming** problem.

Now consider the **gradient flow**

$$\dot{\theta}^i = -g^{ij}(\theta) \partial_j V(\theta) = -g^{ij}(\theta) c_j, \tag{8.15}$$

where $\dot{\theta}^i = \frac{\mathrm{d}}{\mathrm{d}t} \theta^i$, which moves a point $\theta$ in the opposite direction to the gradient (see Equation (2.45)) of $V(\theta)$ and hence decreases its value. Indeed, if a curve $\theta(t)$ obeys the above differential equation, we have

$$\frac{\mathrm{d}}{\mathrm{d}t} V(\theta(t)) = -g^{ij}(\theta(t)) c_i c_j \leq 0.$$

The use of a discrete solution to this differential equation in finding the minimum of $V$ is one variant of the Karmarkar's interior point method, and is called the method of affine projections. Rewriting this in terms of the dual coordinates we have

$$\dot{\eta}_i = -c_i, \tag{8.16}$$

and hence its trajectory is a $\nabla^{(-1)}$-geodesic. In other words, the solution to the interior point method follows a $\nabla^{(-1)}$-geodesic.

Other important examples of gradient flows on the dually flat space are derived from functions of the form

$$U(\theta) = \psi(\theta) + c_i \theta^i. \tag{8.17}$$

The gradient flow of $U$ is represented as

$$\dot{\theta}^i = -g^{ij}(\theta) \partial_j U(\theta) = -g^{ij}(\theta) \{ \eta_j(\theta) + c_j \}, \tag{8.18}$$

or equivalently as

$$\dot{\eta}_i = -(\eta_i + c_i), \tag{8.19}$$

whose trajectory is a $\nabla^{(-1)}$-geodesic, too. In the particular case when $c_i = -\eta_{Qi}$ for a given point $Q$, the gradient flow minimizes the divergence $D(\theta \parallel Q)$ along $\eta_i(t) = \eta_{Qi} + (\eta_i(0) - \eta_{Qi}) e^{-t}$. In addition, we can see from the Pythagorean relation for $D$ that, when the range of $\theta$ is restricted to a $\nabla^{(1)}$-autoparallel submanifold $M$, the gradient flow of $D\left(\theta \parallel Q\right)$ streams into the $\nabla^{(-1)}$-projection of $Q$ onto $M$ along a $\nabla_M^{(-1)}$-geodesic in $M$, where $\nabla_M^{(-1)}$ is the projection of $\nabla^{(-1)}$ onto $M$ with respect to $g$ (see §1.9). This type of gradient flow is often useful in computing the maximum likelihood estimate[1] for an exponential family (e.g. Amari et al. [30]).

There is an attempt to view such gradient flows as **completely integrable dynamical systems**. See Nakamura [163, 164] and Fujiwara and Amari [91]. This relates to other problems such as the QR decomposition of matrices, and the attempt to solve combinatorial problems dynamically by embedding a representation of the problem in a continuous domain (Brockett [57, 58]). Such attempts have the potential to develop and unify completely integrable dynamical systems and the geometry of dual connections.

## 8.2 Neuro-manifolds and nonlinear systems

Consider a nonlinear system which transforms an input signal $x$ into an output signal $y$, and suppose that this system is parameterizable by the finite-dimensional parameter $\xi$. Then the input-output relation may be written as

$$y = f(x; \xi). \tag{8.20}$$

If the system is stochastic, or if there is noise in the system, then the input-output relation is stochastic, and the probability distribution on the output $y$ for an input $x$ may be written as the conditional probability

$$p(y \mid x; \xi). \tag{8.21}$$

If, in addition, the probability distribution of the input signal is given by $q(x)$, then the distribution of the entire system may be written in terms of $\xi$ as $q(x)p(y \mid x; \xi)$. Hence it is possible to investigate using the techniques of information geometry the structure of the manifold created by this nonlinear system.

A typical example of such a nonlinear system is the multilayer perceptron consisting of stochastic neurons, where $\xi$ denotes the parameters of modifiable weights of connection. The space of all such perceptrons is called the neuro-manifold of perceptrons. It is equipped with the Riemannian metric given by the Fisher information.

Another example is a recurrently connected neural network model called the **Boltzmann machine** (Ackley et al. [2]). This is a fully connected network of $n$ stochastic neurons. Each stochastic neuron $N_i$ $(i = 1, \cdots, n)$ has the state $x_i$ which is either 0 or 1, and this state is communicated to the other neurons as the output of this neuron. The neuron $N_i$ takes the outputs $x_j$ of the other neurons $N_j$ $(j \neq i)$ and computes the linear combination

$$u_i = \sum_{j \neq i} w_{ij} x_j - h_i. \tag{8.22}$$

---

[1]Note that the divergence $D$ in the present context turns out $D^{(e)}$, the dual of the Kullback divergence $D^{(m)}$, when $\nabla^{(1)}$ and $\nabla^{(-1)}$ are the e- and m-connections, respectively.

## 8.2. NEURO-MANIFOLDS AND NONLINEAR SYSTEMS

Here $w_{ij}$ is a measure of the strength of the influence which neuron $N_j$ has on neuron $N_i$, and is called the weight of the synaptic connection, while $h_i$ is called the threshold of $N_i$. Let us assume that $w_{ij} = w_{ji}$ and that $w_{ii} = 0$. Each neuron $N_i$ stochastically determines whether in the next time step it will be in the excited state of $x_i = 1$ or the dormant state of $x_i = 0$ using the probability

$$\Pr\{x_i = 1\} = \frac{\exp\{u_i\}}{1 + \exp\{u_i\}}. \tag{8.23}$$

Let us also assume that at each time step only one neuron updates its state.

Let $\boldsymbol{x} = (x_1, \cdots, x_n)$ represent the state of the network. The state changes stochastically through the interaction of the neurons. Clearly, this forms a Markov chain on the $2^n$ element state space $X = \{\boldsymbol{x}\}$. The stationary distribution of this chain is easily computed to be

$$\begin{aligned} p(\boldsymbol{x}) &= Z^{-1}\exp\{-E(\boldsymbol{x})\} \quad \text{where} & (8.24)\\ E(\boldsymbol{x}) &= -\frac{1}{2}\sum w_{ij}x_ix_j + \sum h_ix_i, & (8.25) \end{aligned}$$

where $Z \stackrel{\text{def}}{=} \sum_x \exp\{-E(\boldsymbol{x})\}$.

Let us consider the Boltzmann machine as a device which generates its state according to the stationary distribution given in Equation (8.24). Generalizing, we may divide the state into observable neurons and hidden ones, and given a particular input, consider the queried states as behaving stochastically according to a conditional distribution.

The set of all stationary probability distribution corresponding to Boltzmann machines forms an exponential family $M$ parameterized by $\theta = (w_{ij}, h_i)$. Since a given stationary distribution corresponds to a particular Boltzmann machine, $M$ may be considered to be the manifold of all Boltzmann machines, where it is now possible to introduce the geometric structure of metric and dual connections (Amari et al. [30]). Let $S$ be the manifold of all probability distributions on the state space $X = \{\boldsymbol{x}\}$. This forms a $(2^n - 1)$-dimensional dually flat space with respect to the e-, m-connections, and $M$ is a $\frac{1}{2}n(n+1)$-dimensional e-autoparallel submanifold embedded in this space. $M$ itself is also a dually flat subspace.

The problem of approximating the given distribution $q(\boldsymbol{x})$ of $S$ using $M$, and further, to do so adaptively by learning the parameter $\xi$ is a topic discussed within the field of neural networks. An elegant solution is provided by geometry of dual connections. However, when hidden units are included, the space of distributions is no longer dually flat. In this case the EM algorithm (Dempster et al. [79]) from statistics may be usefully applied, and in addition, it is possible to provide this algorithm with the differential geometrical foundations. See Amari [16] and Byrne [63]. An interesting topic for future development is the analysis of the dynamics of networks not in equilibrium. Yet another would be the analysis of composite neural networks consisting of local expert networks. See the guide to the bibliography at the end of this book for several recent topics related to neural networks and information geometry, including independent component analysis (ICA), the natural gradient method and learning around singularities.

## 8.3 Lie groups and transformation models in information geometry

A Lie group is a manifold with an algebraic structure, and it has a long history of careful and extensive study. However, here too there exists the structure of dual connections, and from this point of view new possibilities for development may be considered. This viewpoint allows an analysis, for example, of Lie groups as completely integrable dynamical systems, and also of the differential equations of the Lax type which have attracted attention from many areas of application. In this section, we discuss a family of probability distributions which forms a group.

We begin by discussing a family of probability distributions whose structure admits characterization as a Lie group. Let $S = \{p(x;\xi) \mid \xi \in \Xi\}$ be a family of distributions for a random variable $x$ parameterized by $\xi$, and suppose that the parameter space $\Xi$ forms a Lie group with $\xi = [\xi^i]$ as its local coordinate system. We denote the group operation by $(\xi, \xi') \mapsto \xi \cdot \xi'$ and the identity element of $\Xi$ by $e$. In addition, suppose that $\Xi$ acts on the range $\mathcal{X}$ of the random variable $x$ from the left, and let us denote the action of $\xi$ on $x$ by $\xi \circ x$. When for all $\xi, \xi' \in \Xi$ the probability distributions in $S$ satisfies

$$p(x;\xi')\,\mathrm{d}x = p(\xi \circ x\,;\,\xi \cdot \xi')\,\mathrm{d}(\xi \circ x) \qquad (8.26)$$

or equivalently

$$\int_A p(x;\xi')\,\mathrm{d}x = \int_{\xi \circ A} p(x\,;\,\xi \cdot \xi')\,\mathrm{d}x \qquad (8.27)$$

for any measurable set $A \subset \mathcal{X}$, we say that $S$ is a transformation model which admits the group structure of $\Xi$. Then since

$$p(x;\xi) = p(k(\xi,x)\,;\,e)\left|\frac{\partial k(\xi,x)}{\partial x}\right|, \qquad (8.28)$$

where

$$k(\xi,x) \stackrel{\mathrm{def}}{=} \xi^{-1} \circ x, \qquad (8.29)$$

the distribution is completely determined by the distribution at the identity element and the group structure. Let us give an example.

**Example 8.1 (Location-scale model)** *Let $\xi = (\mu, \sigma)$ with $\sigma > 0$, and consider the affine transformation on the real line $\mathcal{X} = \mathbb{R}$ defined by*

$$\xi \circ x = \sigma x + \mu \quad \text{and} \quad k(\xi,x) = \xi^{-1} \circ x = \frac{x-\mu}{\sigma}. \qquad (8.30)$$

## 8.3. LIE GROUPS AND TRANSFORMATION MODELS

The operator $\xi$ translates a probability distribution by $\mu$ and scales its variance by a factor of $\sigma$. The range $\Xi$ of $\xi$ forms a Lie group acting transitively on $\mathbb{R}$ with the operations

$$(\mu_1, \sigma_1) \cdot (\mu_2, \sigma_2) = (\mu_1 + \mu_2 \sigma, \, \sigma_1 \sigma_2) \tag{8.31}$$

$$(\mu, \sigma)^{-1} = (-\frac{\mu}{\sigma}, \frac{1}{\sigma}) \tag{8.32}$$

and the identity element $e = (0, 1)$. Note that this group has a matrix representation:

$$(\mu, \sigma) \mapsto \begin{pmatrix} \sigma & \mu \\ 0 & 1 \end{pmatrix}.$$

A distribution family $\{p(x; \xi) \mid \xi \in \Xi\}$ which admits this group structure is called a *location-scale model*, and letting

$$p(x; e) = f(x) \tag{8.33}$$

we have

$$p(x; \xi) = \frac{1}{\sigma} f\left(\frac{x - \mu}{\sigma}\right). \tag{8.34}$$

Next, let $\boldsymbol{x} = (x_1, \ldots, x_N) \in \mathcal{X}_N \stackrel{\text{def}}{=} \mathbb{R}^N$ and suppose that $x_1, \ldots, x_N$ are independently distributed according to the same distribution. Then we have

$$p(\boldsymbol{x}; \xi) = \frac{1}{\sigma^N} f\left(\frac{x_1 - \mu}{\sigma}\right) \cdots f\left(\frac{x_N - \mu}{\sigma}\right). \tag{8.35}$$

In this case, the group $\Xi$ acts on $\mathbb{R}^N$ by

$$k(\xi, \boldsymbol{x}) = \xi^{-1} \circ \boldsymbol{x} = \left(\frac{x_1 - \mu}{\sigma}, \ldots, \frac{x_N - \mu}{\sigma}\right). \tag{8.36}$$

When $N \geq 2$, the isotropy group $\{\xi \mid \xi \circ \boldsymbol{x} = \boldsymbol{x}\}$ of each point $\boldsymbol{x} = (x_1, \ldots, x_N)$ consists only of the identity $e$ unless $x_1 = \cdots = x_N$, which means that $\Xi$ acts freely on $\mathcal{X}_N^\dagger \stackrel{\text{def}}{=} \mathbb{R}^N \setminus \{(x, \ldots, x) \mid x \in \mathbb{R}\}$.

Let us now investigate the natural geometric structures possessed by such statistical models which admit group structure. In particular, it would be interesting to see the relationship between the geometric structure and the Lie group structure, and also between the geometric structure and the distribution $f(x) = p(x; e)$ at the identity element.

Let $S = \{p(x; \xi) \mid \xi \in \Xi\}$ be a statistical model which admits the group structure of $\Xi$, and consider the Fisher metric $g$ and the $\alpha$-connection $\nabla^{(\alpha)}$ on $S$. Then these geometric quantities are uniquely determined by their values at the origin $e$; we state this more precisely in the theorem below.

**Theorem 8.1** *The components of $g$ and $\nabla^{(\alpha)}$ satisfy*

$$g_{ij}(\xi) = B_i^\ell(\xi) B_j^m(\xi) g_{\ell m}(e) \quad \text{and} \tag{8.37}$$

$$\Gamma_{ij,k}^{(\alpha)}(\xi) = B_i^\ell(\xi) B_j^m(\xi) B_k^s(\xi) \Gamma_{\ell m,s}^{(\alpha)}(e)$$
$$+ C_{ij}^\ell(\xi) B_k^m(\xi) g_{\ell m}(e), \tag{8.38}$$

where

$$B_i^\ell(\xi) = \left. \frac{\partial}{\partial \xi'^i} (\xi^{-1} \cdot \xi')^\ell \right|_{\xi'=\xi} \quad \text{and} \tag{8.39}$$

$$C_{ij}^\ell(\xi) = \left. \frac{\partial^2}{\partial \xi'^i \partial \xi'^j} (\xi^{-1} \cdot \xi')^\ell \right|_{\xi'=\xi}. \tag{8.40}$$

*In other words, $g$ and $\nabla^{(\alpha)}$ are invariant under the left action of the group $\Xi$ on itself.*

On the other hand, the Fisher metric and the $\alpha$-connections at the origin are determined by both $k(\xi, x)$, which represents the action of the group $\Xi$ on $\mathcal{X}$, and also the shape of $f(x)$. These may be computed by introducing a coordinate system $[x^a]$ of $\mathcal{X}$ and using

$$\partial_i \log p(x; e) = r_i^a(x) \partial_a \log f(x) + \partial_a r_i^a(x) \tag{8.41}$$

and

$$\partial_i \partial_j \log p(x; e) = r_i^a(x) r_j^b(x) \partial_a \partial_b \log f(x) + s_{ij}^a(x) \partial_a \log f(x)$$
$$- \{\partial_a r_i^a(x)\} \{\partial_b r_j^b(x)\} + \partial_a s_{ij}^a(x), \tag{8.42}$$

where

$$r_i^a(x) = \left. \frac{\partial}{\partial \xi^i} k^a(\xi, x) \right|_{\xi=e} \quad \text{and} \tag{8.43}$$

$$s_{ij}^a(x) = \left. \frac{\partial^2}{\partial \xi^i \partial \xi^j} k^a(\xi, x) \right|_{\xi=e}. \tag{8.44}$$

From the invariance shown in Theorem 8.1, we see that the scalar curvature of the Fisher metric is constant on $S$. In particular, every 2-dimensional statistical model which admits a group structure, a location-scale model for instance, turns out to be a space of constant curvature (see §8.4) with respect to (the Riemannian connection of) the Fisher metric. The corresponding result in the case of normal distributions was found by Amari in 1959 by means of a direct calculation.

We also note that an existence condition for (exact) ancillary statistic (*i.e.* statistic whose distribution does not depend on the parameter $\xi$) is related to the group invariance of a statistical model. To illustrate this, go back to the previous example and let

$$F(\boldsymbol{x}) = \left(\frac{x_1 - \hat{\mu}}{\hat{\sigma}}, \ldots, \frac{x_N - \hat{\mu}}{\hat{\sigma}}\right),$$

where

$$\hat{\mu}(\boldsymbol{x}) = \frac{1}{N}\sum_{t=1}^{N} x_t \quad \text{and} \quad \hat{\sigma}(\boldsymbol{x}) = \left(\frac{1}{N}\sum_{t=1}^{N}(x_t - \hat{\mu})^2\right)^{\frac{1}{2}}.$$

Then $F : \mathcal{X}_N^\dagger \to \mathbb{R}^N$ is invariant under the group action and is an ancillary statistic for every location-scale model. The orbits $\Xi \circ \boldsymbol{x} = \{\xi \circ \boldsymbol{x} \mid \xi \in \Xi\}$ for $\boldsymbol{x} \in \mathcal{X}_N^\dagger$ are completely indexed by $F(\boldsymbol{x})$; i.e. $\Xi \circ \boldsymbol{x} = \Xi \circ \boldsymbol{y}$ iff $F(\boldsymbol{x}) = F(\boldsymbol{y})$. On the other hand, the position of $\boldsymbol{x}$ in the orbit $\Xi \circ \boldsymbol{x}$ is specified by the statistic $G(\boldsymbol{x}) \stackrel{\text{def}}{=} (\hat{\mu}(\boldsymbol{x}), \hat{\sigma}(\boldsymbol{x}))$, which is equivariant in the sense that $G(\boldsymbol{x}) = G(\boldsymbol{y})$ implies $G(\xi \circ \boldsymbol{x}) = G(\xi \circ \boldsymbol{y})$. See Barndorff-Nielsen et al. [42] for the detail.

## 8.4 Mathematical problems posed by information geometry

The differential geometry of dual connections arose from the consideration of the natural geometric structure of a family of probability distributions. However, dual connections also naturally emerge from affine differential geometry, in particular the study of hypersurfaces in an affine space as begun by Blaschke [49]. Recently Nomizu, Kurose, etc. have been researching this topic, and the interest among mathematicians for the field of dualistic differential geometry continues to deepen. We briefly touch on this topic.

We first observe how a pair of dual connections arises in the framework of affine geometry, following Nomizu and Simon [168]. Suppose that an $n$-dimensional manifold $M$ is mapped into the $(n+1)$-dimensional vector space $\mathbb{R}^{n+1}$:

$$f : M \to \mathbb{R}^{n+1}. \tag{8.45}$$

Here we assume that $f$ is an immersion and that the image $f(M)$ forms a hypersurface within $\mathbb{R}^{n+1}$. Take a (local) coordinate system $[\xi^i]$ of $M$ and let $\boldsymbol{e}_i = \partial_i$ be its natural basis. Then each $\boldsymbol{e}_i$ is naturally mapped to the vector $\tilde{\boldsymbol{e}}_i = \partial_i f$ in the tangent space of the hypersurface by $f$, or more precisely, by the induced mapping $f_* = \mathrm{d}f$ from the tangent spaces of $M$ to the tangent spaces of $f(M)$ in $\mathbb{R}^{n+1}$:

$$(f_*)_p : T_p(M) \to T_{f(p)}(f(M)) \subset \mathbb{R}^{n+1}. \tag{8.46}$$

Now suppose in addition that we are given for each point $p$ of $M$ a nonzero vector $\boldsymbol{n}_p \in \mathbb{R}^{n+1}$ which is not contained in the tangent space $T_{f(p)}(f(M))$. Then $\boldsymbol{n} : p \mapsto \boldsymbol{n}_p$ defines a mapping from $M$ into $\mathbb{R}^{n+1}$ as well as $f$, and $\{(e_i)_p, \boldsymbol{n}_p\}$ forms a basis of $\mathbb{R}^{n+1}$ at each point $p$. We call such a pair $(f, \boldsymbol{n})$

an **affine immersion** of $M$. Since $\mathbb{R}^{n+1}$ is flat, it allows parallel translation and differentiation. We may express the change in the tangent vector $\tilde{\bm{e}}_j$ as we move the point $f(p)$ in the direction of $\tilde{\bm{e}}_i$ on $f(M)$ by the partial derivative $\partial_i \tilde{\bm{e}}_j = \partial_i \partial_j f$. Expanding this vector with respect to the basis $\{\tilde{\bm{e}}_i, \bm{n}\}$ we obtain

$$\partial_i \tilde{\bm{e}}_j = \Gamma_{ij}^k \tilde{\bm{e}}_k + g_{ij} \bm{n}. \tag{8.47}$$

Now let us consider the components $\{\Gamma_{ij}^k\}$, which ignore the change in the direction of $\bm{n}$, as determining the coefficients of an affine connection on the original space $M$. Then this introduces the covariant derivative

$$\nabla_{\bm{e}_i} \bm{e}_j = \Gamma_{ij}^k \bm{e}_k \tag{8.48}$$

on $M$. Note that this connection is necessarily symmetric (torsion-free). In addition, let us introduce the metric defined by

$$\langle \bm{e}_i, \bm{e}_j \rangle = g_{ij}. \tag{8.49}$$

Although this metric $g_{ij}$ is symmetric, it is not in general positive definite. Here we only assume that the matrix $[g_{ij}]$ is nonsingular everywhere on $M$. Reflecting that the decomposition $\mathbb{R}^{n+1} = T_{f(p)}(f(M)) \oplus \text{span}(\bm{n}_p)$ induces a projection from $\mathbb{R}^{n+1}$ onto the tangent space $T_{f(p)}(f(M))$ at each point $p$, we may observe that $\nabla$ is a projection of the natural connection of $\mathbb{R}^{n+1}$ and $g$ is the corresponding embedding curvature (see §1.9). This $g$ is also called the **second fundamental form** or the **affine fundamental form** of the immersion.

Let us consider the dual connection $\nabla^*$ of $\nabla$ with respect to $g$, which is uniquely defined by Equation (3.1) as in the case of positive metric. When the vectors $\{\partial_i \bm{n}, i = 1, \cdots, n\}$ consist only of the components in the tangent direction of $f(M)$, or in other words, when they are written using $n^2$ functions $\left\{s_i^j\right\}$ as

$$\partial_i \bm{n} = s_i^j \tilde{\bm{e}}_j, \tag{8.50}$$

we say that the affine immersion $(f, \bm{n})$ is **equiaffine**. Then it may be seen that this property is necessary and sufficient for the dual connection $\nabla^*$ to be symmetric. Noting that the components of the dual connection is represented as

$$\Gamma_{ij,k}^* = \Gamma_{ij,k} + T_{ijk}, \tag{8.51}$$

where $T$ is the covariant tensor of degree 3 defined by

$$T_{ijk} = (\nabla_{\bm{e}_i} g)_{jk} = \partial_i g_{jk} - \Gamma_{ij,k} - \Gamma_{ik,j}, \tag{8.52}$$

we see that the equiaffine property is also equivalent to the symmetry of $T$.

When $(f, \bm{n})$ is equiaffine, the dual connection may also be obtained in a similar manner to Equation (8.47) within the dual linear space $(\mathbb{R}^{n+1})^*$ of $\mathbb{R}^{n+1}$. An element $\bm{y}^*$ of $(\mathbb{R}^{n+1})^*$ is an $\mathbb{R}$-valued linear function on $\mathbb{R}^{n+1}$, and we denote

## 8.4. PROBLEMS POSED BY INFORMATION GEOMETRY

the value $\boldsymbol{y}^*(\boldsymbol{x})$ of the function for the argument $\boldsymbol{x} \in \mathbb{R}^{n+1}$ by $\langle \boldsymbol{x}, \boldsymbol{y}^* \rangle$. Now for each point on $M$ we may choose an element $\boldsymbol{n}^*$ in $(\mathbb{R}^{n+1})^*$ which satisfy

$$\langle \boldsymbol{n}, \boldsymbol{n}^* \rangle = 1 \quad \text{and} \tag{8.53}$$
$$\langle \tilde{\boldsymbol{e}}_i, \boldsymbol{n}^* \rangle = 0, \quad \forall i = 1, \cdots, n. \tag{8.54}$$

We call this $\boldsymbol{n}^* : M \to (\mathbb{R}^{n+1})^*$ a conormal vector field. Then $\{\partial_i \boldsymbol{n}^*, \boldsymbol{n}^*\}$ constitute a basis of $(\mathbb{R}^{n+1})^*$, and the dual connection may be obtained by the expansion

$$\partial_j \partial_i \boldsymbol{n}^* = \Gamma_{ji}^{*k} \partial_k \boldsymbol{n}^* + s_{ij} \boldsymbol{n}^*. \tag{8.55}$$

Note, in addition, that the coefficients $s_{ij}$ are related to $s_i^j$ in Equation (8.50) by $s_{ij} = s_i^k g_{kj}$.

Lauritzen [138] in a more intuitive manner, takes two mutually dual linear spaces and investigates the fundamental cause of the dualistic structure. Let $M$ be an $n$-dimensional manifold, $V$ an $m$-dimensional $(m > n)$ linear space, and $V^*$ its dual. We denote $\boldsymbol{y}^*(\boldsymbol{x})$ by $\langle \boldsymbol{x}, \boldsymbol{y}^* \rangle$ for $\boldsymbol{x} \in V$ and $\boldsymbol{y}^* \in V^*$ as above. Now suppose that we are given immersions of $M$ into $V$ and $V^*$:

$$f : M \to V \tag{8.56}$$
$$k : M \to V^*. \tag{8.57}$$

Then $f(M)$ and $k(M)$ form $n$-dimensional curved subspaces within the $m$-dimensional spaces $V$ and $V^*$, respectively. In addition, the tangent space $T_p(M)$ of $M$ maps naturally to the tangent spaces of the images $f(M)$ in $V$ and $k(M)$ in $V^*$.

Consider a point $p$ in $M$. Now map the basis $\boldsymbol{e}_i = \partial_i$ of $T_p(M)$ to the tangent spaces of $f(M)$ and $k(M)$ by the induced mappings $f_*$ and $k_*$, and let $\tilde{\boldsymbol{e}}_i = \partial_i f$ and $\tilde{\boldsymbol{e}}_i^* = \partial_i k$ denote the results, respectively. Let

$$g_{ij} \stackrel{\text{def}}{=} \langle \tilde{\boldsymbol{e}}_i, \tilde{\boldsymbol{e}}_j^* \rangle, \tag{8.58}$$

and assume that $g_{ij} = g_{ji}$. Then an indefinite inner product on $M$ is defined by $\langle \boldsymbol{e}_i, \boldsymbol{e}_j \rangle = g_{ij}$. In addition, let us define two connections by

$$\Gamma_{ij,k} = \langle \nabla_{\boldsymbol{e}_i} \boldsymbol{e}_j, \boldsymbol{e}_k \rangle = \langle \partial_i \tilde{\boldsymbol{e}}_j, \tilde{\boldsymbol{e}}_k^* \rangle \quad \text{and} \tag{8.59}$$
$$\Gamma_{ij,k}^* = \langle \boldsymbol{e}_k, \nabla_{\boldsymbol{e}_i}^* \boldsymbol{e}_j \rangle = \langle \tilde{\boldsymbol{e}}_k, \partial_i \tilde{\boldsymbol{e}}_j^* \rangle. \tag{8.60}$$

Then these provide $M$ with the dualistic structure $(g, \nabla, \nabla^*)$.

Now let us consider the case when $M = \{p(x; \xi)\}$ is a family of probability distributions on a set $\mathcal{X}$. Let $V$ be a linear space formed by functions on $\mathcal{X}$ satisfying certain regularity conditions, and $V^*$ be the set of signed measures on $\mathcal{X}$. Note that these spaces are infinite dimensional unless $\mathcal{X}$ is a finite set. The inner product of an element $a$ of $V$ and an element $\mu$ of $V^*$ is given by the integral

$$\langle a, \mu \rangle = \int a(x) \mathrm{d}\mu(x). \tag{8.61}$$

We see then that the geometry of the e-, m-connections and the Fisher metric on $M$ is obtained by the mappings

$$f : \xi \mapsto \log p(x, \xi) \qquad (8.62)$$
$$k : \xi \mapsto p(x, \xi) \mathrm{d}x. \qquad (8.63)$$

The dualistic geometry of general $\pm \alpha$-connections may also be seen in this framework by Equations (2.60) and (2.61).

It is known that an arbitrary Riemannian space may be realized as a submanifold of a Euclidean space of sufficiently large dimension. Since this is the case, it is natural to ask whether there is a space into which an $n$-dimensional space equipped with a dualistic structure may be embedded. Although hypersurfaces in $\mathbb{R}^{n+1}$ admit dualistic structures through Equation (8.47), it is not the case that all spaces with dualistic structures be realized in this manner. Kurose [133, 134] and Dillen, Nomizu and Vrancken [80] have studied this problem of realization, and have obtained the following results.

Let $M$ be a simply connected $n(\geq 2)$-dimensional manifold on which a dualistic structure $(g, \nabla, \nabla^*)$ is given, and assume that both $\nabla$ and $\nabla^*$ are symmetric. We say that $(M, g, \nabla)$ is **$\alpha$-conformally flat** if there exist a coordinate system $[\xi^i]$ and a function $\varphi$ on $M$ such that

$$\Gamma_{ij,k} = \frac{1+\alpha}{2}(\partial_k \varphi)g_{ij} - \frac{1-\alpha}{2}\{(\partial_i \varphi)g_{jk} + (\partial_j \varphi)g_{ik}\}. \qquad (8.64)$$

In particular, the $(-1)$-conformally flatness is a property for the connection $\nabla$ alone and is independent of $g$. Note that $(M, g, \nabla)$ is $\alpha$-conformally flat if and only if $(M, g, \nabla^*)$ is $(-\alpha)$-conformally flat. Next, we say that $(M, g, \nabla)$ has **constant curvature** $c$ if the curvature tensor $R$ of $\nabla$ satisfies

$$R(X, Y)Z = c\{\langle Y, Z \rangle X - \langle X, Z \rangle Y\} \qquad (8.65)$$

for any vector fields $X, Y$ and $Z$. When $\nabla$ is the Riemannian connection of $g$, this definition turns out equivalent to the classical one that the Riemannian manifold $(M, g)$ has a constant sectional curvature (see, for instance, Chap.V of Kobayashi and Nomizu [122], vol.I). Obviously $(M, g, \nabla, \nabla^*)$ is dually flat if and only if $(M, g, \nabla)$ (or $(M, g, \nabla^*)$ equivalently) has constant curvature 0. In general, $(M, g, \nabla)$ has constant curvature $c$ if and only if so does $(M, g, \nabla^*)$.

**Theorem 8.2** *([134], [80]) The following conditions are mutually equivalent.*

(i) $(g, \nabla)$ *can be realized by an affine immersion* $(f, \boldsymbol{n})$ *of $M$ into $\mathbb{R}^{n+1}$ through Equation (8.47).*

(ii) $(M, g, \nabla)$ *is 1-conformally flat.*

(iii) $(M, \nabla^*)$ *is $(-1)$-conformally flat.*

**Theorem 8.3** *([133]) The following conditions are mutually equivalent. ($n \geq 3$ is assumed in (i) (ii) $\Rightarrow$ (iii).)*

## 8.4. PROBLEMS POSED BY INFORMATION GEOMETRY

(i) Both $(g, \nabla)$ and $(g, \nabla^*)$ can be realized by affine immersions.

(ii) $(M, g, \nabla)$ (or equivalently $(M, g, \nabla^*)$) is $(\pm 1)$-conformally flat.

(iii) $(M, g, \nabla)$ (or equivalently $(M, g, \nabla^*)$) has constant curvature $c$ for some $c \in \mathbb{R}$.

Kurose [134] has also shown that a divergence $D$ is canonically defined on a space of constant curvature $c$ and that it satisfies the following modification of the Pythagorean relation; given three points $p, q, r$ in $M$, if the $\nabla$-geodesic connecting $p$ and $q$ and the $\nabla^*$-geodesic connecting $q$ and $r$ are orthogonal at $q$, then

$$D(p \| r) = D(p \| q) + D(q \| r) - cD(p \| q) D(q \| r). \tag{8.66}$$

When $c \neq 0$, this may be rewritten as

$$(1 - cD(p \| r)) = (1 - cD(p \| q))(1 - cD(q \| r)), \tag{8.67}$$

which indicates a structure similar to "spherical geometry."

These results are closely related to the discussion in §2.6 and §3.6. Suppose that $M = \{p_\xi\}$ is an $n$-dimensional $\alpha$-family for $\alpha \neq 1$ which is represented as Equation (2.75):

$$\ell^{(\alpha)}(x; \xi) = \sum_{\lambda=0}^{n} \theta^\lambda(\xi) F_\lambda(x),$$

and let $\eta_\lambda(\xi) \stackrel{\text{def}}{=} \int F_\lambda(x) \ell^{(-\alpha)}(x; \xi) \, dx$. Then we have

$$\partial_i \partial_j \theta^\lambda = \Gamma_{ij}^{(\alpha)k} \partial_k \theta^\lambda - \frac{1-\alpha^2}{4} g_{ij} \theta^\lambda, \quad \text{and} \tag{8.68}$$

$$\partial_i \partial_j \eta_\lambda = \Gamma_{ij}^{(-\alpha)k} \partial_k \eta_\lambda - \frac{1-\alpha^2}{4} g_{ij} \eta_\lambda, \tag{8.69}$$

where $\partial_i = \frac{\partial}{\partial \xi^i}$. These equations imply that $(M, g, \nabla^{(\pm \alpha)})$ satisfies the condition (i) in Theorem 8.3 and hence has constant curvature $c$ by (i)$\Rightarrow$(iii), while another observation leads to $c = \frac{1-\alpha^2}{4}$. The corresponding Kurose's divergence turns out to be the $\alpha$-divergence; see Theorem 3.16 in §3.6.

We have seen that the realizability of $(g, \nabla)$ by an affine immersion of codimension 1 does not imply that of $(g, \nabla^*)$. This "breakdown" of duality is also indicated in Equation (8.55) where the second fundamental form $s_{ij}$ is different from $g_{ij}$. Recently Matsuzoe [149] has studied immersions of codimension 2 of the form

$$\partial_i \tilde{e}_j = \Gamma_{ij}^k \tilde{e}_k + t_{ij} f + g_{ij} n, \tag{8.70}$$

where $f$ and $n$ are mappings from $M$ into $\mathbb{R}^{n+2}$, and $\{\tilde{e}_1, \cdots, \tilde{e}_n, f, n\}$ are assumed to form a basis of $\mathbb{R}^{n+2}$ at every point on $M$. He proved that $(g, \nabla)$ is realized by such an immersion of codimension 2 if and only if so is $(g, \nabla^*)$. He also showed that a necessary and sufficient condition for the realizability is given by the conformally-projectively flatness: we say that $(g, \nabla)$ is **conformally-projectively flat** if there exist a coordinate system $[\xi^i]$ and two functions $\varphi$ and $\psi$ on $M$ such that

$$\Gamma_{ij,k} = (\partial_k \varphi) g_{ij} - \{(\partial_i \psi) g_{jk} + (\partial_j \psi) g_{ik}\}. \tag{8.71}$$

Conformal dualistic geometry, in which dualistic structures are transformed conformally, was developed by Okamoto, Amari and Takeuchi [175] from the theory of sequential estimation. In this variant of the estimation problem the number of observations are sequentially decided based on the information contained in what is already known. In a manner similar to the importance of curvature in statistical inference, conformal curvature plays an important role in sequential estimation. Conformal dualistic geometry also plays an important role within mathematics.

We see, then, that in addition to its applications, information geometry poses interesting mathematical problems. We enumerate several of these below.

(i) Find the condition(s) necessary for a given $n$-dimensional space with a dualistic structure to be realizable as an $n$-dimensional submanifold in $\mathbb{R}^m$. Also, find the condition(s) necessary to be realizable as a pair of $n$-dimensional submanifolds in two $m$-dimensional mutually dual linear spaces $V$ and $V^*$.

(ii) Find the condition(s) necessary for an $n$-dimensional space with a dualistic structure to be realizable as a submanifold of an $m$-dimensional dually flat space.

(iii) Let $(M, g)$ be a Riemannian space. We say that $(M, g)$ can be flattened if there exist a pair of affine connections $\nabla$ and $\nabla^*$ such that $(M, g, \nabla, \nabla^*)$ is dually flat. Show whether this is always possible. If not, find the invariant which characterizes those spaces which may be flattened.

(iv) Develop the theory of infinite-dimensional spaces with dualistic structures. Two examples are: the geometry of the set consisting of all (or most) absolutely continuous probability density functions $p(x)$ on $\mathbb{R}$ (as given, this is not a manifold), and the geometry of the infinite-dimensional space of stochastic processes. See §2.5, §4.8.2 and §5.2. Recently a nice mathematical foundation based on the theory of Orlicz space has been given to the first example by Pistone and his coworkers [189, 97, 188].

(v) Analyze the global structure of spaces with dualistic structures.

(vi) Extend the notion of dualistic structure to the case when a Finsler metric is given instead of a Riemannian metric, and analyze spaces with such structures. This relates to non-regular families of probability distributions for which the central limit theorem does not work, and corresponds to stable distributions other than normal distributions. See Amari [8].

# Guide to the Bibliography

There are a lot of good textbooks on general differential geometry. Some of them are Kobayashi and Nomizu [122], Spivak [202], Lang [137] and Helgason [106]. **Chapter 1** of the present book is a digest of some elementary parts of these textbooks.

Although there does not yet exist a standard text covering the entirety of information geometry, the fundamental ideas and applications to the higher-order asymptotic theory of statistical inference are found in Amari [9], Barndorff-Nielsen [41], Murrey and Rice [153] and Kass and Vos [119]. The last two are expository textbooks. In addition, Amari et al. [23] and Dodson (ed.) [81] are collections of papers focusing on information geometry, and Amari [15, 17], Barndorff-Nielsen [44] and Kass [118] are expository papers.

Historically speaking, Rao's paper [190] was the one which first suggested the idea of considering the Fisher information as a Riemannian metric on the space formed by a family of probability distributions. This Riemannian structure, the Fisher metric in our terminology, was studied further by many researchers, among which are Akin [4], James [114], Atkinson and Mitchell [34], Skovgaard [201], Oller [176], Oller and Cuadras [178] and Oller and Corcuera [177].

The $\alpha$-connections were introduced in [65] by Chentsov. Although somewhat difficult to read, it is an extremely interesting book, in which the Fisher metric and the $\alpha$-connections were characterized by the invariance under Markovian morphisms. The invariance of geometrical structures on statistical manifolds was discussed also by Lauritzen [139], Campbell [64], Dawid [76, 77], Picard [187], Li [142], Corcuera and Giummole [68] and Burbea [60].

The statistical meaning of curvature was first clarified by Efron [83]. This work became a trigger for the later developments of information geometry; see the historical remarks in §2.4. We should also note that the theory of exponential families is an important background of information geometry. The book [37] of Barndorff-Nielsen is a comprehensive work on this subject. These preceding works inspired Amari to pursue a new idea of information geometry in [5, 6], where the $\alpha$-connections were introduced and proved to be useful in the asymptotic theory of statistical estimation. One can see that several manifestations of the $\pm\alpha$-duality already appeared and played important roles in these papers. The essence of the duality was elucidated by Nagaoka and Amari [161]. This paper introduced the notion of dual connections, developed the general theory of dually flat spaces, and applied it to the geometry of $\alpha$-connections.

**Sections 2.6, 3.1, 3.3, 3.4** and **3.6** of the present book are mostly based on this paper. Although these results are now well known through [9] and later publications, the paper itself was never published in a major journal. The editors or reviewers of some major journals were reluctant to accept new ideas of combining statistics and differential geometry. Theorems 3.18 and 3.19 were shown in the master thesis [155] of Nagaoka, but were not dealt with in [161] nor [9] for the reason that the authors could find little significance in replacing the definitions of expectation and variance with their $\alpha$-versions. The attempt made by Curado and Tsallis [74] and Tsallis et al. [208] has encouraged the authors to revive the results.

The dualistic geometry of general divergences (contrast functions) was developed by Eguchi [84, 85, 87], to which **§3.2** is essentially indebted. The yoke geometry of Barndorff-Nielsen [39, 40, 41] is a natural generalization of the divergence geometry. Important examples of divergences on statistical models are given by the $f$-divergences which were introduced by Csiszár [70]. He made several pioneering works related to information geometry such as [71, 72, 73]. A comprehensive study of the $f$-divergences is found in Vajda [214].

There are slightly different or more general frameworks for information geometry. Barndorff-Nielsen [38, 41] used the observed Fisher information and the observed $\alpha$-connections instead of our definitions, which he called the expected Fisher information and the expected $\alpha$-connections, to construct the observed geometry. Barndorff-Nielsen and Jupp [47] studied a symplectic structure of a dualistic manifold. Critchley et al. [69] proposed the preferred point geometry, and Zhu and Rohwer [229, 230] investigated a Bayesian information geometry. See also Burbea and Rao [61], which treats a situation where the entropy $H(p)$ in Equation (3.57) is replaced with a more general form of functionals.

Statistics has constantly been the primary field for applications of information geometry. **Sections 4.1–4.6** are mostly based on Amari [5, 6, 7, 9], while the source of **§4.7** is Kumon and Amari [132] and Amari [9]. The idea of local exponential family bundles described in **§4.8.1** was discussed by Amari [11] and further pursued by Barndorff-Nielsen and Jupp [46]. The subject of **§4.8.2** — the theory of estimating functions for semi-parametric and non-parametric models using dual connections on fiber bundles — was developed by Amari and Kumon [29] and Amari and Kawanabe [27, 28]. This theory was applied to ICA (Independent Component Analysis) by Amari and Cardoso [24], Amari [20, 21] and Kawanabe and Murata [120].

Besides these, there are lots of works related to statistical inference and differential geometry. The book [9] contains a comprehensive list of references, which would be helpful for those interested in the applications to statistics. In addition, many new developments were made after the publication of this book, among which are as follows. As mentioned in **§4.5**, the higher-order asymptotics of statistical estimation and prediction were further discussed by Eguchi and Yanagimoto [88], Komaki [123] and Kano [116]. The relationship between sequential estimation and conformal geometry was established by Okamoto et al. [175]. Geometrical aspects of the EM algorithm (Dempster et al. [79]) were studied by Csiszár and Tusnády [73], Amari [16] and Byrne [63], while Matsuyama

[147, 148] proposed the $\alpha$-EM algorithm to accelerate convergence. Myung et al. [154] applied a geometrical idea to give a foundation to Rissanen's MDL and to generalize it. Conformal geometry was used by Amari and Wu [32] to improve the kernel support vector machine. Amari [22] studied an invariant decomposition of higher-order correlations in order to elucidate the correlational structure of an ensemble of neurons. See also Eguchi [86], Kass [117], Pace and Salvan [181] and Vos [216, 217, 218, 219]. The last author pointed out that the dually flat structure of an exponential family is applicable to a quasi-likelihood in the theory of generalized linear model.

In **Chapter 5** we have surveyed some topics from differential geometrical studies of dynamical systems and time series, chiefly based on Amari [12] and Ohara et al. [172, 171, 173]. Related works are found in Brockett [56], Kumon [131], Ravishanker et al. [193], Komaki [124], Zang et al. [228], Brigo et al. [55], Kulhavy [128, 129, 130] and Xu [223]. See also Sekine [195, 196] for geometry of some infinite-dimensional spaces consisting of continuous-time stochastic processes.

The geometrical framework connecting multiterminal information theory and statistical inference was given in the papers of Amari and Han [13, 26, 99, 100]. **Chapter 6** is an overview of this subject which is also indebted to such information theoretical works as Han [98], Ahlswede and Brunashev [3] and Shalaby and Papamarcou [197]. A comprehensive survey is found in [100].

In **Chapter 7** we have discussed some ideas for extending the framework of information geometry to quantum systems. Among the materials of the quantum information geometry, the quantum relative entropy has been studied so far mainly in mathematical physics and operator algebras; see *e.g.* Umegaki [212], Lindblad [144], Araki [33], Uhlmann [209], and Ohya and Petz [174]. Reflecting that its classical counterpart (i.e. the Kullback divergence) is one of the key concepts in the classical information geometry, it seems a natural idea to use the relative entropy to define a geometrical structure on the quantum state space. Ingarden et al. [112] pointed out that the relative entropy induces a Riemannian metric on the quantum state space. Nagaoka [158] introduced the dualistic structure from the relative entropy and showed that it is dually flat just like the classical e-, m-connections on the space of probability distributions. Petz [183] and Petz and Toth [186] studied the same geometrical structure from a different point of view, elucidated its relation to the Bogoliubov inner product (or the canonical correlation), and derived the Cramér-Rao inequality in this geometry which is a special case of our Theorem 7.3. The quantum version of Csiszár's $f$-divergence treated in **§7.2** was introduced by Petz [182] (see also [174]), extending the approach of [33] where the relative entropy was defined via the relative modular operator. The quantum $\alpha$-divergence and the induced dualistic structure were introduced by Hasegawa [101] and further studied by himself and Petz [102, 185]. These geometrical structures may be regarded as formal quantum analogues of the $\alpha$-connections and the Fisher metric, and are discussed in **§7.2** of the present book in the light of the general theory of dual connections developed in Chapter 3.

We should note, however, that it is by no means clear whether the geometrical structure induced from a divergence is relevent, as in the classical case, to statistical problems such as parameter estimation and hypothesis testing. Statistical inference problems on quantum states were actively studied by Helstrom, Holevo, Belavkin, Yuen, Lax, Kennedy and others around the 1970s; see the books of Helstrom [107] and Holevo [110] and their lists of references. We cannot find the relative entropy nor any divergence in these works, but instead there are some analogues of the score functions (i.e. the derivatives of the logarithmic likelihood) such as the symmetric logarithmic derivatives (SLDs) and the right logarithmic derivatives (RLDs). Nagaoka [158] showed that a dualistic structure can be induced from the SLDs and elucidated some relations to the parameter estimation of quantum states. He also pointed out that the e-connection in this structure has a nonvanishing torsion, which forms a remarkable contrast to the dualistic structure induced from a divergence. The geometry of relative entropy and that of SLD were unified as the geometry of generalized covariance in Nagaoka [160]. The content of §7.3 and the first three theorems in §7.4 are mostly based on [160] except that Theorem 7.1 is new. The notion of quasi-classical model and the propositions (I) through (VI) in §7.4 are basically due to Nagaoka [156], whereas Young [226] obtained similar results to (I)–(IV). See also Nagaoka and Fujiwara [162].

Let us take a brief look at other approaches related to quantum extensions of information geometry, although we do not intend to give a comprehensive list of references. The monotonicity of a Riemannian metric on the quantum state space with respect to completely positive maps was studied by Morozowa and Chentsov [152], Petz [184] and Lesniewski and Ruskai [140]. A mathematically rigorous foundation was given to the $\alpha$-connections on the infinite-dimensional quantum state space by Gibilisco and Isola [96] as a noncommutative extension of the work of Gibilisco and Pistone [97] mentioned below. A connection on a fiber bundle on the state space was studied by Uhlmann [210, 211], Dąbrowski and Jadczyk [75], Hübner [111], among others, in connection with the Berry phase, and its relation to the geometry of SLD was discussed by Fujiwara [89, 90] and Matsumoto [146]. The future of the quantum information geometry will essentially depend on the development of the theory of quantum statistical inference. For recent studies on the parameter estimation of quantum states, see Nagaoka [157, 159], Fujiwara and Nagaoka [92, 93, 94], Matsumoto [145, 146] and Hayashi [104, 105]. The theory of hypothesis testing of quantum states is also important. For example, a statistical meaning of the quantum relative entropy is elucidated by the quantum version of Stein's lemma, which was proved by Hiai and Petz [109] and Ogawa and Nagaoka [169]. See also Wootters [222], Braunstein and Caves [51, 52], and Brody and Hughston [59] for geometrical study of quantum statistical inference.

Besides quantum mechanics, there are several topics in physics related to information geometry. Geometrical aspects of statistical physics were discussed by Balian [36] and Streater [204]. Recently, the Tsallis entropy [207, 74, 208] has been highlighted in the field of statistical physics. This subject seems to have close relation to the theory of $\alpha$-connections discussed in **Sections 2.6 and 3.6**. See also Tanaka [206], where the information geometry was applied to elucidation of the method of mean field approximation.

As pointed out in §8.1, a dually flat structure naturally arises when a smooth convex function is arbitrarily given. This observation enables us to apply the framework of information geometry to several new fields. Note that the idea of introducing a divergence $D$ from a convex function $\psi$ via Equation (8.8) was already studied by Bregmann [53] without noticing the inherent dualistic geometrical structure. See Azoury and Warmuth [35] for a related subject. The application to linear programming is found in Amari's unpublished note [10]. See also Tanabe and Tsuchiya [205]. The idea was extended to semidefinite programming problems by Ohara [170]. Gradient flows on dually flat spaces and their relations to completely integrable systems were discussed by Nakamura [163, 164] and Fujiwara and Amari [91]. See also Onishi and Imai [179, 180], where $\nabla$- and $\nabla^*$-Voronoi diagrams in computational geometry were studied.

The gradient of a function depends on the choice of the underlying Riemannian metric as in Equations (2.45) and (2.46), and when a gradient flow is applied to an optimization process, it is often important to choose a "good" metric in order to achieve a good performance. The learning method using the gradient flow based on such a natural metric as the Fisher metric on a statistical model or the invariant metric on a Lie group is called the natural gradient method, which is applicable to wide range of learning and optimization problems. The invariant Riemannian metric on the general linear group $GL(n)$ was used for the natural gradient method in ICA by Amari [18, 19]. This approach was extended to linear dynamical systems for the purpose of multiterminal deconvolution problem by Zhang et al. [228]. See also Edelman et al. [82] for geometry of gradient algorithms on Grassmann and Stiefel manifolds. The natural gradient method for multilayer perceptrons, which works much better than the conventional backpropagation method, was studied by Yang and Amari [224] and Amari et al. [31], for which Rattray et al. [191, 192] proved the optimality in learning by means of a statistical-physical approach. Some extensions of the natural gradient method are found in Amari et al. [25] and Kivinen and Warmuth [121].

Neural networks bring up several interesting problems related to both statistics and geometry, some of which were briefly explained in §8.2. Neuro-manifolds of Boltzmann machines and higher-order neurons were discussed in Amari [14] and Amari et al. [30]. The EM-algorithm and the natural gradient method mentioned above are important tools for learning procedures of neural networks; see in particular Amari et al. [16, 31], Byrne [63], Rattray et al. [191, 192], and Yang and Amari [224]. Study on neuro-manifolds has revealed an interesting fact that a hierarchical system such as multilayer perceptrons, ARMA models, Gaussian mixtures, etc. in general includes singularities at which the Fisher information degenerates. Singularities have serious influence on learning speed and model selection. Watanabe [220] has given an algebraic-geometrical foundation to it. Fukumizu and Amari [95] shows an interesting behavior of learning dynamics at singularities. These are new subjects to be studied further.

Transformation models on which some Lie groups are acting have many special properties, which were extensively studied by Barndorff-Nielsen et al.

[42, 43]. In §**8.3** we have briefly touched upon the dualistic structure of such transformation models, although the subject has not fully been investigated yet.

The notion of dual connections naturally arises in the framework of affine differential geometry as seen in §**8.4**. Mathematical studies on this subject are found in Nomizu and Pinkall [166], Nomizu and Simon [168], Nomizu and Sasaki [167], Dillen et al. [80], Kurose [133, 134, 135], Abe [1], Lauritzen [138], Matsuzoe [149], Noguchi [165], Li et al. [141] and Uohashi et al. [213]. On the other hand, Shima [198, 199, 200] studied a Hessian manifold which is a Riemannian manifold with the metric given by the second derivatives of a convex potential function $\psi(\theta)$ in a local coordinate system as in Equation (3.34). This is essentially equivalent to the notion of dually flat space. He investigated this kind of manifolds including their global structures from purely mathematical points of view. Infinite-dimensional statistical manifolds were studied by Kambayashi [115], von Friedrich [215] and Lafferty [136], and a mathematical foundation was given to this problem in the framework of Orlicz space by Pistone and his group in [97, 188, 189].

# Bibliography

[1] N. Abe. Affine immersions and conjugate connections. *Tensor, N.S.*, 55:276–280, 1994.

[2] D.H. Ackley, G.E. Hinton, and T.J. Sejnowski. A learning algorithm for boltzmann machines. *Cognitive Science*, 9:147–169, 1985.

[3] R. Ahlswede and M. Burnashev. On minimax extimation in the presence of side information about remote data. *The Annals of Statistics*, 18:141–171, 1990.

[4] E. Akin. *The Geometry of Population Genetics*. Lecture Notes in Biomathematics 31. Springer-Verlag, 1979.

[5] S. Amari. Theory of information spaces—a geometrical foundation of statistics. POST RAAG Report 106, 1980.

[6] S. Amari. Differential geometry of curved exponential families—curvature and information loss. *The Annals of Statistics*, 10:357–385, 1982.

[7] S. Amari. Geometrical theory of asymptotic ancillarity and conditional inference. *Biometrika*, 69:1–17, 1982.

[8] S. Amari. Finsler geometry of families of non-regular probability distributions. RIMS Kokyuroku 538, Research Institute for Mathematical Sciences, Kyoto University, 1984. (in Japanese).

[9] S. Amari. *Differential-Geometrical Methods in Statistics*. Lecture Notes in Statistics 28. Springer-Verlag, 1985.

[10] S. Amari. Convex problems and differential geometry. unpublished note (in Japanese), 1987.

[11] S. Amari. Differential geometrical theory of statistics. In *[23], Chapter II*, pages 19–94, 1987.

[12] S. Amari. Differential geometry of a parametric family of invertible linear systems—Riemannian metric, dual affine connections and divergence. *Mathematical Systems Theory*, 20:53–82, 1987.

[13] S. Amari. Fisher information under restriction of Shannon information. *Annals of the Institute of Statistical Mathematics*, 41:623–648, 1989.

[14] S. Amari. Dualistic geometry of the manifold of higher-order neurons. *Neural Networks*, 4:443–451, 1991.

[15] S. Amari. Information geometry. *Applied Mathematics*, pages 37–56, 1992. (in Japanese).

[16] S. Amari. Information geometry of EM and em algorithms for neural networks. *Neural Networks*, 8:1379–1408, 1995.

[17] S. Amari. Information geometry. *Contemporary Mathematics*, 203:81–95, 1997.

[18] S. Amari. Natural gradient works efficiently in learning. *Neural Computation*, 10:251–276, 1998.

[19] S. Amari. Natural gradient for over- and under-complete bases in ICA. *Neural Computation*, 11:1875–1883, 1999.

[20] S. Amari. Superefficiency in blind source separation. *IEEE Transactions on Signal Processing*, 47:936–944, 1999.

[21] S. Amari. Estimating functions of independent component analysis for temporally correlated signals. *Neural Computation*, 12:1155–1179, 2000.

[22] S. Amari. Information geometry on hierarchical decomposition of stochastic interactions. submitted.

[23] S. Amari, O.E. Barndorff-Nielsen, R.E. Kass, S.L. Lauritzen, and C.R. Rao. *Differential Geometry in Statistical Inference*. IMS Lecture Notes: Monograph Series 10. Institute of Mathematical Statistics, Hayward, California, 1987.

[24] S. Amari and J.-F. Cardoso. Blind source separation—semiparametric statistical approach. *IEEE Transactions on Signal Processing*, 45(11):2692–2700, 1997.

[25] S. Amari, T.P. Chen, and A.Cichocki. Nonholonomic orthogonal learning algorithms for blind source separation. *Neural Computation*, 12:1133–1154, 2000.

[26] S. Amari and T.S. Han. Statistical inference under multi-terminal rate restrictions—a differential geometrical approach. *IEEE Transactions on Information Theory*, IT-35:217–227, 1989.

[27] S. Amari and M. Kawanabe. Information geometry of estimating functions in semiparametric statistical models. *Bernoulli*, 3:29–54, 1997.

[28] S. Amari and M. Kawanabe. Estimating functions in semiparametric statistical models. In I.V. Basawa, V.P. Godambe, and R.L. Taylor, editors, *IMS Selected Proceedings of the Symposium on Estimating Functions*, volume 32, pages 65–81, 1998.

[29] S. Amari and M. Kumon. Estimation in the presence of infinitely many nuisance parameters—geometry of estimating functions. *The Annals of Statistics*, 16:1044–1068, 1988.

[30] S. Amari, K. Kurata, and H. Nagaoka. Information geometry of Boltzmann machines. *IEEE Transactions on Neural Networks*, 3:260–271, 1992.

[31] S. Amari, H. Park, and K. Fukumizu. Adaptive method of realizing natural gradient learning for multilayer perceptrons. *Neural Computation*. to appear.

[32] S. Amari and S. Wu. Improving support vector machine classifiers by modifying kernel function. *Neural Networks*, 12:783–789, 1999.

[33] H. Araki. Relative entropy of states of von neumann algebras. *Publications of the Research Institute for Mathematical Sciences, Kyoto University*, 11:809–833, 1976.

[34] C. Atkinson and A.F. Mitchell. Rao's distance measure. *Sankhyā: The Indian Journal of Statistics, Ser. A*, 43:345–365, 1981.

[35] K.S. Azoury and M.K. Warmuth. Relative loss bounds for on-line density estimation with the exponential family of distributions. submitted.

[36] R. Balian, Y. Alhassid, and H. Reinhardt. Dissipation in many-body systems: A geometric approach based on information theory. *PHYSICS REPORTS (Review Section of Physics Letters)*, 131(1,2):1–146, 1986.

[37] O.E. Barndorff-Nielsen. *Information and Exponential Families in Statistical Theory*. Wiley, New York, 1978.

[38] O.E. Barndorff-Nielsen. Likelihood and observed geometries. *The Annals of Statistics*, 14:856–873, 1986.

[39] O.E. Barndorff-Nielsen. Differential geometry and statistics: Some mathematical aspects. *Indian Journal of Mathematics*, 29, Ramanujan Centenary Volume:335–350, 1987.

[40] O.E. Barndorff-Nielsen. On some differential geometric concepts and their relations to statistics. In *[81]*, pages 53–90, 1987.

[41] O.E. Barndorff-Nielsen. *Parametric Statistical Models and Likelihood*. Lecture Notes in Statistics 50. Springer-Verlag, 1988.

[42] O.E. Barndorff-Nielsen, P. Blæsild, and P.S. Eriksen. *Decomposition and Invariance of Measures, and Statistical Transformation Models.* Lecture Notes in Statistics 58. Springer-Verlag, 1989.

[43] O.E. Barndorff-Nielsen, P. Blæsild, J.L. Jensen, and B. Jørgensen. Exponential transformation models. *Proceedings of the Royal Society of London*, A379:41–65, 1982.

[44] O.E. Barndorff-Nielsen, R.D. Cox, and N. Reid. The role of differential geometry in statistical theory. *International Statistics Review*, 54:83–96, 1986.

[45] O.E. Barndorff-Nielsen and E.B.V. Jensen, editors. *Geometry in Present Day Science.* World Scientific, 1999.

[46] O.E. Barndorff-Nielsen and P.E. Jupp. Approximating exponential models. *Annals of the Institute of Statistical Mathematics*, 41:247–267, 1989.

[47] O.E. Barndorff-Nielsen and P.E. Jupp. Yokes and symplectic structures. *Journal of Statistical Planning and Inference*, 63:133–146, 1997.

[48] R. Bhatia. *Matrix Analysis.* Springer-Verlag, 1997.

[49] W. Blaschke. *Vorlesungen über Differentialgeometrie II, Affine Differentialgeometrie.* Springer, Berlin, 1923.

[50] P. Bloomfield. An exponential model for the spectrum of a scalar time series. *Biometrika*, 60:217–226, 1973.

[51] S.L. Braunstein and C.M. Caves. Statistical distance and the geometry of quantum states. *Physical Review Letters*, 72:3439–3442, 1994.

[52] S.L. Braunstein and C.M. Caves. Geometry of quantum states. In V.P. Belavkin, O. Hirota, and R.L. Hudson, editors, *Quantum Communications and Measurement*, pages 21–30. Plenum Press, New York, 1995.

[53] L.M. Bregman. The relaxation method of finding the common point of convex sets and its application to the solution of problems in convex programming. *USSR Computational Mathematics and Physics*, 7:200–217, 1967.

[54] D. Brigo. Diffusion processes, manifolds of exponentail densities, and nonlinear filtering. In *[45]*, pages 75–96, 1999.

[55] D. Brigo, B. Hanzon, and F.L. Gland. A differential geometric approach to nonlinear filtering: the projection filter. *IRISA Report*, 1995.

[56] R.W. Brockett. Some geometric questions in the theory of linear systems. *IEEE Transactions on Automatic Control*, 21:449–455, 1976.

[57] R.W. Brockett. Dynamical systems that sort lists, diagonalize matrices and solve linear programming problems. In *Proceedings of 27th IEEE Conference on Decision and Control*, pages 799–. IEEE, 1988.

[58] R.W. Brockett. Least squares matching problems. *Linear Algebra and Its Applications*, 122–124:761–777, 1989.

[59] D.C. Brody and L.P. Hughston. *The Geometry of Statistical Physics*. Imperial College Press/ World Scientific Pub. to appear.

[60] J. Burbea. Informative geometry of probability spaces. *Expositiones Mathematicae*, 4:347–378, 1986.

[61] J. Burbea and C.R. Rao. Entropy differential metric, distance and divergence measures in probability spaces: A unified approach. *Journal of Multivariate Analysis*, 12:575–596, 1982.

[62] J.P. Burg. Maximum entropy spectral analysis. In *Proceedings of the 37th Meeting of the Society of Exploration Geophysicists*, 1967. Also in *Modern Spectrum Analysis*.

[63] W. Byrne. Alternating minimization and Boltzmann machine learning. *IEEE Transactions on Neural Networks*, 3:612–620, 1992.

[64] L.L. Campbell. The relation between information theory and the differential geometric approach to statistics. *Information Sciences*, 35:199–210, 1985.

[65] N.N. Chentsov (Čencov). *Statistical Decision Rules and Optimal Inference*. American Mathematical Society, Rhode Island, U.S.A., 1982. (Originally published in Russian, Nauka, Moscow, 1972).

[66] H. Chernoff. A measure of asymptotic efficiency for tests of a hypothesis based on a sum of observations. *Annals of Mathematical Statistics*, 23:493–507, 1952.

[67] B.S. Clarke and A.R. Barron. Information theoretic asymptotics of Bayes methods. *IEEE Transactions on Information Theory*, 36:453–471, 1990.

[68] I.M. Corcuera and F. Giummolè. A characterization of monotone and regular divergences. *Annals of the Institute of Statistical Mathematics*, 50:433–450, 1998.

[69] F. Critchley, P. Marriott, and M. Salmon. Preferred point geometry and statistical manifolds. *The Annals of Statistics*, 21:1197–1224, 1993.

[70] I. Csiszár. Information type measures of difference of probability distributions and indirect observations. *Studia Scientiarum Mathematicarum Hungarica*, 2:299–318, 1967.

[71] I. Csiszár. On topological properties of $f$-divergence. *Studia Scientiarum Mathematicarum Hungarica*, 2:329–339, 1967.

[72] I. Csiszár. $I$-divergence geometry of probability distributions and minimization problems. *The Annals of Probability*, 3:146–158, 1975.

[73] I. Csiszár and G. Tusnády. Information geometry and alternating minimization procedures. In E.F. Dedewicz et al., editor, *Statistics and Decisions, Supplement Issue No.1*, pages 205–237. R. Oldenbourg Verlag, Munich, 1984.

[74] E.M.F. Curado and C. Tsallis. Generalized statistical mechanics: connection with thermodynamics. *Journal of Physics A*, 24:L69–L72, 1991.

[75] L. Dąbrowski and A. Jadczyk. Quantum statistical holonomy. *Journal of Physics A*, 22:3167–3170, 1989.

[76] A.P. Dawid. Discussion to Efron's paper. *The Annals of Statistics*, 3:1231–1234, 1975.

[77] A.P. Dawid. Further comments on a paper by Bradley Efron. *The Annals of Statistics*, 5:1249, 1977.

[78] A. Dembo and O. Zeitouni. *Large Deviations Techniques and Applications, 2nd ed.* Springer, New York, 1998.

[79] A.P. Dempster, N.M. Laird, and D.B. Rubin. Maximum likelihood from incomplete data via the EM algorithm. *Journal of the Royal Statistical Society*, B39:1–38, 1977.

[80] F. Dillen, K. Nomizu, and L. Vrancken. Conjugate connections and Radon's theorem in affine differential geometry. *Monatshefte für Mathematik*, 109:221–235, 1990.

[81] C.T.J. Dodson, editor. *Geometrization of Statistical Theory*. Proceedings of the GST Workshop. ULDM Publications, Department of Mathematics, University of Lancaster, 1987.

[82] A. Edelman, T. Arias, and S.T. Smith. The geometry of algorithms with orthogonality constraints. *SIAM Journal on Matrix Analysis and Applications*, 20:303–353, 1998.

[83] B. Efron. Defining the curvature of a statistical problem (with application to second order efficiency) (with discussion). *The Annals of Statistics*, 3:1189–1242, 1975.

[84] S. Eguchi. Second order efficiency of minimum contrast estimators in a curved exponential family. *The Annals of Statistics*, 11:793–803, 1983.

[85] S. Eguchi. A characterization of second order efficiency in a curved exponential family. *Annals of the Institute of Statistical Mathematics*, 36:199–206, 1984.

[86] S. Eguchi. A geometric look at nuisance parameter effect of local powers in testing hypothesis. *Annals of the Institute of Statistical Mathematics*, 43:245–260, 1991.

[87] S. Eguchi. Geometry of minimum contrast. *Hiroshima Mathematical Journal*, 22(3):631–647, 1992.

[88] S. Eguchi and T. Yanagimoto. Asymptotic improvement of maximum likelihood estimators using relative entropy risk. Research Memorandum 665, The Institute of Statistical Mathematics, Tokyo, 1998.

[89] A. Fujiwara. *A geometrical study in quantum information systems*. PhD thesis, University of Tokyo, 1995.

[90] A. Fujiwara. Geometry of quantum information systems. In *[45]*, pages 35–48, 1999.

[91] A. Fujiwara and S. Amari. Gradient systems in view of information geometry. *Physica D*, 80:317–327, 1995.

[92] A. Fujiwara and H. Nagaoka. Quantum Fisher metric and estimation for pure state models. *Physics Letters A.*, 201:119–124, 1995.

[93] A. Fujiwara and H. Nagaoka. Coherency in view of quantum estimation theory. In K. Fujikawa and Y.A. Ono, editors, *Quantum Coherence and Decoherence*, pages 303–306. Elsevier Science B. V., 1996.

[94] A. Fujiwara and H. Nagaoka. An estimation theoretical characterizatoin of coherent states. *Journal of Mathematical Physics*, 40:4227–4239, 1999.

[95] K. Fukumizu and S. Amari. Local minima and plateaus in hierarchical structures of multilayer perceptrons. *Neural Networks*. to appear.

[96] P. Gibilisco and T. Isola. Connections on statistical manifolds of density operators by geometry of noncommutative $l^p$-spaces. *Infinite Dimensional Analysis, Quantum Probability and Related Topics*, 2(1):169–178, 1999.

[97] P. Gibilisco and G. Pistone. Connections on non-parametric statistical manifolds by Orlicz space geometry. *Infinite Dimensional Analysis, Quantum Probabilities and Related Topics*, 1(2):325–347, 1998.

[98] T.S. Han. Hypothesis testing with multiterminal data compression. *IEEE Transactions on Information Theory*, 33(6):759–772, 1987.

[99] T.S. Han and S. Amari. Parameter estimation with multiterminal data compression. *IEEE Transactions on Information Theory*, 41(6):1802–1833, 1995.

[100] T.S. Han and S. Amari. Statistical inference under multiterminal data compression. *IEEE Transactions on Information Theory*, 44(6):2300–2324, 1998.

[101] H. Hasegawa. $\alpha$-divergence of the non-commutative information geometry. *Reports on Mathematical Physics*, 33:87–93, 1993.

[102] H. Hasegawa and D. Petz. Non-commutative extension of information geometry ii. In O. Hirota, A.S. Holevo, and C.M. Caves, editors, *Quantum Communication, Computing and Measurement*, pages 109–118. Plenum Publishing, 1997.

[103] J.H. Havrda and F. Chavat. Qualification methods of classification processes: Concepts of structural $\alpha$-entropy. *Kybernetica*, 3:30–35, 1967.

[104] M. Hayashi. A linear programming approach to attainable Cramér-Rao type bound. In O. Hirota, A.S. Holevo, and C.M. Caves, editors, *Quantum Communication, Computing and Measurement*. Plenum Publishing, 1997.

[105] M. Hayashi. Asymptotic estimation theory for a finite dimensional pure state model. *Journal of Physics A*, 31:4633–4655, 1998.

[106] S. Helgason. *Differential Geometry, Lie Groups, and Symmetric Spaces*. Academic, New York, 1978.

[107] C.W. Helstrom. *Quantum Detection and Estimation Theory*. Academic Press, New York, 1976.

[108] M. Henmi and R. Kobayashi. Hooke's law in statistical manifolds and divergences. *Nagoya Journal of Mathematics*. to appear.

[109] F. Hiai and D. Petz. The proper formula for relative entropy and its asymptotics in quantum probability. *Communications in Mathematical Physics*, 143:99–114, 1991.

[110] A.S. Holevo. *Probabilistic and Statistical Aspects of Quantum Theory*. North-Holland, 1982.

[111] M. Hübner. Computation of Uhlmann's parallel transport for density matrices and the Bures distance on three-dimensional Hilbert space. *Physics Letters A*, 179:226–230, 1993.

[112] R.S. Ingarden, H. Janyszek, A. Kossakowski, and T. Kawaguchi. Information geometry of quantum statistical systems. *Tensor, N.S.*, 37:105–111, 1982.

[113] R.S. Ingarden, Y. Sato, K. Sugawa, and M. Kawaguchi. Information thermodynamics and differential geometry. *Tensor, N.S.*, 33:347–353, 1979.

[114] A.T. James. The variance information manifold and the function on it. In P.K. Krishnaiah, editor, *Multivariate Analysis*, pages 157–169. Academic Press, 1973.

[115] T. Kambayashi. Statistical manifold of infinite dimension. *Transactions of the Japan Society for Industrial and Applied Mathematics*, 4(3):211–228, 1994. (in Japanese).

[116] Y. Kano. Third-order efficientcy implies fourth-order efficiency. *Journal of the Japan Statistical Society*, 26:101–117, 1996.

[117] R.E. Kass. Canonical parameterization and zero parameter effect curvature. *Journal of the Royal Statistical Society*, B46:86–92, 1984.

[118] R.E. Kass. The geometry of asymptotic inference (with discussions). *Statistical Science*, 4:188–234, 1989.

[119] R.E. Kass and P.W. Vos. *Geometrical Foundations of Asymptotic Inference*. John Wiley, New York, 1997.

[120] K. Kawanabe and N. Murata. Semiparametric estimation in ICA and information geometry. in preparation.

[121] L. Kivinen and M.K. Warmuth. Exponentiated gradient versus gradient descent for linear predictors. *Information and Computation*, 132:1–63, 1997.

[122] S. Kobayashi and K. Nomizu. *Foundations of Differential Geometry, I, II*. Interscience, New York, 1963, 1969.

[123] F. Komaki. On asymptotic properties of predictive distributions. *Biometrika*, 83(2):299–313, 1996.

[124] F. Komaki. An estimating method for parametric spectral densities of Gaussian time series. *Journal of Time Series Analysis*, 20:31–50, 1999.

[125] F. Komaki. Least informative prior for prediction. in preparation.

[126] K. Kraus. *States, Effects, and Operations*. Lecture Notes in Physics 190. Springer-Verlag, 1983.

[127] R. Kubo, M. Toda, and N. Hashitsume. *Statistical Physics II, Nonequilibrium Statistical Mechanics*. Springer Series in Solid State Sciences. Springer-Verlag, 1991.

[128] R. Kulhavy. A Bayes-closed approximation of recursive non-linear estimation. *International Journal of Adaptive Control and Signal Processing*, 4:271–285, 1990.

[129] R. Kulhavy. Recursive nonlinear estimation: A geometric view of recursive estimation provides a deeper insight into the problem of its approximation under memory limitation. *Automatica*, 26(3):545–555, 1990.

[130] R. Kulhavý. *Recursive Nonlinear Estimation: A Geometric Approach.* Lecture Notes in Control and Information Sciences 216. Springer-Verlag, 1996.

[131] M. Kumon. Identification of nonminimum-phase transfer functions using a higher-order spectrum. *Annals of the Institute of Statistical Mathematics*, 44(2):239–260, 1992.

[132] M. Kumon and S. Amari. Geometrical theory of higher-order asymptotics of test, interval estimator and conditional inference. *Proceedings of the Royal Society of London*, A387:429–458, 1983.

[133] T. Kurose. Dual connections and affine geometry. *Mathematische Zeitshrift*, 203:115–121, 1990.

[134] T. Kurose. On the divergence of 1-conformally flat statistical manifolds. *The Tôhoku Mathematical Journal*, 46:427–433, 1994.

[135] T. Kurose. 1-conformally flat statistical manifolds and their realization in affine space. *Fukuoka University Science Reports*, 29(2):209–219, 1999.

[136] J.D. Lafferty. The density manifold and configuration space quantization. *Transactions of the AMS*, 305:699–741, 1988.

[137] S. Lang. *Differential Manifolds.* Springer, New York, 1985.

[138] S. Lauritzen. Conjugate connection in statistical theory. In *[81]*, pages 33–51, 1987.

[139] S. Lauritzen. Statistical manifolds. In *[23]*, pages 163–216, 1987.

[140] A. Lesniewski and M.B. Ruskai. Monotone Riemannian metrics and relative entropy on non-commutative probability spaces. Mathematical Physics Preprint Archive mp_arc 99-59, 1999.

[141] A.-M. Li, U. Simon, and G. Zhao. *Global Affine Differential Geometry of Hypersurfaces.* Walter de Gruyter, Berlin, New York, 1993.

[142] H. Li. Determination of geometrical structures on statistical experiments, probability and statistics. In A. Badrkian, P-A Meyer, and J-A Yan, editors, *Proceedings of the Wuhan Meeting*, pages 133–148. World Scientific, 1993.

[143] E.H. Lieb. Convex trace functions and the Wigner-Yanase-Dyson conjecture. *Advances in Mathematics*, 11:267–288, 1973.

[144] G. Lindblad. Completely positive maps and entropy inequalities. *Communications in Mathematical Physics*, 40:147–151, 1975.

[145] K. Matsumoto. A new approach to the Cramér-Rao type bound of the pure state model. Technical Report METR 96-09, Dept. of Math. Eng. and Inform. Phys, Univ. of Tokyo, 1996.

[146] K. Matsumoto. *A geometrical approach to quantum estimation theory.* PhD thesis, University of Tokyo, 1998.

[147] Y. Matsuyama. The $\alpha$-EM algorithm: A block connectable generalized learning tool for neural networks. In J. Mira et al., editor, *Biological and Artificial Computation: From Neuroscince to Technology*, Lecture Notes in Computer Science 1240, pages 483–492. Springer Verlag, 1997.

[148] Y. Matsuyama. The $\alpha$-EM algorithm and its basic properties. *Transactions of the Institute of Electronics, Information and Communication Engineers, D-I*, J82:1347–1358, 1999. (in Japanese).

[149] H. Matsuzoe. On realization of conformally-projectively flat statistical manifolds and the divergences. *Hokkaido Mathematical Journal*, 27:409–421, 1998.

[150] T. Matumoto. Any statistical manifold has a cotrast function –on the $C^3$-functions taking the minimum at the diagonal of the product manifold–. *Hiroshima Mathematical Journal*, 23:327–332, 1993.

[151] A.F.S. Mitchell. The information matrix, skewness tensors and $\alpha$-connections for the general multivariate elliptic distribution. *Annals of the Institute of Statistical Mathematics*, 41:289–304, 1989.

[152] E.A. Morozowa and N.N. Chentsov. Markov invariant geometry on state manifolds. *Itogi Nauki i Tehniki*, 36:69–102, 1990. (in Russian).

[153] M.K. Murrey and J.W. Rice. *Differential Geometry and Statistics*. Chapman, 1993.

[154] I.J. Myung, V. Balasubramanian, and M.A. Pitt. Counting probability distributions: Differential geometry and model selection. submitted.

[155] H. Nagaoka. *Foundations of statistical geometry and applications to robust estimation.* Master thesis, Dept. of Math. Eng. and Instr. Phys, University of Tokyo, 1982. (in Japanese).

[156] H. Nagaoka. On the Fisher information of a quantum statistical model. In *Proceedings of 10th Symposium on Information Theory and Its Applications*, pages 241–246. Society of Information Theory and Its Applications in Japan, 1987. (in Japanese).

[157] H. Nagaoka. A new approach to Cramér-Rao bounds for quantum state estimation. Technical Report IT 89-42, The Institute of Electronics, Information and Communication Engineers, Tokyo, 1989.

[158] H. Nagaoka. On the parameter estimation problem for quantum statistical models. In *Proceedings of 12th Symposium on Information Theory and Its Applications*, pages 577–582. Society of Information Theory and Its Applications in Japan, 1989.

[159] H. Nagaoka. A generalization of the simultaneous diagonalizatoin of Hermitian matrices and its relation to quantum estimation theory. *Transactions of the Japan Society for Industrial and Applied Mathematics*, 1:305–318, 1991. (in Japanese).

[160] H. Nagaoka. Differential geometrical aspects of quantum state estimation and relative entropy. In V.P. Belavkin, O. Hirota, and R.L. Hudson, editors, *Quantum Communications and Measurement*, pages 449–452. Plenum Press, New York, 1995. (An extended version is found in METR 94-14, Dept. of Math. Eng. and Instr. Phys, Univ. of Tokyo).

[161] H. Nagaoka and S. Amari. Differential geometry of smooth families of probability distributions. Technical Report METR 82-7, Dept. of Math. Eng. and Instr. Phys, Univ. of Tokyo, 1982.

[162] H. Nagaoka and A. Fujiwara. Autoparallelity of a quantum statistical manifold. Technical Report UEC-IS-2000-4, Grad. School of Inform. Sys., The Univ. of Electro-Communications, 2000.

[163] Y. Nakamura. Completely integrable gradient systems on the manifolds of Gaussian and multinomial distributions. *Japan Journal of Industrial and Applied Mathematics*, 10(2):179–189, 1993.

[164] Y. Nakamura. A Tau-function for the finite Toda molecule and information spaces. *Contemporary Mathematics*, 179:205–211, 1994.

[165] M. Noguchi. Geometry of statistical manifolds. *Differential Geometry and its Applications*, 2:197–222, 1992.

[166] K. Nomizu and O. Pinkall. On the geometry of affine immersions. *Mathematische Zeitshrift*, 195:165–178, 1987.

[167] K. Nomizu and T. Sasaki. *Affine Differential Geometry*. Cambridge University Press, 1994.

[168] K. Nomizu and U. Simon. Notes on conjugate connections. In F. Dillen and L. Verstraelen, editors, *Geometry and topology of submanifolds*, volume IV, pages 152–173. World Scientific, 1992.

[169] T. Ogawa and H. Nagaoka. Strong converse and Stein's lemma in quantum hypothesis testing. *IEEE Transactions on Information Theory*. to appear.

[170] A. Ohara. Information geometric analysis of an interior point method for semidefinite programming. In *[45]*, pages 49–74, 1999.

[171] A. Ohara and S. Amari. Differential geometric structures of stable state feedback systems with dual connections. *Kybernetika*, 30(4):369–386, 1994.

[172] A. Ohara and T. Kitamori. Geometric structures of stable state feedback systems. *IEEE Transactions on Automatic Control*, 38(10):1579–1583, 1993.

[173] A. Ohara, N. Suda, and S. Amari. Dualistic differential geometry of positive definite matrices and its applications. *Linear Algebra and its Applications*, 247:31–53, 1996.

[174] M. Ohya and D. Petz. *Quantum entropy and its use*. Springer-Verlag, 1993.

[175] I. Okamoto, S. Amari, and K. Takeuchi. Asymptotic theory of sequential estimation procedures for curved exponential families. *The Annals of Statistics*, 19:961–981, 1991.

[176] J.M. Oller. Information metric for extreme value and logistic probability distributions. *Sankhyā: The Indian Journal of Statistics, Ser. A*, 49:17–23, 1987.

[177] J.M. Oller and J.M. Corcuera. Intrinsic analysis of statistical estimation. *The Annals of Statistics*, 23:1562–1581, 1995.

[178] J.M. Oller and C.M. Cuadras. Rao's distance for negative multinomial distributions. *Sankhyā: The Indian Journal of Statistics, Ser. A*, 47:75–83, 1985.

[179] K. Onishi and H. Imai. Voronoi diagram in statistical parametric space by Kullback-Leibler divergence. In *Proceedings of 13th ACM Symposium on Computational Geometry*, pages 463–465, 1997.

[180] K. Onishi and H. Imai. Riemannian computational geometry – Voronoi diagram and Delaunay-type triangulation in dually flat space –. submitted.

[181] L. Pace and A. Salvan. The geometric structure of the expected/observed likelihood expansions. *Annals of the Institute of Statistical Mathematics*, 46(4):649–666, 1994.

[182] D. Petz. Quasi-entropies for finite quantum systems. *Reports on Mathematical Physics*, 23:57–65, 1986.

[183] D. Petz. Geometry of canonical correlation on the state space of a quantum system. *Journal of Mathematical Physics*, 35:780–795, 1994.

[184] D. Petz. Monotone metrics on matrix spaces. *Linear Algebra and Its Applications*, 244:81–96, 1996.

[185] D. Petz and H. Hasegawa. On the Riemannian metric of $\alpha$-entropies of density matrices. *Letters in Mathematical Physics*, 38(2):221–225, 1996.

[186] D. Petz and G. Toth. The Bogoliubov inner product in quantum statistics. *Letters in Mathematical Physics*, 27:205–216, 1993.

[187] D.B. Picard. Statistical morphisms and related invariance properties. *Annals of the Institute of Statistical Mathematics*, 44(1):45–61, 1992.

[188] G. Pistone and M.P. Rogantin. The exponential statistical manifold: Mean parameters, orthogonality, and space transformation. *Bernoulli*, 5:721–760, 1999.

[189] G. Pistone and C. Sempi. An infinite-dimensional geometric structure on the space of all the probability measures equivalent to a given one. *The Annals of Statistics*, 23:1543–1561, 1995.

[190] C.R. Rao. Information and accuracy attainable in the estimation of statistical parameters. *Bulletin of the Calcutta Mathematical Society*, 37:81–91, 1945.

[191] M. Rattray and D. Saad. Analysis of natural gradient descent for multilayer neural networks. *Physical Review E*, 59(4):4523–4532, 1999.

[192] M. Rattray, D. Saad, and S. Amari. Natural gradient descent for on-line learning. *Physical Review Letters*, 81:5461–5464, 1999.

[193] N. Ravishanker, E.L. Melnick, and C. Tsai. Differential geometry of ARMA models. *Journal of Time Series Analysis*, 11(3):259–275, 1990.

[194] A. Rényi. On measures of entropy and information. In *Proceedings of the 4th Berkeley Symposium on Mathematical Statistics and Probability*, volume 1, pages 547–561. University of California Press, 1961.

[195] J. Sekine. The Hilbert-Riemannian structure of equivalent Gaussian measures associated with the Fisher information. *Osaka Journal of Mathematics*, 32:71–95, 1995.

[196] J. Sekine. Information geometry for symmetric-diffusions. *Potential Analysis*. to appear.

[197] H.M.H. Shalaby and A. Papamarcou. Multiterminal detection with zero-rate data compression. *IEEE Transactions on Information Theory*, 38(2):254–267, 1992.

[198] H. Shima. Homogeneous Hessian manifolds. *Annales de l'Institut Fourier*, 30:91–128, 1980.

[199] H. Shima. Vanishing theorems for compact Hessian manifolds. *Annales de l'Institut Fourier*, 36(3):183–205, 1986.

[200] H. Shima. Geometry of Hessian domains. *Differential Geometry and its Applications*, 7:277–290, 1997.

[201] L.T. Skovgaard. A Riemannian geometry of the multivariate normal model. *Scandinavian Journal of Statistics*, 11:211–233, 1984.

[202] M. Spivak. *A comprehensive Introduction to Differential Geometry, 2nd ed.* Publish or Perish, Boston, 1979.

[203] W.F. Stinespring. Positive functions on $C^*$-algebras. *Proceedings of the American Mathematical Society*, 6:211–216, 1955.

[204] R.F. Streater. Statistical dynamics and information geometry. *Contemporary Mathematics*, 203:117–131, 1997.

[205] K. Tanabe and T. Tsuchiya. New geometrical theory of linear programming. *SUURIKAGAKU (Mathematical Sciences)*, pages 32–37, 1988. (in Japanese).

[206] T. Tanaka. Information geometry of mean field approximation. *Neural Computation*. to appear.

[207] C. Tsallis. Possible generalization of Boltzmann-Gibbs statistics. *Journal of Statistical Physics*, 52:479–, 1988.

[208] C. Tsallis, R.S. Mendes, and A.R. Plastino. The role of constraints within generalized nonextensive statistics. *Physica A*, 261:534–554, 1998.

[209] A. Uhlmann. Relative entropy and the wigner-yanase-dyson-lieb concavity in an interpolation theory. *Communications in Mathematical Physics*, 54:21–32, 1977.

[210] A. Uhlmann. Parallel transport and "quantum holonomy" along density operators. *Reports on Mathematical Physics*, 24:229–240, 1986.

[211] A. Uhlmann. Density operators as an arena for differential geometry. *Reports on Mathematical Physics*, 33:253–263, 1993.

[212] H. Umegaki. Conditional expectation in an operator algebra, iii. *Kôdai Mathematical Seminar Reports*, 11:51–64, 1959.

[213] K. Uohashi, A. Ohara, and T. Fujii. 1-conformally flat statistical submanifolds. *Osaka Journal of Mathematics*, 37(2), 2000. to appear.

[214] I. Vajda. *Theory of Statistical Inference and Information*. Kluwer Academic Publishers, 1989.

[215] T. von Friedrich. Die Fisher-information und symplektische strukturen. *Mathematische Nachrichten*, 152:273–296, 1991.

[216] P.W. Vos. Fundamental equations for statistical submanifolds with applications to the Bartlett correction. *Annals of the Institute of Statistical Mathematics*, 41(3):429–450, 1989.

[217] P.W. Vos. A geometric approach to detecting influential curves. *The Annals of Statistics*, 19(3):1570–1581, 1991.

[218] P.W. Vos. Geometry of $f$-divergence. *Annals of the Institute of Statistical Mathematics*, 43(3):515–537, 1991.

[219] P.W. Vos. Quasi-likelihood or extended quasi-likelihood? an information-geometric approach. *Annals of the Institute of Statistical Mathematics*, 47(1):49–64, 1995.

[220] S. Watanabe. Algebraic analysis for non-identifiable learning machines. *Neural Computation*. to appear.

[221] B.C. Wei. *Exponential family Nonlinear Models*. Lecture Notes in Statistics 130. Springer-Verlag, 1998.

[222] W. K. Wootters. Statistical distance and Hilbert space. *Physical Review D*, 23:357–362, 1981.

[223] D. Xu. Differential geometrical structures related to forecasting error variance ratios. *Annals of the Institute of Statistical Mathematics*, 43:621–646, 1991.

[224] H.H. Yang and S. Amari. Complexity issues in natural gradient descent method for training multilayer perceptrons. *Neural Computation*, 10:2137–2157, 1998.

[225] S. Yoshizawa and K. Tanabe. Dual differential geometry associated with the Kullback-Leibler information on the Gaussian distributions and its 2-parameter deformations. *Science University of Tokyo Journal of Mathematics*, 35:113–137, 1999.

[226] T.Y. Young. Asymptotically efficient approaches to quantum-mechanical parameter estimation. *Information Sciences*, 9:25–42, 1975.

[227] H.P. Yuen and M. Lax. Multiple-parameter quantum estimation and measurement of nonselfadjoint observables. *IEEE Transactions on Information Theory*, 19:740–750, 1973.

[228] L. Zhang, S. Amari, and A. Cichocki. Semiparametric model and super efficiency in blind deconvolution. submitted.

[229] H. Zhu. Bayesian geometric theory of learning algorithms. In *Proceedings of International Conference of Neural Networks (ICNN'97)*, volume 3, pages 1041–1044, 1997.

[230] H. Zhu and R. Rohwer. Measurements of generalisation based on information geometry. In S.W. Ellacott, J.C. Mason, and I.J.Anderson, editors, *Mathematics of Neural Networks: Models, Algorithms and Applications*, pages 394–398. Kluwer Academic Publishers, Boston, 1997.

# Index

acceptance region, 101
additivity
— of the Fisher metric, 31
— of the Kullback divergence, 58
affine connection, 13
affine coordinate system, 18
$\alpha$- —, 33
affine differential geometry, 175
affine fundamental form, 176
affine immersion, 176
affine parameter, 21
affine projections
method of —, 169
affine subspace, 21
affine transformation, 18
$\alpha$-affine coordinate system, 33
$\alpha$-affine manifold, 46
$\alpha$-autoparallel, 33
$\alpha$-conformally flat, 178
$\alpha$-connection, 33
— on the quantum state space, 149, 151
— on the system space, 119
$\alpha$-coordinate system, 121
$\alpha$-divergence, 57, 71, 179
— on the system space, 122
quantum —, 152
$\alpha$-family, 48, 179
$\alpha$-flat, 33
$\alpha$-model, 126
$\alpha$-parallel, 33
$\alpha$-representation, 46
— on the quantum state space, 153
$\alpha$-spectrum, 120
$\alpha$-stochastic realization, 126
$\alpha$-version of Cramér-Rao inequality, 74
ancillary fiber, 112
ancillary statistic, 100, 174
AR model, 117
ARMA model, 118
asymptotically efficient estimator, 32, 84, 93
asymptotic Cramér-Rao inequality, 84, 93
autocorrelations, 121

autoparallel, 20
$\alpha$- —, 33
autoparallel curve, 21

base manifold, 108
bias-corrected estimator, 94
Bloomfield exponential model, 118
Bogoliubov inner product, 155
Boltzmann machine, 170
Boltzmann-Gibbs distribution, 68

$C^\infty$ curve, 6
$C^\infty$ diffeomorphism, 3, 5
$C^\infty$ differentiable manifold, 3
$C^\infty$ function, 4
$C^\infty$ mapping, 5
canonical correlation, 155
canonical divergence, 61
Pythagorean relation for —, 62
triangular relation of —, 62
canonical parameters, 34
chain rule
— of the Fisher metric, 31
— of the Kullback divergence, 57
completely integrable dynamical system, 170
completely positive map, 154
conformally flat
$\alpha$- —, 178
conformally-projectively flat, 180
conjugate connection, 51
connection, 13
$\alpha$- —, 33, 119, 149, 151
e- —, exponential —, 35, 157
m- —, mixture —, 36, 157
— coefficients, 13
affine —, 13
conjugate —, 51
dual —, 51
Levi-Civita —, 24
metric —, 23
Riemannian —, 24
symmetric —, 19
torsion-free —, 19

conormal vector field, 177
consistent estimator, 83
constant curvature
    space of —, 174, 178
contrast function, 54
convex analysis, 167
coordinate functions, 2
coordinate system, 1
    affine —, 18
    dual —, 59
    Euclidean —, 24
    normal —, 77
coordinate transformation, 3
coordinates, 1
cotangent space, 41
covariant derivative, 13, 15, 16
CP map, 154
Cramér-Rao inequality, 31, 43
    $\alpha$-version of —, 74
    asymptotic —, 84, 93
    quantum —, 160, 162
Cramér's theorem, 69
critical region, 101
cumulant generating function, 69
curvature tensor (field), 18
curve, 6
    autoparallel —, 21
    parallel translation along a —, 15
    vector field along a —, 14
curved exponential family, 36, 87

deficiency of a test, 105
denormalization, 47
density operator, 145
diffeomorphism, 3, 5
differentiable manifold, 1
differential
    — of a function, 42
    — of a mapping, 7
dimension, 1
divergence, 53, 54, 61
    $\alpha$- —, 57, 71, 122
    $f$- —, 56
    $(g, \nabla)$- —, 61
    $(g, \nabla)$- — of a curve, 64
    canonical —, 61
    dual —, 55
    Kullback —, 57, 67
    Kurose's —, 72, 179
    quantum $\alpha$- —, 152
    quantum $f$- —, 151
dual connection, 51
dual coordinate system, 59
dual divergence, 55
dual foliation, 75, 138
dual parameters, 65

dualistic structure, 51
    induced —, 52
dually flat space, 58

e-connection, 35
    — on the quantum state space, 157
e-representation, 41
    — on the quantum state space, 155
efficient estimator, 32, 84
efficient score function, 113
efficient test, 102
Efron curvature, 106
Einstein's convention, 5
embedding curvature, 23
entropy, 66, 122
    quantum relative —, 152
    relative —, 57
    von Neumann —, 153
equiaffine immersion, 176
estimating equation, 109
estimating function, 109
estimating submanifold, 88
estimation, 82
estimator
    asymptotically efficient —, 32, 84
    bias-corrected —, 94
    consistent —, 83
    efficient —, 32, 84
    first-order efficient —, 32, 84
    locally unbiased —, 43
    maximum likelihood —, 84, 87, 93, 96
    second-order efficient —, 96
    unbiased —, 31, 82
Euclidean coordinate system, 24
expectation parameters, 65
exponent of error, 136
exponential connection, 35
    — on the quantum state space, 157
exponential family, 34, 85
    curved —, 87
exponential map, 77
exponential representation, 41
    — on the quantum state space, 155

$f$-divergence, 56
    quantum —, 151
$\mathcal{F}$-multilinearity, 9
fiber, 108
    ancillary —, 112
    information —, 112
    nuisance —, 112
fiber bundle, 107, 108, 130
first-order efficient estimator, 32, 84, 93
first-order uniformly most powerful test, 102
Fisher information, 28
Fisher information matrix, 28

# INDEX

Fisher metric, 29
    — on the quantum state space, 149, 151, 156
    — on the system space, 119
flat, 18
    $\alpha$- —, 33
    dually —, 58
foliation, 75, 138

$(g, \nabla)$-divergence, 61
    — of a curve, 64
Gaussian times series, 117
generalized covariance, 155
geodesics, 21
gradient, 42
    natural — method, 185
gradient flow, 169

Hellinger distance, 57
higher-order asymptotic theory
    — of estimation, 84, 94
    — of tests, 100
Hilbert bundle, 109

ICA, 113
impulse response, 116
independent component analysis, 113
induced dualistic structure, 52
information
    Fisher —, 28
    Kullback-Leibler —, 57
information fiber, 112
information loss, 98
information metric, 29
information theory, 133
invariance
    — of the Fisher metric and the $\alpha$-connection, 38

Jeffreys' prior, 44
joint convexity
    — of $f$-divergence, 56
    — of quantum $f$-divergence, 154

Karmarkar's interior point method, 169
Kubo-Mori inner product, 155
Kullback divergence, 57, 67
    quantum equivalent of —, 152
Kullback-Leibler information, 57
Kurose's divergence, 72, 179

large deviations, 68, 69, 136, 139
Legendre transformations, 60
level of a test, 101
Levi-Civita connection, 24
Lie group, 172
likelihood function, 84

linear programming, 169
linear system, 115
    minimal phase —, 116
    stable —, 116, 127
local exponential family bundle, 107
locally unbiased estimator, 43
location-scale model, 109, 172
logarithmic moment generating function, 69

MA model, 117
m-connection, 36
    — on the quantum state space, 157
m-representation, 40
    — on the quantum state space, 148
manifold, 3
    differentiable —, 1, 3
    Riemannian —, 11
marginal distributions, 137
maximum entropy
    principle of —, 68
    principle of —, 125
maximum likelihood estimator, 84, 87, 93, 96
measurement
    quantum —, 146
    simple —, 147
metric
    Fisher —, 29, 119, 149, 151, 156
    information —, 29
    Riemannian —, 11
metric connection, 23
minimal phase system, 116
mixed coordinate system, 76, 137
mixed state, 146
mixture connection, 36
    — on the quantum state space, 157
mixture family, 35
mixture parameters, 35
mixture representation, 40
    — on the quantum state space, 148
model, 26
modifed Pythagorean relation, 72, 179
monotonicity
    — of $f$-divergence, 56
    — of quantum $f$-divergence, 154
    — of the Fisher metric, 31
multilinear mapping, 8
multiterminal information source, 133
multiterminal information theory, 133

$\nabla$-projection, 63
natural basis, 7
natural gradient method, 185
natural parameters, 34
normal coordinate system, 77
nuisance fiber, 112

nuisance parameter, 109
nuisance space, 112
nuisance tangent space, 111

observable
    quantum —, 147
observed point, 86
operator convex function, 154

parallel, 15, 17
    $\alpha$- —, 33
parallel translation, 15
parametric model, 26
points, 1
POM, 146
positive-operator-valued measure, 146
potential
    — of Legendre transformation, 60
power function, 101
power spectrum, 116
principle of maximum entropy, 68, 125
probability distributions, 25
probability operator-valued measure, 146
projection
    $\nabla$- —, 63
    — of a connection, 22
pure state, 146
Pythagorean relation
    — for canonical divergence, 62
    modified —, 72, 179

quantum $\alpha$-divergence, 152
quantum Cramér-Rao inequality, 160, 162
quantum equivalent of the Kullback
    divergence, 152
quantum $f$-divergence, 151
quantum measurement, 146
quantum observable, 147
quantum relative entropy, 152
quantum state, 145
quasiclassical, 163

relative entropy, 57
    quantum —, 152
relative modular operator, 150
representation of a tangent vector
    0- —, 44
    $\alpha$- —, 46, 153
    e- —, exponential —, 41, 155
    m- —, mixture —, 40, 148
Riemann-Christoffel curvature tensor
    (field), 18
Riemannian connection, 24
Riemannian manifold, 11
Riemannian metric, 11

Sanov's theorem, 69, 139
score function, 109
second fundamental form, 176
second-order efficient estimator, 96
simple measurement, 147
single-letterization, 143
SLD, 156
stable feedback, 127
stable linear system, 116, 127
stable matrix, 127
state
    quantum —, 145
statistical inference, 81
    — for multiterminal information, 133
statistical manifold, 28
statistical model, 25, 26, 81
stochastic realization, 125
    $\alpha$- —, 126
submanifold, 10
sufficient statistic, 29, 56, 86
    invariance with respect to a —, 38
symmetric connection, 19
symmetric logarithmic derivative, 156
symmetrized inner product, 155
system space, 117

tangent space, 5, 7
tangent vector, 5, 7
tensor field, 8
    — of type $(q, r)$, 8
    components of a —, 9
    curvature —, 18
    Riemann-Christoffel curvature —, 18
    torsion —, 18
test, 82, 100
    efficient —, 102
    first-order uniformly most powerful —,
    102
    uniformly most powerful —, 101
time series, 115
torsion tensor (field), 18
torsion-free connection, 19
transfer function, 116
triangular relation, 62, 77
true distribution, 26

unbiased estimator, 31, 82
underlying distribution, 26
uniformly most powerful test, 101

vector field, 8
    — along a curve, 14
    conormal —, 177
von Neumann entropy, 153

0-rate encoding, 135
0-representation, 44